STUDENT'S SOLUTIONS MANUAL
MILTON LOYER

Penn State University

to accompany

ELEMENTARY STATISTICS
NINTH EDITION

Mario F. Triola

Dutchess Community College

PEARSON

Addison
Wesley

Boston   San Francisco   New York
London   Toronto   Sydney   Tokyo   Singapore   Madrid
Mexico City   Munich   Paris   Cape Town   Hong Kong   Montreal

Reproduced by Pearson Addison-Wesley from electronic files supplied by the author.

Copyright © 2004 Pearson Education, Inc.
Publishing as Pearson Addison-Wesley, 75 Arlington Street, Boston, MA 02116

ISBN 0-321-12217-8

3 4 5 VHG 06 05 04

TABLE OF CONTENTS

PREFACE

This manual contains the solutions to the odd-numbered exercises for each section of the textbook Elementary Statistics, Ninth Edition, by Mario Triola, and the solutions for all of the end-of-chapter review and cumulative review exercises of that text. To aid in the comprehension of calculations, worked problems typically include intermediate steps of algebraic and/or computer/calculator notation. When appropriate, additional hints and comments are included and prefaced by NOTE.

Many statistical problems are best solved using particular formats. Recognizing and following these patterns promote understanding and develop the capacity to apply the concepts to other problems. This manual identifies and employs such formats whenever practicable.

For best results, read the text carefully before attempting the exercises, and attempt the exercises before consulting the solutions. This manual has been prepared to provide a check and extra insights for exercises that have already been completed and to provide guidance for solving exercises that have already been attempted but have not been successfully completed.

I would like to thank Mario Triola for writing an excellent elementary statistics book and for inviting me to prepare this solutions manual.

Chapter 1

Introduction to Statistics

1-2 Types of Data

1. Parameters, since 87 and 13 refer to the entire population.

3. Statistic, since 0.65 refers to the selected sample.

5. Discrete, since annual salary cannot assume any value over a continuous span – as an annual salary of $98,765.4321, for example, is not a physical possibility. The annual presidential salary is typically stated in multiples of $1000 - and it cannot be paid in US currency in units smaller than a cent.

7. Discrete, since the calculated percent cannot be any value over a continuous span. Because the number having guns in their homes must be an integer, the possible values for the percent are limited to multiples of 100/1059.

9. Ratio, since differences are meaningful and zero height has a natural meaning.

11. Interval, since differences are meaningful but ratios are not. Refer to exercise #21.

13. Ordinal, since the ratings give relative position in a hierarchy.

15. Ratio, since differences are meaningful and zero "yes" responses has a natural meaning.

17. a. The sample is the ten adults who were asked for an opinion.
    b. There is more than one possible answer. The population the reporter had in mind may have been all the adults living in that particular city.
    No. People living in that city but who do not pass by that corner (shut-ins, those with no business in that part of town, etc.) will not be represented in the sample. Furthermore, it was not specified that the reported selected the ten adults in an unbiased manner - e.g., he may have (consciously or subconsciously) selected ten friendly-looking people.

19. a. The sample is the 1059 randomly selected adults.
    b. The population is all the adults in the entire country. NOTE: While it is not specified that the adults were selected from within USA, this is the logical assumption.
    Yes. If the adults in the sample were randomly selected, there are no biases that would prevent the sample from being representative. NOTE: This is a sample of adults, not of homes. The appropriate inference is that about 39% of the adults live in a home with a gun, not that there is a gun in about 39% of the homes. Since adults were selected at random, homes with more adults are more likely to be represented in the survey – and that bias would prevent the survey from being representative of all homes.

21. Temperature ratios are not meaningful because a temperature of 0° does not represent the absence of temperature in the same sense that $0 represents the absence of money. The zero temperature in the example (whether Fahrenheit or Centigrade) was determined by a criterion other than "the absence of temperature."

23. This is an example of ordinal data. It is not interval data because differences are not meaningful – i.e., the difference between the ratings +4 and +5 does not necessarily represent the same differential in the quality of the food as the difference between the ratings 0 and +1.

1-3 Critical Thinking

NOTE: The alternatives given in exercises #1 and #3 are not the only possible answers.

1. Males weigh more than females – and truck drivers tend to be male, while about half the adults who do not drive trucks are female.

3. There are more minority drivers than white drivers in Orange County – if the rates of speeding were the same for both types of drivers, you would expect the more numerous type of driver to be issued more tickets.

5. The study was financed by those with a vested interest in promoting chocolate, who would be more likely to report only the favorable findings. While chocolate contains an antioxidant associated with decreased risk of heart disease and stroke, it may also contain other ingredients associated with an increased risk of those or other health problems.

7. No. The sample was self-selected and therefore not necessarily representative of the general population (or even of the constituencies of the various women's groups contacted).

9. Telephone directories do not contain unlisted numbers, and consumer types electing not be listed would not be represented in the survey. In addition, sometimes numbers appear in directories more than once (e.g., under the husband's name and under the wife's name), and consumer types electing multiple listings would be over-represented in the survey. Also excluded from the survey would be those without telephones (for economic, religious, or other reasons), transient persons (both frequent movers and those who just happened to have moved) not in the current directory, and persons whose living arrangements (group housing, extended families, etc.) do not include a phone in their own name.

11. Motorcyclists that died in crashes in which helmets may have saved their lives could not be present to testify.

13. No. While 0.94 is the average of the 29 different brands in the survey, it is not a good estimate of the average of all the cigarettes smoked for at least two significant reasons. First, the 29 brands in the survey are not representative of all brands on the market – they include only 100mm filtered brands that are not menthol or light. Secondly, some brands are much more popular than others. Even if every brand on the market were in the survey, the average would not be a good estimate of all the cigarettes smoked – the values of the popular brands will carry more weight when the amounts of each brand that are actually smoked are taken into consideration.

15. Most people born after 1945 are still living. This means the sample would include only those who did not reach their normal life expectancy and would not be representative.

NOTE for exercises #17-20: Multiplying or dividing by 1 will not change a value. Since 100% is the same as 1, multiplying or dividing by 100% will not change a value – but remember that the percent sign is a necessary unit (just like inches or pounds) that cannot be ignored.

17. a. $17/25 = .68 = .68 \times 100\% = 68\%$
    b. $35.2\% = 35.2\%/100\% = .352$
    c. $57\% = 57\%/100\% = .57; .57 \times 1500 = 855$
    d. $.486 = .486 \times 100\% = 48.6\%$

19. a. There is not enough accuracy in the reported percent to determine an exact answer. While $.52 \times 1038 = 539.76$ suggests an answer of 540, x/1038 rounds to 52% for $535 \leq x \leq 544$. The actual number could be anywhere from 535 to 544 inclusive.
    b. $52/1038 = .05 = 5\%$

21. There is not enough accuracy in the reported percents to determine an exact answer. It is tempting to conclude that
    $.08 \cdot (1875) = 150$ admit to committing a campus crime
    $.62 \cdot (150) = 93$ say they did so under the influence of alcohol or drugs.
    But x/1875 rounds to 8% for $141 \leq x \leq 159$, and so the number admitting to a crime could be anywhere from 141 to 159 inclusive. Similarly, the number that say they committed the crime under the influence of alcohol or drugs could be anywhere from 87 to 99 inclusive – since 87/141 and 99/159 both round to 62%.

23. Assuming that each of the 20 individual subjects is ultimately counted as a success or not (i.e., that there are no "dropouts" or "partial successes"), the success rates in fraction form must be one of 0/20, 1/20, 2/20,..., 19/20, 20/20. In percentages, these rates are multiples of 5 (0%, 5%, 10%,..., 95%, 100%), and values such as 53% and 58% are not mathematical possibilities.

25. The answers will vary according to student interests. Here is one example.
    a. Should our school consider a dress code?
    b. Should our school consider a dress code outlawing lewd or unsanitary practices?
    c. Should our school consider a dress code limiting personal freedom and diverting attention from more important educational issues?

1-4 Design of Experiments

1. Experiment, since the effect of an applied treatment is measured.

3. Observational study, since specific characteristics are measured on unmodified subjects.

5. Retrospective, since the data are collected by going back in time.

7. Cross-sectional, since the data are collected at one point in time.

9. Convenience, since the sample was simply those who happen to pass by.

11. Random, since each adult has an equal chance of being selected.
    NOTE: This is a complex situation. The above answer ignores the fact that there is not a 1 to 1 correspondence between adults and phone numbers. An adult with more than one number will have a higher chance of being selected. An adult who shares a phone number with other adults will have a lower chance of being selected. And, of course, adults with no phone numbers have no chance of being selected at all.

13. Cluster, since the population of interest (assumed to be all students at The College of Newport) was divided into classes which were randomly selected in order to interview all the students in each selected class.
    NOTE: Ideally the division into classes should place each student into one and only one class (e.g., if every student must take exactly one PE class each semester, select the PE classes at random). In practice such divisions are often made in ways that place some students in none of the classes (e.g., by selecting from all 2 pm M-W-F classes) or in more than one of the classes (e.g., by selecting from all the classes offered in the college). With careful handling, imperfect divisions should not significantly affect the results.

15.   Systematic, since every fifth element in the population (assumed to be all drivers passing the checkpoint during its operation) was sampled.

17.   Stratified, since the population of interest (assumed to be all workers) was divided into 3 subpopulations from which the actual sampling was done.

19.   Cluster, since the population of interest (assumed to be all cardiac patients) were divided into hospitals which were randomly selected in order to survey all such patients at each hospital.

21.   Yes, this is a random sample because each of the 1000 tablets (assumed to be the population) has an equal chance of being selected. Yes, this is a simple random sample because every possible group of 50 tablets has the same chance of being chosen.

23.   No, this is not a random sample because each element in the population does not have an equal chance of being selected. If the population of interest is all the residents of the city, then those who do not pass that street corner have no chance of being selected. Even if the population of interest is all those who pass that street corner, the reporter's personal preferences would likely prevent each person from having an equal chance of being selected – e.g., the reporter might not want to stop a woman with three noisy children, or a man walking very fast and obviously in a hurry to get somewhere. Since the sample is not random, it cannot be a simple random sample.

25.   Yes, this is a random sample because each person in Orange County has an equal chance of being selected. No, this is not a simple random sample because every possible sample of 200 Orange County residents does not have the same chance of being chosen – a sample with 75 men and 45 women, for example, is not possible. The survey will include a simple random sample of Orange County men and a simple random sample of Orange County women, but not a simple random sample of Orange County adults.

27.   Assume that "students who use this book" refers to a particular semester (or other point in time). Before attempting any sampling procedure below, two preliminary steps are necessary: (1) obtain from the publisher a list of each school that uses the book and (2) obtain from each school the number of students enrolled in courses requiring the text. Note that using only the number of books sold to the school by the publisher will miss those students purchasing used copies turned in from previous semesters. "Students" in parts (a)-(e) below is understood to mean students using this book.
   a. Random - conceptually (i.e., without actually doing so) make a list of all N students by assigning to each school one place on the list for each of its students; pick 100 random numbers from 1 to N -- each number identifies a particular school and a conceptual student; for each identified school select a student at random (or more than one student, if that school was identified more than once). In this manner every student has the same chance of being selected -- in fact, every group of 100 has the same chance of being selected, and the sample is a simple random sample.
   b. Systematic - place the n schools in a list, pick a random number between 1 and (n/10) to determine a starting selection, select every $(n/10)^{th}$ school; at each of the 10 systematically selected schools, randomly (or systematically) select and survey 10 students.
   c. Convenience - randomly (or conveniently) select and survey 100 students from your school.
   d. Stratified - divide the schools into categories (e.g., 2 year, 4 year private, 4 year public, proprietary) and select 1 school from each category; at each school randomly (or by a stratified procedure) select and survey the number of students equal to the percent of the total students that are in that category.
   e. Cluster - select a school at random and survey all the students (or cluster sample again by grouping the students in classes and selecting a class at random in which to survey

all the students); repeat (i.e., select another school at random) as necessary until the desired sample of 100 is obtained.

29. No. Even though every state has the same chance of being selected, not every voter has the same chance of being selected. Once a state is selected, any particular registered voter in a small state is competing against fewer peers and has a higher chance of being selected than any particular registered voter in a large state.

31. a. The two samples are self-selected, and the participants were not randomly assigned to their respective groups. It appears that the researchers asked a group of drivers without cell phones to participate, and then approached a group of drivers who already had cell phones. It could be that drivers who self-select to have cell phones are anxious types that would have more accidents anyway.
     b. Those with cell phones were specifically asked to use the phones while driving. This is not likely to produce the same results as "normal" usage – not to mention the fact that knowing they are part of a study will likely put them on their best behavior and not represent their typical conduct in such a situation.

Review Exercises

1. No, the responses cannot be considered representative of the population of the United States for at least the following reasons.
    a. Only AOL subscribers were contacted. Those who are not AOL subscribers, not to mention those with no Internet connections at all, will not be represented. This will introduce a significant bias – since non-Internet users might to be the less affluent, less educated and less business-oriented.
    b. The sample was self-selected. The responses represent only those who had enough time and/or interest in the question. Typically, those with strong feelings one way or the other are more likely to respond and will be over-represented in the results.

2. Let N be the total number of full-time students and n be the desired sample size.
    a. Random. Obtain a list of all N full-time students, number the students from 1 to N, select n random numbers from 1 to N, and poll each student whose number on the list is one of the random numbers selected.
    b. Systematic. Obtain a list of all N full-time students, number the students from 1 to N, let m be the largest integer less than the fraction N/n, select a random number between 1 and m, begin with the student whose number is the random number selected, and poll that student and every mth student thereafter.
    c. Convenience. Select a location (e.g., the intersection of major campus walkways) by which most of the students usually pass, and poll the first n full-time students that pass.
    d. Stratified. Obtain a list of all N full-time students and the gender of each, divide the list by gender, and randomly select and poll n/2 students from each gender.
    e. Cluster. Obtain a list of all the classes meeting at a popular time (e.g., 10 am Monday), estimate how many of the classes would be necessary to include n students, select that many of the classes at random, and poll all of the students in each selected class.

3. a. Ratio, since differences are valid and there is a meaningful 0.
    b. Ordinal, since there is a hierarchy but differences are not valid.
    c. Nominal, since the categories have no meaningful inherent order.
    d. Interval, since differences are valid but the 0 is arbitrary.

4. a. Discrete, since the number of shares held must be an integer.
    NOTE: Even if partial shares are allowed (e.g., 5½ shares), the number of shares must be some fractional value and not any value on a continuum -- e.g., a person could not own π shares.
    b. Ratio, since differences between values are consistent and there is a natural zero.
    c. Stratified, since the set of interest (all stockholders) was divided into subpopulations (by states) from which the actual sampling was done.
    d. Statistic, since the value is determined from a sample and not the entire population.
    e. There is no unique correct answer, but the following are reasonable possibilities.
    (1) The proportion of stockholders holding above that certain number of shares (which would vary from company to company) that would make them "influential." (2) The proportion of stockholders holding below that certain number of shares (which would vary from company to company) that would make them "insignificant." (3) The numbers of shares (and hence the degree of influence) held by the largest stockholders.
    f. There are several possible valid answers. (1) The results would be from a self-selected group (i.e., those who chose to respond) and not necessarily a representative group. (2) If the questionnaire did not include information on the numbers of shares owned, the views of small stockholders (who are probably less knowledgeable about business and stocks) could not be distinguished from those of large stockholders (whose views should carry more weight).

5. a. Systematic, since the selections are made at regular intervals.
    b. Convenience, since those selected were the ones who happened to attend.
    c. Cluster, since the stockholders were organized into groups (by stockbroker) and all the stockholders in the selected groups were chosen.
    d. Random, since each stockholder has the same chance of being selected.
    e. Stratified, since the stockholders were divided into subpopulations from which the actual sampling was done.

6. a. Blinding occurs when those involved in an experiment (either as subjects or evaluators) do not know whether they are dealing with a treatment or a placebo. It might be used in this experiment by (a) not telling the subjects whether they are receiving Sleepeze or the placebo and/or (b) not telling any post-experiment interviewers or evaluators which subjects received Sleepeze and which ones received the placebo. Double-blinding occurs when neither the subjects nor the evaluators know whether they are dealing with a treatment or a placebo.
    b. The data reported will probably involve subjective assessments (e.g., "On a scale of 1 to 10, how well did it work?") that may be subconsciously influenced by whether the subject was known to have received Sleepeze or the placebo.
    c. In a completely randomized block design, subjects are assigned to the groups (in this case to receive Sleepeze or the placebo) at random.
    d. In a rigorously controlled block design, subjects are assigned to the groups (in this case to receive Sleepeze or the placebo) in such a way that the groups are similar with respect to extraneous variables that might affect the outcome. In this experiment it may be important to make certain each group has approximately the same age distribution, degree of insomnia, number of males, number users of alcohol and/or tobacco, etc.
    e. Replication involves repeating the experiment on a sample of subjects large enough to ensure that atypical responses of a few subjects will not give a distorted view of the true situation.

Cumulative Review Exercises

NOTE: Throughout the text intermediate mathematical steps will be shown as an aid to those who may be having difficulty with the calculations. In practice, most of the work can be done continuously on calculators and the intermediate values are unnecessary. Even when the calculations cannot be done continuously, DO NOT WRITE AN INTERMEDIATE VALUE ON YOUR PAPER AND THEN RE-ENTER IT IN THE CALCULATOR. That practice can introduce round-off errors and copying errors. Store any intermediate values in the calculator so that you can recall them with infinite accuracy and without copying errors.

1. $$\frac{169.1+144.2+179.3+178.5+152.6+166.8+135.0+201.5+175.2+139.0}{10} = \frac{1638.5}{10}$$
$$= 163.85$$

2. $$\frac{98.20 - 98.60}{.62} = \frac{-.40}{.62} = -.645$$

3. $$\frac{98.20 - 98.60}{0.62/\sqrt{106}} = \frac{-.40}{.0602} = -6.642$$

4. $$\left[\frac{(1.96)(15)}{2}\right]^2 = \left[\frac{29.4}{2}\right]^2 = [14.7]^2 = 216.09$$

5. $$\sqrt{\frac{(5-7)^2 + (12-7)^2 + (4-7)^2}{3-1}} = \sqrt{\frac{(-2)^2 + (5)^2 + (-3)^2}{2}} = \sqrt{\frac{4 + 25 + 9}{2}} = \sqrt{\frac{38}{2}}$$
$$= \sqrt{19} = 4.359$$

6. $$\frac{(183 - 137.09)^2}{137.09} + \frac{(30 - 41.68)^2}{41.68} = \frac{(45.91)^2}{137.09} + \frac{(-11.68)^2}{41.68} = \frac{2107.7281}{137.09} + \frac{136.4224}{41.68}$$
$$= 15.375 + 3.273 = 18.647$$

7. $$\sqrt{\frac{10(513.27) - 71.5^2}{10(9)}} = \sqrt{\frac{5132.7 - 5112.25}{90}} = \sqrt{\frac{20.45}{90}} = \sqrt{.2272} = .477$$

8. $$\frac{8(151,879) - (516.5)(2176)}{\sqrt{8(34,525.75) - 516.5^2}\sqrt{8(728,520) - 2176^2}} = \frac{1215032 - 1123904}{\sqrt{9433.75}\sqrt{1093184}} = \frac{91128}{1015522} = .897$$

9. $0.95^{500} = 7.27E\text{-}12 = 7.27\cdot10^{-12}$
$= .00000000000727$; moving the decimal point left 12 places

NOTE: Calculators and computers vary in their representation of such numbers. This manual assumes they will be given in scientific notation as a two-decimal number between 1.00 and 9.99 inclusive followed by an indication (usually E for *exponent* of the multiplying power of ten) of how to adjust the decimal point to obtain a number in the usual notation (rounded to three significant digits).

10. $8^{14} = 4.40E+12 = 4.40\cdot10^{12}$
$= 4,400,000,000,000$; moving the decimal point right 12 places

11. $9^{12} = 2.82E+11 = 2.82\cdot10^{11}$
$= 282,000,000,000$; moving the decimal point right 11 places

12. $.25^{17} = 5.82E\text{-}11 = 5.82\cdot10^{-11}$
$= .0000000000582$; moving the decimal point left 11 places

Chapter 2

Describing, Exploring, and Comparing Data

2-2 Frequency Distributions

1. Subtracting the first two consecutive lower class limits indicates that the class width is 100 - 90 = 10. Since there is a gap of 1.0 between the upper class limit of one class and the lower class limit of the next, class boundaries are determined by increasing or decreasing the appropriate class limits by (1.0)/2 = 0.5. The class boundaries and class midpoints are given in the table below.

pressure	class boundaries	class midpoint	frequency
90 - 99	89.5 - 99.5	94.5	1
100 - 109	99.5 - 109.5	104.5	4
110 - 119	109.5 - 119.5	114.5	17
120 - 129	119.5 - 129.5	124.5	12
130 - 139	129.5 - 139.5	134.5	5
140 - 149	139.5 - 149.5	144.5	0
150 - 159	149.5 - 159.5	154.5	1
			40

NOTE: Although they often contain extra decimal points and may involve consideration of how the data were obtained, class boundaries are the key to tabular and pictorial data summaries. Once the class boundaries are obtained, everything else falls into place. Here the first class width is readily seen to be 99.5 - 89.5 = 10.0 and the first midpoint is (89.5 + 99.5)/2 = 94.5. In this manual, class boundaries will typically be calculated first and then used to determine other values. In addition, the sum of the frequencies is an informative number used in many subsequent calculations and will be shown as an integral part of each table.

3. Since the gap between classes as presented is 1.0 the appropriate class limits are increased or decreased by (1.0)/2 = .05 to obtain the class boundaries and the following table.

cholesterol	class boundaries	class midpoint	frequency
0 - 199	-0.5 - 199.5	99.5	13
200 - 399	199.5 - 399.5	299.5	11
400 - 599	399.5 - 599.5	499.5	5
600 - 799	599.5 - 799.5	699.5	8
800 - 999	799.5 - 999.5	899.5	2
1000 - 1199	999.5 - 1199.5	1099.5	0
1200 - 1399	1199.5 - 1399.5	1299.5	1
			40

The class width is 199.5 - (-0.5) = 200; the first midpoint is (-0.5 + 199.5)/2 = 99.5.

5. The relative frequency for each class is found by dividing its frequency by 40, the sum of the frequencies. NOTE: As before, the sum is included as an integral part of the table. For relative frequencies, this should always be 1.000 (i.e., 100%) and serves as a check for the calculations.

pressure	relative frequency
90 - 99	.025
100 - 109	.100
110 - 119	.425
120 - 129	.300
130 - 139	.125
140 - 149	.000
150 - 159	.025
	1.000

7. The relative frequency for each class is found by dividing its frequency by 40, the sum of the frequencies. NOTE: In #5, the relative frequencies were expressed as decimals; here they are expressed as percents. The choice is arbitrary.

cholesterol	relative frequency
0 - 199	.325
200 - 399	.275
400 - 599	.125
600 - 799	.200
800 - 999	.050
1000 -1199	.000
1200 -1399	.025
	1.000

9. The cumulative frequencies are determined by repeated addition of successive frequencies to obtain the combined number in each class and all previous classes. NOTE: Consistent with the emphasis that has been placed on class boundaries, we choose to use upper class boundaries in the "less than" column. Conceptually, pressures occur on a continuum and the integer values reported are assumed to be the nearest whole number representation of the precise measure. An exact pressure of 99.7, for example, would be reported as 100 and fall in the second class. The values in the first class, therefore, are better described as being "less than 99.5" (using the upper class boundary) than as being "less than 100." This distinction becomes crucial in the construction of pictorial representations in the next section. In addition, the fact that the final cumulative frequency must equal the total number (i.e, the sum of the frequency column) serves as a check for calculations. The sum of cumulative frequencies, however, has absolutely no meaning and is not included.

(#9) pressure	cumulative frequency
less than 99.5	1
less than 109.5	5
less than 119.9	22
less than 129.5	34
less than 139.5	39
less than 149.5	39
less than 159.5	40

11. The cumulative frequencies are determined by repeated addition of successive frequencies to obtain the combined number in each class and all previous classes. NOTE: Consistent with the emphasis that has been placed on class boundaries, we choose to use upper class boundaries in the "less than" column.

(#11) cholesterol	cumulative frequency
less than 199.5	13
less than 399.5	24
less than 599.5	29
less than 799.5	37
less than 999.5	39
less than 1199.5	39
less than 1399.5	40

13. The relative frequencies are determined by dividing the given frequencies by 200, the sum of the given frequencies. The fact that the sum of the relative frequencies is 1.000 provides a check to the arithmetic. For a fair die, we expect each relative frequency to be close to 1/6 = .167. As these relative frequencies all fall between .135 and .210, they do not appear to differ significantly from the values expected for a fair die.

(#13) outcome	relative frequency
1	.135
2	.155
3	.210
4	.200
5	.140
6	.160
	1.000

15. For a lower class limit of 0 for the first class and a class width of 50, the frequency distribution is given at the right.
NOTE: The class limits for the first class are 0-49 and not 0-50.

weight (lbs)	frequency
0 - 49	6
50 - 99	10
100 - 149	10
150 - 199	7
200 - 249	8
250 - 299	2
300 - 349	4
350 - 399	3
400 - 449	3
450 - 499	0
500 - 549	1
	54

17. The separate frequency distributions are given below.

male head circumference	frequency
34.0 - 35.9	2
36.0 - 37.9	0
38.0 - 39.9	5
40.0 - 41.9	29
42.0 - 43.9	14
	50

female head circumference	frequency
34.0 - 35.9	1
36.0 - 37.9	3
38.0 - 39.9	14
40.0 - 41.9	27
42.0 - 43.9	5
	50

It appears that the head circumferences tend to be larger for baby boys than for baby girls.

19. The separate relative frequency distributions are given below, to the right of the figure giving the actual frequencies. The relative frequencies were obtained by dividing the actual frequencies for each gender by the total frequencies for that gender.

M	age	F	male ages	relative frequency	female ages	relative frequency
11	19-28	8	19-28	.099	19-28	.205
43	29-38	18	29-38	.387	29-38	.462
31	39-48	4	39-48	.279	39-48	.103
22	49-58	7	49-58	.198	49-58	.179
4	59-68	2	59-68	.036	59-68	.051
111		39		1.000		1.000

Both genders have more runners in the 29-38 age group than in any other ten year spread. But the second most populous category is the one above that for the males, and the one below that for the females. It appears that the male runners tend to be slightly older than the female runners.

21. Assuming that "start the first class at 200 lb" refers to the first lower class limit produces the frequency table given at the right.

weight (lbs)	frequency
200 - 219	6
220 - 239	5
240 - 259	12
260 - 279	36
280 - 299	87
300 - 319	28
320 - 339	0
340 - 359	0
360 - 379	0
380 - 399	0
400 - 419	0
420 - 439	0
440 - 459	0
460 - 479	0
480 - 499	0
500 - 519	1
	175

In general, an outlier can add several rows to a frequency table. Even though most of the added rows have frequency zero, the table tends to suggest that these are possible valid values – thus distorting the reader's mental image of the distribution.

2-3 Visualizing Data

1. The answer depends upon what is meant by "the center." Since the ages seem to range from about 10 to about 70, 40 could be called the center value - in the sense that it is half way between the lowest and highest points on the horizontal axis that represent observed data. Since the ages are concentrated at the lower end of the scale, an age near 24 could be called the center value - in the sense that about ½ of the 131 ages appear to be below 24 and about ½ of the ages appear to be above 24. Or an age near 26 could be called the center value - in the sense that a fulcrum placed under 26 on the horizontal would appears to nearly "balance" the histogram.

3. The percentage younger than 30 is about $(1+38+38+16)/131 = 93/131 = .710 = 71.0\%$.

5. Since, the shading representing Group A covers about 40% of the "pie," the approximate percentage of people with Group A blood is 40%. If the chart is based on a sample of 500 people, approximately $(.40)(500) = 200$ people had Group A blood.

7. Obtain the relative frequencies by dividing each frequency by the total frequency for each sample. The two relative frequency histograms are given, using the same scale, below at the right. Each sample distribution spreads out in both directions from a "typical" (or most frequently occurring value), but the faculty data appears to be shifted to the left by about one interval. Since each interval represents 3 years, the faculty cars are about 3 years newer than the student cars. In addition, the majority of the student car ages fall below the class with the highest frequency, while faculty car ages tend to occur above their most populous class.
    NOTE: The class boundaries in the histograms are -0.5, 2.5, 5.5, etc. Assuming the car ages were obtained by subtracting the model year from the calendar year, that boundary scheme is consistent with the new models being introduced ½ year "early" and sold at a steady rate during the calendar year. In truth, this data is difficult to present in a graph that accounts for the underlying continuity of the ages. The figures given below are probably the best for the level of this text.

STUDENTS

age	frequency	relative frequency
0- 2	23	.106
3- 5	33	.152
6- 8	63	.290
9-11	68	.313
12-14	19	.088
15-17	10	.046
18-20	1	.005
21-23	0	.000
	217	1.000

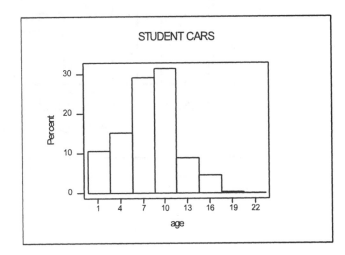

FACULTY

age	frequency	relative frequency
0- 2	30	.197
3- 5	47	.309
6- 8	36	.237
9-11	30	.197
12-14	8	.053
15-17	0	.000
18-20	0	.000
21-23	1	.007
	152	1.000

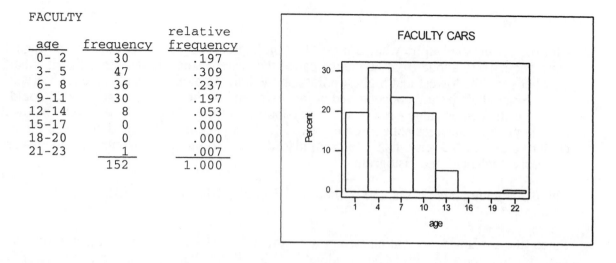

9. See the figure below. The class boundaries are -0.5,49.5,99.5,149.5,...549.5. The bars extend from class boundary to class boundary. For visual simplicity, the horizontal axis has been labeled 0,50,150,etc. The "center" of the weights (i.e., the "balance point" of the histogram) appears to be about 190.

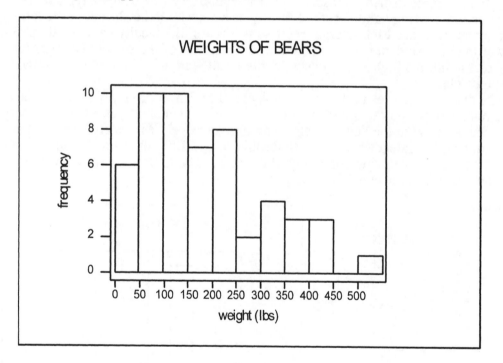

11. See the figure below. The frequencies are plotted above the class midpoints, and "extra" midpoints are added so that both polygons begin and end with a frequency of zero. The females are represented by the solid line, and the males by the dashes.

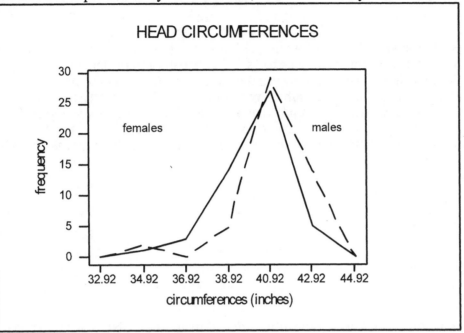

Both polygons have approximately the same shape, but the male circumferences appear to be slightly larger.

13. See the figures below. Since ages are reported as of the last birthday, the class boundaries are 19,29,39,49,59,69 – i.e., the first class in each figure includes all ages from 19.000 to 28.999.

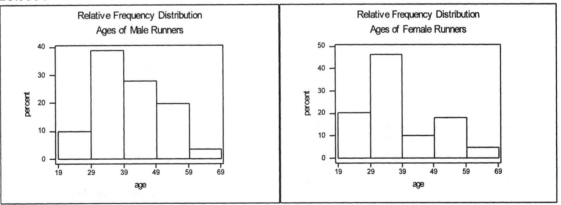

The two histograms are almost identical, except for the "dip" at age 39-49 for the females. It could be that physical or family conditions that typically occur about that age for females are not conducive to running marathons. Because of that dip, it appears that the male runners are slightly older than their female counterparts.

15. The original numbers are listed by the row in which they appear in the stem-and-leaf plot.

```
stem leaves        original numbers
  20 0005          200, 200, 200, 205
  21 69999         216, 219, 219, 219, 219
  22 2233333       222, 222, 223, 223, 223, 223, 223
  23
  24 1177          241, 241, 247, 247
```

17.   The dotplot is constructed using the original scores as follows. Each space represents 1 unit.

```
        :               :       :
        :               :      ::
      --:----:------T-----:--:------T----------:----:----T--
       200      210     220      230        240     250
```

19.   The expanded stem-and-leaf plot below on the left is one possibility. NOTE: The text claims that stem-and-leaf plots enable us to "see the distribution of data and yet keep all the information in the original list." Following the suggestion to round the nearest inch not only loses information but also uses subjectivity to round values exactly half way between. Since always rounding such values "up" creates a slight bias, many texts suggest rounding toward the even digit – so that 33.5 becomes 34, but 36.5 becomes 36. The technique below of using superscripts to indicate the occasional decimals is both mathematically clear and visually uncluttered.

```
stem | leaves
   3 | 6  7
   4 | 0  0  1  3  3⁵
   4 | 6  6  7  8  8  9
   5 | 0  2  2⁵  3  3  4
   5 | 7³  7⁵  8  9  9  9
   6 | 0  0⁵  1  1  1⁵  2  3  3  3  3⁵  4  4  4
   6 | 5  5  6⁵  7  7⁵  8⁵
   7 | 0  0⁵  2  2  2  2  3  3⁵
   7 | 5  6⁵
```

21.   See the figure below, with bars arranged in order of magnitude. Networking appears to be the most effective job-seeking approach.

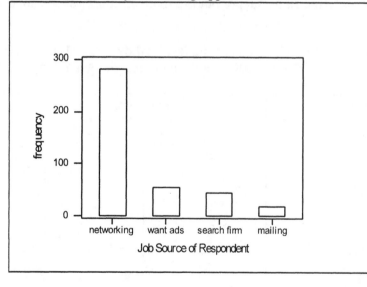

23.   See the figure at the right. The sum of the frequencies is 50. The relative frequencies are:
$23/50 = 46\%$ $9/50 = 18\%$
$12/50 = 24\%$ $6/50 = 12\%$.
The corresponding central angles are:
$(.46)360° = 165.6°$
$(.18)360° = 64.8°$
$(.24)360° = 86.4°$
$(.12)360° = 43.2°$.
To be complete, the figure needs to be titled with the name of the quantity being measured.

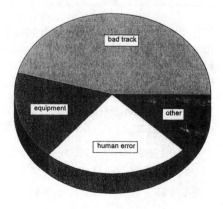

Causes of Train Derailments

25.   The scatter diagram is given below. The figure should have a title, and each axis should
       be labeled both numerically and with the name of the variable. An "x" marks a single
       occurrence, while numbers indicate multiple occurrences at a point. Cigarettes high in tar
       also tend to be high in CO. The points cluster about a straight line from (0,0) to (18,18),
       indicating that the mg of CO tends to be about equal to the mg of tar.

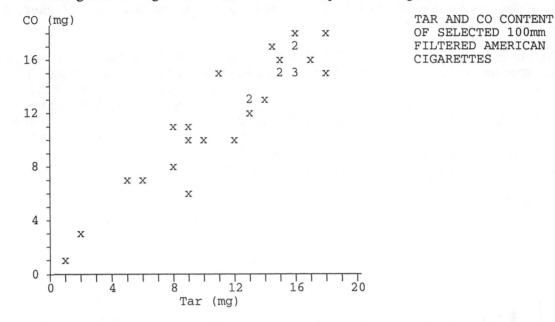

27.   The time series graph is given below. An investor could use this graph to project where
       the market might be next year and invest accordingly. The flat portion at the right might
       be suggesting that the market has reached a point at which it will level off for a while –
       and that the rapid growth of the last several years may be at end.

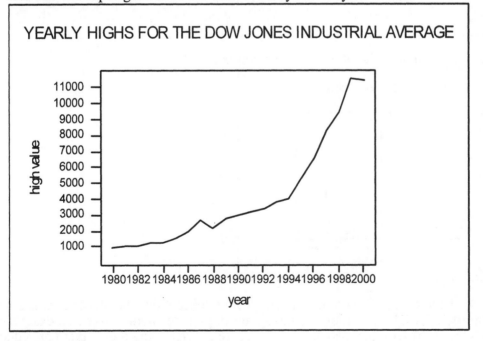

29.   According to the figure, 422,000 started and 10,000 returned.
       10,000/422,000 = 2.37%

31.   The figure indicates the number of men had just dropped to 37,000 on November 9 when
       the temperature was 16°F (-9°C), and had just dropped to 24,000 on November 14 when
       the temperature was -6°F (-21°C). The number who died during that time, therefore, was
       37,000 - 24,000 = 13,000.

33.   NOTE: Exercise #20 of section 2-2 dealt with frequency table representations of this data
       and specified a class width of 20. Using a class width with an odd number of units of
       measure allows the class midpoint to have the same number of decimal places as the
       original data and often produces more appealing visual representations. Here we use a
       class width of 25 – with class midpoints of 200, 225, etc. – and so the figures will differ
       from ones made using the previous classes. The histogram bars extend from class
       boundary to class boundary (i.e., from 187.5 to 212.5 for the first class), but for
       convenience the labels have been placed at the class midpoints.

   a. The histogram with the outlier is as shown below.

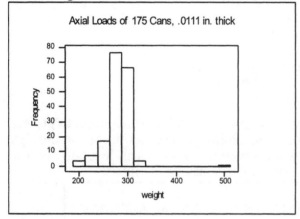

   b. The histogram without the outlier is as shown below.

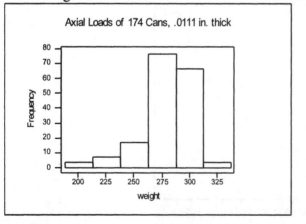

   c. The basic shape of the histogram does not change, except that a distant piece has been
      "broken off." NOTE: The redrawn histogram in part (b) should not be an exact copy of
      the one in part (a) with the distant bar erased. Since removing the distant bar reduces
      the effective width of the figure significantly, the rule that the height should be
      approximately 3/4 of the width requires either making the remaining bars wider (and
      keeping the figure's height) or making them shorter (and keeping the figure's reduced
      width). To do otherwise produces a figure too tall for its width – and one that tends to
      visually overstate the differences between classes.

2-4 Measures of Center

NOTE: As it is common in mathematics and statistics to use symbols instead of words to represent quantities that are used often and/or that may appear in equations, this manual employs symbols for the measures of central tendency as follows:

mean = $\bar{x}$ mode = M
median = $\tilde{x}$ midrange = m.r.

Also, this manual generally follows the author's guideline of presenting means, medians and midranges accurate to one more decimal place than found in the original data. The mode, the only measure which must be one of the original pieces of data, is presented with the same accuracy as the original data. This manual will, however, recognize the following two exceptions to these guidelines.
(1) When there are an odd number of data, the median will be one of the original values and will not be reported to one more decimal place – so the median of 1,2,3,4,5 would be reported as 3.
(2) When the mean falls exactly halfway between two values that meet the guidelines, an extra decimal place ending in 5 will be used – so the mean of 1,2,3,3 would be reported as 9/4=2.25.

1. Arranged in order, the 6 scores are: 0 0 0 176 223 548
   a. $\bar{x} = (\Sigma x)/n = (947)/6 = 157.8$ c. M = 0
   b. $\tilde{x} = (0 + 176)/2 = 88.0$ d. m.r. = (0 + 548)/2 = 274.0
   Whether or not there is a problem with scenes of tobacco use in animated children's films depends on the scene. If a likeable character uses tobacco without any criticism or ill effects, this would suggest to children that such usage is appropriate and acceptable. In most children's films, however, tobacco usage is typically associated with evil and undesirable characters and situations – suggesting that tobacco usage is not appropriate or acceptable for decent persons.

NOTE: The median is the middle score when the scores are arranged in order, and the midrange is halfway between the first and last score when the scores are arranged in order. It is therefore usually helpful to begin by placing the scores in order. This will not affect the mean, and it may also aid in identifying the mode. In addition, no measure of central tendency can have a value lower than the smallest score or higher than the largest score – remembering this helps to protect against gross errors, which most commonly occur when calculating the mean.

3. Arranged in order, the 16 scores are:
   .03 .07 .09 .13 .13 .17 .24 .30 .39 .43 .43 .44 .45 .47 .47 .48
   a. $\bar{x} = (\Sigma x)/n = (4.72)/16 = .295$ c. M = .13,.43,.47 [tri-modal]
   b. $\tilde{x} = (.30 + .39)/2 = .345$ d. m.r. = (.03 + .48)/2 = .255
   No; this sample means weights each brand of cereal equally and does not take into account which of the cereals have higher (or lower) rates of consumption.

5. Arranged in order, the 15 scores are:
   .12 .13 .14 .16 .16 .16 .17 .17 .17 .18 .21 .24 .24 .27 .29
   a. $\bar{x} = (\Sigma x)/n = (2.81)/15 = .187$ c. M = .16,.17 [bi-modal]
   b. $\tilde{x} = .17$ d. m.r. = (.12 + .29)/2 = .205
   Yes; these values appear to be significantly above the allowed maximum.

7. Arranged in order, the 20 scores are:
   15 17 17 17 17 17 17 18 18 18 18 18 19 19 19 19 20 21 21 21
   a. $\bar{x} = (\Sigma x)/n = (366)/20 = 18.3$ c. M = 17
   b. $\tilde{x} = (18 + 18)/2 = 18.0$ d. m.r. = (15 + 21)/2 = 18.0
   The sample results appear to be very consistent – i.e., there appears to be little variation from person to person. This suggests that there is little variation in the population, and that the sample mean should be a good estimate of the population mean – in the sense that other samples would likely produce similar results that vary little from this sample.

9. Arranged in order, the scores are as follows.
    JV: 6.5 6.6 6.7 6.8 7.1 7.3 7.4 7.7 7.7 7.7
    Pr: 4.2 5.4 5.8 6.2 6.7 7.7 7.7 8.5 9.3 10.0

Jefferson Valley	Providence
$n = 10$	$n = 10$
$\bar{x} = (\Sigma x)/n = (71.5)/10 = 7.15$	$\bar{x} = (\Sigma x)/n = (71.5)/10 = 7.15$
$\tilde{x} = (7.1 + 7.3)/2 = 7.20$	$\tilde{x} = (6.7 + 7.7)/2 = 7.20$
$M = 7.7$	$M = 7.7$
m.r. $= (6.5 + 7.7)/2 = 7.10$	m.r. $= (4.2 + 10.0)/2 = 7.10$

    Comparing only measures of central tendency, one might suspect the two sets are identical.
    The Jefferson Valley times, however, are considerably less variable.
    NOTE: This is the reason most banks have gone to the single waiting line. While it
    doesn't make service faster, it makes service times more equitable by eliminating the "luck
    of the draw" – i.e., ending up by pure chance in a fast or slow line and having unusually
    short or long waits.

11. Arranged in order, the scores are as follows.
        McDonald's: 92 118 128 136 153 176 192 193 240 254 267 287
    Jack in the Box: 74 109 109 190 229 255 270 300 328 377 428 481

McDonald's	Jack in the Box
$n = 12$	$n = 12$
$\bar{x} = (\Sigma x)/n = (2236)/12 = 186.3$	$\bar{x} = (\Sigma x)/n = (3150)/12 = 262.5$
$\tilde{x} = (176 + 192)/2 = 184.0$	$\tilde{x} = (255 + 270)/2 = 262.5$
$M = $ [none]	$M = 109$
m.r. $= (92 + 287)/2 = 189.5$	m.r. $= (74 + 481)/2 = 277.5$

    McDonald's appears to be faster. Yes; the difference appears to be significant.

13. The following values were obtained for the head circumference data, where x_i indicates the
    i^{th} score from the ordered list.

males	females
$n = 50$	$n = 50$
$\bar{x} = (\Sigma x)/n = 2054.9/50 = 41.10$	$\bar{x} = (\Sigma x)/n = 2002.4/50 = 40.05$
$\tilde{x} = (x_{25}+x_{26})/2$	$\tilde{x} = (x_{25}+x_{26})/2)/2$
$\quad = (41.1+41.1)/2 = 41.10$	$\quad = (40.2+40.2 = 40.20$

    No; on the basis of the means and medians alone, there does not appear to be a significant
    difference between the genders.

15. The following values were obtained for Boston rainfall, where x_i indicates the ith score
    from the ordered list.

Thursday	Sunday
$n = 52$	$n = 52$
$\bar{x} = (\Sigma x)/n = 3.57/52 = .069$	$\bar{x} = (\Sigma x)/n = 3.52/52 = .068$
$\tilde{x} = (x_{26}+x_{27})/2 = (.00 + .00)/2 = .000$	$\tilde{x} = (x_{26}+x_{27})/2 = (.00 + .00)/2 = .000$

    If "it rains more on weekends" refers to the amount of rain, the data do not support the
    claim. The amount of rainfall appears to be virtually the same for Thursday and Sunday.
    If "it rains more on weekends" refers to the frequency of rain (regardless of the amount),
    then the proportions of days on which there was rain would have to be compared.

17.   The x values below are the class midpoints from the given frequency table.

x	f	x·f
44.5	8	356.0
54.5	44	2398.0
64.5	23	1483.5
74.5	6	447.0
84.5	107	9041.5
94.5	11	1039.5
104.5	1	104.5
	200	14870.0

$$\overline{x} = (\Sigma x \cdot f)/n$$
$$= (14870.0)/200$$
$$= 74.35$$

NOTE: The mean time was calculated to be 74.35 minutes. According to the rule given in the text, this value should be rounded to one decimal place. The text describes how many decimal places to present in an answer, but not the actual rounding process. When the figure to be rounded is <u>exactly</u> half-way between two values (i.e., the digit in the position to be discarded is a 5, and there are no further digits because the calculations have "come out even"), there is no universally accepted rounding rule. Some authors say to always round up such a value; others correctly note that always rounding up introduces a consistent bias, and that the value should actually be rounded up half the time and rounded down half the time. And so some authors suggest rounding toward the even value (e.g., .65 becomes .6 and .75 becomes .8), while others simply suggest flipping a coin. In this manual, answers <u>exactly</u> half-way between will be reported without rounding (i.e., stated to one more decimal than usual).

19.   The x values below are the class midpoints from the given frequency table.

x	f	x·f
43.5	25	1087.5
47.5	14	665.0
51.5	7	360.5
55.5	3	166.5
59.5	1	59.5
	50	2339.0

$$\overline{x} = (\Sigma x \cdot f)/n$$
$$= (2339.0)/50$$
$$= 46.8$$

The mean speed of 46.8 mi/hr of those ticketed by the police is more than 1.5 times the posted speed limit of 30 mi/hr. NOTE: This indicates nothing about the mean speed of *all* drivers, a figure which may or may not be higher than the posted limit.

21.   a. Arranged in order, the original 54 scores are:

    26   29   34   40   46   48   60   62   64   65   76   79   80   86   90
    94 105 114 116 120 125 132 140 140 144 148 150 150 154 166
    166 180 182 202 202 204 204 212 220 220 236 262 270 316 332
    344 348 356 360 365 416 436 446 514
    $\overline{x} = (\Sigma x)/n = (9876)/54 = 182.9$

   b. Trimming the highest and lowest 10% (or 5.4=5 scores), the remaining 44 scores are:

    48   60   62   64   65   76   79   80   86   90   94 105 114 116 120
    125 132 140 140 144 148 150 150 154 166 166 180 182 202 202
    204 204 212 220 220 236 262 270 316 332 344 348 356 360
    $\overline{x} = (\Sigma x)/n = (7524)/44 = 171.0$

   c. Trimming the highest and lowest 20% (or 10.8=11 scores), the remaining 32 scores are:

    79   80   86   90   94 105 114 116 120 125 132 140 140 144 148
    150 150 154 166 166 180 182 202 202 204 204 212 220 220 236
    262 270
    $\overline{x} = (\Sigma x)/n = (5093)/32 = 159.2$

In this case, the mean gets smaller as more scores are trimmed. In general, means can increase, decrease, or stay the same as more scores are trimmed. The mean decreased here because the higher scores were farther from the original mean than were the lower scores.

23. If n=10 values have $\bar{x} = (\Sigma x)/n = 75.0$, then $\Sigma x = 750$.
     a. Since the sum of the 9 given values is 698, the 10th value must be 750-698 = 52.
     b. Generalizing the result in part (a), n-1 values can be freely assigned. The nth value must be whatever is needed (whether positive or negative) to reach the sum $(\Sigma x) = n\cdot\bar{x}$ determined by the given mean.

25. The w values below are the weights – which can sum to any value, since the weighted mean is found by dividing by Σw.

x	w	x·w
65	.15	9.75
83	.15	12.45
80	.23	12.00
90	.15	13.50
92	.40	36.80
	1.00	84.50

$$\bar{x} = (\Sigma x\cdot w)/(\Sigma w)$$
$$= (84.50)/(1.00)$$
$$= 84.50$$

27. Let $\bar{x}_h$ stand for the harmonic mean: $\bar{x}_h = n/[\Sigma(1/x)]$
$$= 2/[1/40 + 1/60]$$
$$= 2/[.0417]$$
$$= 48.0$$

29. R.M.S. $= \sqrt{\Sigma x^2/n}$
$$= \sqrt{[(110)^2 + (0)^2 + (-60)^2 + (12)^2]/4}$$
$$= \sqrt{15844/4}$$
$$= \sqrt{3961}$$
$$= 62.9$$

2-5 Measures of Variation

NOTE: Although not given in the text, the symbol R will be used for the range throughout this manual. Remember that the range is the difference between the highest and the lowest scores, and not necessarily the difference between the last and the first values as they are listed. Since calculating the range involves only the subtraction of 2 original pieces of data, that measure of variation will be reported with the same accuracy as the original data.

1.

x	x $-\bar{x}$	$(x-\bar{x})^2$	x^2
0	-157.83	24910.3089	0
0	-157.83	24910.3089	0
0	-157.83	24910.3089	0
176	18.17	330.1489	30976
223	65.17	4247.1289	49729
548	390.17	152232.6289	300304
947	0.02	231540.8834	381009

$\bar{x} = (\Sigma x)/n = 947/6 = 157.83$

$R = 5480 = 548$

by formula 2-4,
$s^2 = \Sigma(x-\bar{x})^2/(n-1)$
$= 231540.8834/5$
$= 46308.17$
$= 46308.2$

by formula 2-5,
$s^2 = [n(\Sigma x^2) - (\Sigma x)^2]/[n(n-1)]$
$= [6(381009) - (947)^2]/[6(5)]$
$= [13888888889245]/[30]$
$= 46308.17$
$= 46308.2$

$s = \sqrt{46308.17} = 215.2$

The times appear to vary widely.

NOTE: When finding the square root of the variance to obtain the standard deviation, use <u>all</u> the decimal places of the variance, and not the rounded value reported as the answer. The best way to do this is either to keep the value on the calculator display or to place it in the memory. Do <u>not</u> copy down all the decimal places and then re-enter them to find the square root, as that could introduce round-off and/or copying errors.

When using formula 2-4, constructing a table having the first three columns shown above helps to organize the calculations and makes errors less likely. In addition, verify that $\Sigma(x-\bar{x}) = 0$ [except for a possible small discrepancy at the last decimal, due to using a rounded value for the mean] before proceeding – if such is not the case, there is an error and further calculation is fruitless. For completeness, and as a check, both formulas 2-4 and 2-5 were used above. In general, only formula 2-5 will be used throughout the remainder of this manual for the following reasons:

    (1) When the mean does not "come out even," formula 2-4 involves round-off error and/or many messy decimal calculations.
    (2) The quantities Σx and Σx^2 needed for formula 2-5 can be found directly and conveniently on the calculator from the original data without having to construct a table like the one above.

3. preliminary values: $n = 16$, $\Sigma x = 4.72$, $\Sigma x^2 = 1.8144$
    $R = .48 - .03 = .45$
    $s^2 = [n(\Sigma x^2) - (\Sigma x)^2]/[n(n-1)] = [16(1.8144) - (4.72)^2]/[16(15)]$
       $= (6.7520)/240 = .028$
    $s = .168$

No; this value is not likely to be a good estimate for the standard deviation among the amounts in each gram consumed by the population. This value gives equal weight to each cereal in the sample, no matter how popular (or unpopular) it is with consumers. The sample is not representative of the consumption patters.

NOTE: The quantity $[n(\Sigma x^2) - (\Sigma x)^2]$ cannot be less than zero. A negative value indicates that there is an error and that further calculation is fruitless. In addition, remember to find the value for s by taking the square root of the precise value of s^2 showing on the calculator display <u>before</u> it is rounded to one more decimal place than the original data.

5. preliminary values: $n = 15$, $\Sigma x = 2.81$, $\Sigma x^2 = .5631$
    $R = .29 - .12 = .17$
    $s^2 = [n(\Sigma x^2) - (\Sigma x)^2]/[n(n-1)] = [15(.5631) - (2.81)^2]/[15(14)]$
       $= (.5504)/210 = .00262$
    $s = .051$

No; the intent is to lower the mean amount. If all drivers had high concentrations (e.g., between .35 and .30), there would be a low s – but a high level of drunk driving.
NOTE: Following the usual round-off rule of giving answers with one more decimal than the original data produces a variance of .003, which has only one significant digit.
This is an unusual situation occurring because values less than 1.0 become smaller when they are squared. Since the original data had 2 significant digits, we provide 3 significant digits for the variance.

7. preliminary values: $n = 20$, $\Sigma x = 366$, $\Sigma x^2 = 6746$
    $R = 21 - 15 = 6$
    $s^2 = [n(\Sigma x^2) - (\Sigma x)^2]/[n(n-1)] = 20(6746) - (366)^2]/[20(19)] = (964)/380 = 2.5$
    $s = 1.6$

The fact that both R and s are small (relative to the recorded data values) means that reaction times tended to be consistent and showed little variation.

9.   <u>Jefferson Valley</u>   <u>Providence</u>

$n = 10$, $\Sigma x = 71.5$, $\Sigma x^2 = 513.27$ $n = 10$, $\Sigma x = 71.5$, $\Sigma x^2 = 541.09$

$R = 7.7 - 6.5 = 1.2$ $R = 10.0 - 4.2 = 5.8$

$s^2 = [n(\Sigma x^2) - (\Sigma x)^2]/[n(n-1)]$ $s^2 = [n(\Sigma x^2) - (\Sigma x)^2]/[n(n-1)]$

   $= [10(513.27) - (71.5)^2]/[10(9)]$ $= [10(541.09) - (71.5)^2]/[10(9)]$

   $= 20.45/90 = 0.23$ $= 298.65/90 = 3.32$

  $s = 0.48$ $s = 1.82$

Exercise #9 of section 2-4 indicated that the mean waiting time was 7.15 minutes at each bank.   The Jefferson Valley waiting times, however, are considerably less variable.   The range measures the difference between the extremes.   The longest and shortest waits at Jefferson Valley differ by a little over 1 minute (R=1.2), while the longest and shortest waits at Providence differ by almost 6 minutes (R=5.8).   The standard deviation measures the typical difference from the mean.   A Jefferson Valley customer usually receives service within about ½ minute (s=0.48) of 7.15 minutes, while a Providence customer usually receives service within about 2 minutes (s=1.82) of the mean.

11.   <u>McDonald's</u>   <u>Jack in the Box</u>

$n = 12$, $\Sigma x = 2236$, $\Sigma x^2 = 461540$ $n = 12$, $\Sigma x = 3150$, $\Sigma x^2 = 1009962$

$R = 287 - 92 = 195$ $R = 481 - 74 = 407$

$s^2 = [n(\Sigma x^2) - (\Sigma x)^2]/[n(n-1)]$ $s^2 = [n(\Sigma x^2) - (\Sigma x)^2]/[n(n-1)]$

   $= [12(461540) - (2236)^2]/[12(11)]$ $= [12(1009962) - (3150)^2]/[12(11)]$

   $= 538784/132 = 4081.7$ $= .2197044/132 = 16644.3$

  $s = 63.9$ $s = 129.0$

There appears to be about twice the variability among the times at Jack in the Box as there is at McDonald's.   In other words, McDonald's seems to be doing better at producing a uniform system for dealing with all orders.

13.   <u>males</u>   <u>females</u>

$n = 50$, $\Sigma x = 2054.9$, $\Sigma x^2 = 84562.23$ $n = 50$, $\Sigma x = 2002.4$, $\Sigma x^2 = 80323.82$

$s^2 = [n(\Sigma x^2) - (\Sigma x)^2]/[n(n-1)]$ $s^2 = [n(\Sigma x^2) - (\Sigma x)^2]/[n(n-1)]$

   $= [50(84562.23) - (2054.9)^2]/[50(49)]$ $= [50(80323.82) - (2002.4)^2]/[50(49)]$

   $= 5497.49/2450 = 2.24$ $= 6585.24/2450 = 2.69$

  $s = 1.50$ $s = 1.64$

While there was slightly more variability among the female values in this particular sample, there does not appear to be a significant difference between the genders.

15.   <u>Thursday</u>   <u>Sunday</u>

$n = 52$, $\Sigma x = 3.57$, $\Sigma x^2 = 1.6699$ $n = 52$, $\Sigma x = 3.52$, $\Sigma x^2 = 2.2790$

$s^2 = [n(\Sigma x^2) - (\Sigma x)^2]/[n(n-1)]$ $s^2 = [n(\Sigma x^2) - (\Sigma x)^2]/[n(n-1)]$

   $= [52(1.6699) - (3.57)^2]/[52(51)]$ $= [52(2.2790) - (3.52)^2]/[52(51)]$

   $= 74.0899/2652 = .02793$ $= 106.1176/2652 = .04001$

  $s = .167$ $s = .200$

The values are close, although there may be slightly more variation among the amounts for Sundays than among the amounts for Thursdays.

17.

x	f	f·x	f·x²
44.5	8	356.0	15842.00
54.5	44	2398.0	130691.00
64.5	23	1483.5	95685.75
74.5	6	447.0	33301.50
84.5	107	9041.5	764006.75
94.5	11	1039.5	98232.75
104.5	1	104.5	10920.25
	200	14870.0	1148680.00

$s^2 = [n(\Sigma f \cdot x^2) - (\Sigma f \cdot x)^2]/[n(n-1)]$
$= [200(1148680.00) - (1487.0)^2]/[200(199)] = (8619100.00)/29800 = 216.56$
$s = 14.7$

19.

x	f	f·x	f·x²
43.5	25	1087.5	47305.25
47.5	14	665.0	31587.50
51.5	7	360.5	18565.75
55.5	3	166.5	9240.75
59.5	1	59.5	3540.25
	50	2339.0	110239.50

$s^2 = [n(\Sigma f \cdot x^2) - (\Sigma f \cdot x)^2]/[n(n-1)]$
$= [50(110239.50) - (2339.0)^2]/[50(49)] = (41054.00)/2450 = 16.76$
$s = 4.1$

21. Answers will vary. Assuming that the ages range from 25 to 75, the Range Rule of Thumb suggests $s \approx$ range/4 $= (75-25)/4 = 50/4 = 12.5$.

23. Given $\bar{x} = 38.86$ and $s = 3.78$, the Range Rule of Thumb suggests
minimum "usual" value $= \bar{x} - 2s = 38.86 - 2(3.78) = 38.86 - 7.56 = 31.30$
maximum "usual" value $= \bar{x} + 2s = 38.86 + 2(3.78) = 38.86 + 7.56 = 46.42$
Yes, in this context a length of 47.0 cm would be considered unusual.

25. a. The limits 61.1 and 66.1 are 1 standard deviation from the mean. The Empirical Rule for Data with a Bell-shaped Distribution states that about 68% of the heights should fall within those limits.
b. The limits 56.1 and 76.1 are 3 standard deviations from the mean. The Empirical Rule for Data with a Bell-shaped Distribution states that about 99.7% of the heights should fall within those limits.

27. The limits 58.6 and 68.6 are 2 standard deviations from the mean. Chebyshev's Theorem states that there must be at least $1 - 1/k^2$ of the scores within k standard deviations of the mean. Here $k = 2$, and so the proportion of the heights within those limits is at least $1 - 1/2^2 = 1 - 1/4 = 3/4 = 75\%$.

29. The following values were obtained using computer software. Using the accuracy for $\bar{x}$ and s given by the computer, we calculate the coefficient of variation to the nearest .01%.

calories	grams of sugar
n = 16	n = 16
$\bar{x} = 3.7625$	$\bar{x} = .2950$
s = .2217	s = .1677
CV = $s/\bar{x}$	CV = $s/\bar{x}$
= .2217/3.7625	= .1677/.2950
= .0589 = 5.89%	= .5685 = 56.85%

Relative to their mean amounts, there is considerably more variability in the grams of sugar than in the calories. There are two possible explanations:
(1) The amount of calories may be easier to control than the weight of sugar.
(2) The weights of sugar are so small that any slight deviation seems large by comparison.

31.   A standard deviation of s = 0 is possible only when $s^2 = 0$, and $s^2 = \Sigma(x-\bar{x})^2/(n-1) = 0$ only when $\Sigma(x-\bar{x})^2 = 0$. Since each $(x-\bar{x})^2$ is non-negative, $\Sigma(x-\bar{x})^2 = 0$ only when every $(x-\bar{x})^2 = 0$ – i.e., only when every x is equal to $\bar{x}$. In simple terms, zero variation occurs only when all the scores are identical.

33.   The Everlast brand is the better choice. In general, a smaller standard deviation of lifetimes indicates more consistency from battery to battery – signaling a more dependable production process and a more dependable final product. Assuming a bell-shaped distribution of lifetimes, for example, that empirical rule states that about 68% of the lifetimes will fall within one standard deviation of the mean. Here, those limits would be
    for Everlast: 50 ± 2 or 48 months to 52 months
    for Endurance: 50 ± 6 or 44 months to 56 months
While a person might be lucky and purchase a long-lasting Endurance battery, an Everlast battery is much more likely to last for the advertised 48 months.

35.   section 1
    $n = 11, \Sigma x = 201, \Sigma x^2 = 4001$
    $R = 20 - 1 = 19$
    $s^2 = [n(\Sigma x^2) - (\Sigma x)^2]/[n(n-1)]$
       $= [11(4001) - (201)^2]/[11(10)]$
       $= 3610/110 = 32.92$
    $s = 5.7$

    section 2
    $n = 11, \Sigma x = 119, \Sigma x^2 = 1741$
    $R = 19 - 2 = 17$
    $s^2 = [n(\Sigma x^2) - (\Sigma x)^2]/[n(n-1)]$
       $= [11(1741) - (119)^2]/[11(10)]$
       $= 1990/110 = 45.36$
    $s = 6.7$

The range values give the impression that section 1 had more variability than section 2. The range can be misleading because it is based only on the extreme scores. In this case, the lowest score in section 1 was so distinctly different from the others that to include it in any measure trying to give a summary about the section as a whole would skew the results. For the mean, where the value is only one of 11 used in the calculation, the effect is minimal; for the range, where the value is one of only 2 used in the calculation, the effect is dramatic. The standard deviation values give the impression that section 2 had slightly more variability.
NOTE: In this case, section 2 seems considerably more variable (or diverse), and even the standard deviation by itself fails to accurately distinguish between the sections.

37.   The following values were obtained for the smokers in Table 2-1 using computer software. All subsequent calculations use the accuracy of these values.
    $n = 40$ $\bar{x} = 172.5$ $s = 119.5$
    The signal-to-noise ratio is
    $\bar{x}/s = 172.5/119.5 = 1.44$.

38.   The following values were obtained for the smokers in Table 2-1 using computer software. All subsequent calculations use the accuracy of these values.
    $n = 40$ $\bar{x} = 172.5$ $\tilde{x} = 170.0$ $s = 119.5$
    Pearson's index is skewness is
    $I = 3(\bar{x} - \tilde{x})/s = 3(172.5 - 170.0)/119.5 = 3(2.5)/119.5 = .063$
Since $-1.00 < I < 1.00$ for these data, there is not significant skewness.

39.   The most spread, and hence the largest standard deviation, occurs when the scores are evenly divided between the minimum and maximum values – i.e., for ½ the values occurring at the minimum and the other ½ occurring at the maximum. For n=10 scores between 70 and 100 inclusive, the largest possible standard deviation occurs for the values:
    70 70 70 70 70 100 100 100 100 100
    preliminary values: $n = 10$
                $\Sigma x = 850$
                $\Sigma x^2 = 74500$

$s^2 = [n(\Sigma x^2) - (\Sigma x)^2]/[n(n-1)]$
$= [10(74500) - (850)^2]/[10(9)]$
$= (22500)/90 = 250$
$s = 15.8$

NOTE: This agrees with the notion that the "standard deviation" is describing the "typical spread" of the data. Since the mean of the values is $\bar{x} = (\Sigma x)/n = 850/10 = 85$, each of the values is exactly 15 away from the center and any reasonable measure of typical spread should give a value close to 15.

41. For greater accuracy and understanding, we use 3 decimal places and formula 2-4.
    a. the original population

x	x-μ	(x-μ)²
3	-3	9
6	0	0
9	3	9
18	0	18

$\mu = (\Sigma x)/N = 18/3 = 6$

$\sigma^2 = \Sigma(x-\mu)^2/N = 18/3 = 6$
$[\sigma = 2.449]$

b. the nine samples: using $s^2 = \Sigma(x-\bar{x})^2/(n-1)$ [for each sample, n = 2]

sample	$\bar{x}$	s²	s
3,3	3.0	0	0
3,6	4.5	4.5	2.121
3,9	6.0	18.0	4.245
6,3	4.5	4.5	2.121
6,6	6.0	0	0
6,9	7.5	4.5	2.121
9,3	6.0	18.0	4.245
9,6	7.5	4.5	2.121
9,9	9.0	0	0
	54.0	54.0	16.974

mean of the 9 s² values

$(\Sigma s^2)/9 = 54.0/9 = 6$

c. the nine samples: using $\sigma^2 = \Sigma(x-\mu)^2/N$ [for each sample, N = 2]

sample	μ	σ²
3,3	3.0	0
3,6	4.5	2.25
3,9	6.0	9.00
6,3	4.5	2.25
6,6	6.0	0
6,9	7.5	2.25
9,3	6.0	9.00
9,6	7.5	2.25
9,9	9.0	0
	54.0	27.00

mean of the 9 σ² values

$(\Sigma \sigma^2)/9 = 27.00/9 = 3$

d. The approach in (b) of dividing by n-1 when calculating the sample variance gives a better estimate of the population variance. On the average, the approach in (b) gave the correct population variance of 6. The approach in (c) of dividing by n underestimated the correct population variance. When computing sample variances, divide by n-1 and not by n.

e. No. An unbiased estimator is one that gives the correct answer on the average. Since the average value of s^2 in part (b) was 6, which was the correct value calculated for σ^2 in part (a), s^2 is an unbiased estimator of σ^2. Since the average value of s in part (b) is $(\Sigma s)/9 = 16.974/9 = 1.886$, which is not the correct value of 2.449 calculated for σ in part (a), s is not an unbiased estimator of σ.
NOTE: Since the average value of $\bar{x}$ in part (b) is $(\Sigma \bar{x})/9 = 54.0/9 = 6.0$, which is the correct value calculated for μ in part (a), $\bar{x}$ is an unbiased estimator of μ.

2-6 Measures of Relative Standing

1. a. $x - \mu = 160 - 100 = 60$
    b. $60/\sigma = 60/16 = 3.75$
    c. $z = (x - \mu)/\sigma = (160 - 100)/16 = 60/16 = 3.75$
    d. Since Einstein's IQ converts to a z score of $3.75 > 2$, it is considered unusually high.

3. In general, $z = (x - \mu)/\sigma$.
    a. $z_{60} = (60 - 69.0)/2.8 = -3.21$
    b. $z_{85} = (85 - 69.0)/2.8 = 5.71$
    c. $z_{69.72} = (69.72 - 69.0)/2.8 = 0.26$

5. $z = (x - \mu)/\sigma$
    $z_{70} = (70 - 63.6)/2.5 = 2.56$
    Yes, that height is considered unusual since $2.56 > 2.00$.

7. $z = (x - \mu)/\sigma$
    $z_{101} = (101 - 98.20)/0.62 = 4.52$
    Yes, that temperature is unusually high, since $4.52 > 2.00$. It suggests that either the person is healthy but has a very unusual temperature for a healthy person, or that person is sick (i.e., has an elevated temperature attributable to some cause).

9. In general $z = (x - \overline{x})/s$
    psychology: $z_{85} = (85 - 90)/10 = -5/10 = -0.50$
    economics: $z_{45} = (45 - 55)/5 = -10/5 = -2.00$
    The psychology score has the better relative position since $-0.50 > -2.00$.

11. preliminary values: $n = 36$, $\Sigma x = 29.4056$, $\Sigma x^2 = 24.02111984$
    $\overline{x} = (\Sigma x)/n = 29.4056/36 = .81682$
    $s^2 = [n(\Sigma x^2) - (\Sigma x)^2]/[n(n-1)]$
    $\quad = [36(24.02111984) - (29.4056)^2]/[36(35)] = .07100288/1260 = .000056351$
    $s = .007507$
    $z = (x - \overline{x})/s$
    $z_{.7901} = (.7901 - .81682)/.007507 = -3.56$
    Yes; since $-3.56 < -2.00$, .7901 is an unusual weight for regular Coke.

13. Let $b =$ # of scores below x; $n =$ total number of scores.
    In general, the percentile of score x is $(b/n) \cdot 100$.
    The percentile for a cotinine level of 149 is $(17/40) \cdot 100 = 42.5$.

15. Let $b =$ # of scores below x; $n =$ total number of scores.
    In general, the percentile of score x is $(b/n) \cdot 100$.
    The percentile for a cotinine level of 35 is $(6/40) \cdot 100 = 15$.

17. To find P_{20}, $L = (20/100) \cdot 40 = 8$ – a whole number.
    The mean of the 8th and 9th scores, $P_{20} = (44+48)/2 = 46$.

19. To find P_{75}, $L = (75/100) \cdot 40 = 30$ – a whole number.
    The mean of the 30th and 31st scores, $P_{75} = (250+253)/2 = 251.5$.

21. To find P_{33}, $L = (33/100) \cdot 40 = 13.2$, rounded up to 14.
    Since the 14th score is 121, $P_{33} = 121$.

23. To find P_{01}, $L = (1/100) \cdot 40 = 0.4$, rounded up to 1.
    Since the 1st score is 0, $P_{01} = 0$.

NOTE: For exercises 25-36, refer to the ordered list at the right.

#	level
1	2
2	8
3	44
4	62
5	62
6	89
7	94
8	98
9	98
10	112
11	123
12	125
13	126
14	130
15	146
16	149
17	149
18	173
19	175
20	181
21	207
22	223
23	237
24	254
25	264
26	267
27	271
28	280
29	293
30	301
31	309
32	318
33	325
34	384
35	447
36	462
37	531
38	596
39	600
40	920

25. Let b = # of scores below x; n = total number of scores
In general, the percentile of score x is (b/n)·100.
The percentile of score 123 is (10/40)·100 = 25.

27. Let b = # of scores below x; n = total number of scores
In general, the percentile of score x is (b/n)·100.
The percentile of score 271 is (26/40)·100 = 65.

29. To find P_{85}, L = (85/100)·40 = 34 – a whole number.
The mean of the 34th and 35th scores, P_{85} = (384+447)/2 = 415.5.

31. To find Q_1 = P_{25}, L = (25/100)·40 = 10 – a whole number.
The mean of the 10th and 11th scores, P_{25} = (112+123)/2 = 117.5.

33. To find P_{18}, L = (18/100)·40 = 7.2 rounded up to 8.
Since the 8th score is 98, P_{18} = 98.

35. To find P_{58}, L = (58/100)·40 = 23.2 rounded up to 24.
Since the 23rd score is 237, P_{58} = 254.

37. In general, z scores are not affected by the particular unit of measurement that is used. The relative position of a score (whether it is above or below the mean, its rank in an ordered list of the scores, etc.) is not affected by the unit of measurement, and relative position is what a z score communicates. Mathematically, the same units (feet, centimeters, dollars, etc.) appear in both the numerator and denominator of $(x-\mu)/\sigma$ and cancel out to leave the z score unit-free. In fact the z score is also called the *standard score* for that very reason – it is a standardized value that is independent of the unit of measure employed.
NOTE: In more technical language, changing from one unit of measure to another (feet to centimeters, °F to °C, dollars to pesos, etc.) is a linear transformation – i.e., if x is the score in one unit, then (for some appropriate values of a and b) y = ax + b is the score in the other unit. In such cases it can be shown (see exercise 2.5 #36) that
$\mu_y = a\mu_x + b$ and
$\sigma_y = a\sigma_x$.
The new z score is the same the old one, since
$$z_y = (y - \mu_y)/\sigma_y$$
$$= [(ax + b) - (a\mu_x + b)]/a\sigma_x$$
$$= [ax - a\mu_x]/a\sigma_x$$
$$= (x - \mu_x)/\sigma_x$$
$$= z_x$$

39. a. uniform
b. bell-shaped
c. In general, the shape of a distribution of not affected if all values are converted to z scores. The shape of a distribution does not depend on the unit of measurement -- e.g., if the heights of a group of people form a bell-shaped distribution, they will do so whether measured in inches or centimeters. For a histogram, for example, using a different unit of measurement simply requires a re-labeling of the horizontal axis but does not change the shape of the figure. Since z scores are merely a re-labeling to standard units, converting all the scores to z scores will not change the shape of a distribution.

41.   a. The interquartile range is $Q_3 - Q_1$.
        For $Q_3 = P_{75}$, $L = (75/100) \cdot 40 = 30$ – a whole number.
          The mean of the 30th and 31st scores, $Q_3 = (250+253)/2 = 251.5$.
        For $Q_1 = P_{25}$, $L = (25/100) \cdot 40 = 10$ – a whole number.
          The mean of the 10th and 11th scores, $Q_1 = (86+87)/2 = 86.5$.
        The interquartile range is $251.5 - 86.5 = 165$.
      b. The midquartile is $(Q_1 + Q_3)/2 = (251.5+86.5)/2 = 169$.
      c. The 10-90 percentile range is $P_{90} - P_{10}$.
        For P_{90}, $L = (90/100) \cdot 40 = 36$ – a whole number.
          The mean of the 36th and 37th scores, $P_{90} = (289+290)/2 = 289.5$.
        For P_{10}, $L = (10/100) \cdot 40 = 4$ – a whole number.
          The mean of the 4th and 5th scores, $P_{10} = (3+17)/2 = 10$.
        The 10-90 percentile range is $289.5 - 10 = 279.5$
      d. Yes, $Q_2 = P_{50}$ by definition. Yes, they are always equal.
      e. For $Q_2 = P_{50}$, $L = (50/100) \cdot 40 = 20$ – a whole number.
          The mean of the 20th and 21st scores, $Q_2 = (167+173)/2 = 170$.
        No; in this case $170 = Q_2 \neq (Q_1 + Q_3)/2 = 169$, which demonstrates that the
        median does not necessarily equal the midquartile.

43.   a. $D_1 = P_{10}$, $D_5 = P_{50}$, $D_8 = P_{80}$
      b. In each case D_d, $L = (10d/100) \cdot 40 = 4d$ – a whole number.
        $D_d = (x_{4d} + x_{4d+1})/2$ for $d = 1,2,3,4,5,6,7,8,9$
        $D_1 = (x_4 + x_5)/2 = (3+17)/2 = 10.0$
        $D_2 = (x_8 + x_9)/2 = (44+48)/2 = 46.0$
        $D_3 = (x_{12} + x_{13})/2 = (103+112)/2 = 107.5$
        $D_4 = (x_{16} + x_{17})/2 = (130+131)/2 = 130.5$
        $D_5 = (x_{20} + x_{21})/2 = (167+173)/2 = 170.0$
        $D_6 = (x_{24} + x_{25})/2 = (208+210)/2 = 209.0$
        $D_7 = (x_{28} + x_{29})/2 = (234+245)/2 = 239.5$
        $D_8 = (x_{32} + x_{33})/2 = (265+266)/2 = 265.5$
        $D_9 = (x_{36} + x_{37})/2 = (289+290)/2 = 289.5$
      c. In each case $Quintile_q$, $L = (20q/100) \cdot 40 = 8q$ – a whole number.
        $Quintile_q = (x_{8d} + x_{8d+1})/2$ for $q = 1,2,3,4$
        $Quintile_1 = (x_8 + x_9)/2 = (44+48)/2 = 46.0$
        $Quintile_2 = (x_{16} + x_{17})/2 = (130+131)/2 = 130.5$
        $Quintile_3 = (x_{24} + x_{25})/2 = (208+210)/2 = 209.0$
        $Quintile_4 = (x_{32} + x_{33})/2 = (265+266)/2 = 265.5$

2-7 Exploratory Data Analysis (EDA)

The exercises in this section may be done much more easily when ordered lists of the values are available. On the next page appear ordered lists for the data of exercises #1-8. The left-most column gives the ordered ID numbers 1-50. Data sets with more than 50 values (except for the 54 bear lengths in exercise #6) have more than one column – the second column being ordered values 51-100, etc.

NOTE: A boxplot can be misleading if an extreme value artificially extends the whisker beyond the reasonable data. This manual follows the convention of extending the whisker only to the highest and lowest values within $1.5(Q_3 - Q_1)$ of Q_2 – any point more than 1.5 times the width of the box away from the median will be considered an outlier and represented separately. This type of boxplot is discussed in exercise #13. The boxplots given in this manual can be converted to the ones presented in the text by extending the whisker to the most extreme outlier.

The following ordered lists are used for exercises #1-8.

	#1 dig	#2 bud	#3 cal	#4 nic	#5 wgt	#6 lgn	#7 alc	#8 temp	#8 temp	#8 temp
01	0	0.250	3.3	0.1	.870	36.0	0	96.5	98.3	99.1
02	0	0.325	3.5	0.2	.872	37.0	0	96.9	98.3	99.2
03	0	6.400	3.6	0.5	.874	40.0	0	97.0	98.4	99.4
04	0	8.000	3.6	0.6	.882	40.0	0	97.0	98.4	99.4
05	0	10.000	3.6	0.7	.888	41.0	0	97.0	98.4	99.5
06	0	15.000	3.6	0.7	.891	43.0	0	97.1	98.4	99.6
07	1	17.000	3.7	0.8	.897	43.5	0	97.1	98.4	
08	1	18.500	3.7	0.8	.898	46.0	0	97.1	98.4	
09	1	25.000	3.8	0.8	.908	46.0	0	97.2	98.4	
10	2	50.000	3.9	0.8	.908	47.0	0	97.3	98.4	
11	2	55.000	3.9	0.9	.908	48.0	0	97.3	98.4	
12	2	70.000	3.9	1.0	.911	48.0	0	97.3	98.4	
13	2	70.000	4.0	1.0	.912	49.0	0	97.4	98.4	
14	3	72.000	4.0	1.0	.913	50.0	0	97.4	98.4	
15	3	100.000	4.0	1.0	.920	52.0	0	97.4	98.5	
16	4		4.1	1.0	.924	52.5	0	97.5	98.5	
17	4			1.0	.924	53.0	0	97.5	98.5	
18	4			1.0	.933	53.0	0	97.6	98.5	
19	5			1.1	.936	54.0	0	97.6	98.6	
20	5			1.1	.952	57.3	0	97.6	98.6	
21	5			1.1	.983	57.5	0	97.6	98.6	
22	5			1.2		58.0	0	97.6	98.6	
23	5			1.2		59.0	0	97.6	98.6	
24	5			1.2		59.0	0	97.7	98.6	
25	5			1.2		59.0	0	97.8	98.6	
26	6			1.2		60.0	3	97.8	98.6	
27	6			1.3		60.5	4	97.8	98.6	
28	6			1.4		61.0	5	97.8	98.6	
29	7			1.4		61.0	7	97.8	98.6	
30	7					61.5	8	97.9	98.6	
31	7					62.5	13	97.9	98.6	
32	7					63.0	20	97.9	98.6	
33	7					63.0	28	98.0	98.6	
34	8					63.0	33	98.0	98.7	
35	8					63.5	34	98.0	98.7	
36	9					64.0	38	98.0	98.7	
37	9					64.0	39	98.0	98.7	
38	9					64.0	39	98.0	98.7	
39	9					65.0	46	98.0	98.7	
40	9					65.0	51	98.0	98.8	
41						66.5	72	98.0	98.8	
42						67.0	73	98.0	98.8	
43						67.5	74	98.0	98.8	
44						68.5	76	98.0	98.8	
45						70.0	80	98.0	98.8	
46						70.5	88	98.2	98.8	
47						72.0	113	98.2	98.9	
48						72.0	123	98.2	98.9	
49						72.0	142	98.2	99.0	
50						72.0	414	98.2	99.0	
						73.0				
						73.5				
						75.0				
						76.5				

1.   Consider the 40 digits.
      For $Q_1 = P_{25}$, $L = (25/100)\cdot 40 = 10$ – a whole number, use 10.5.
      For $\tilde{x} = Q_2 = P_{50}$, $L = (50/100)\cdot 40 = 20$ – a whole number, use 20.5.
      For $Q_3 = P_{75}$, $L = (75/100)\cdot 40 = 30$ – a whole number, use 30.5.

   $\min = x_1 = 0$
   $Q_1 = (x_{10}+x_{11})2 = (2+2)/2 = 2$
   $Q_2 = (x_{20}+x_{21})2 = (5+5)/2 = 5$
   $Q_3 = (x_{30}+x_{31})2 = (7+7)/2 = 7$
   $\max = x_{40} = 9$

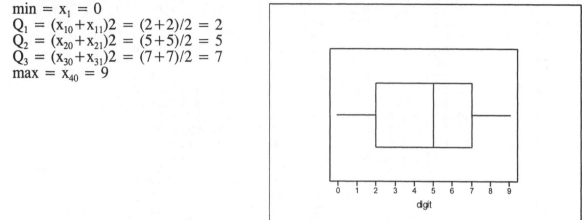

   Since the quartiles (indicated by the vertical bars) divide the figure into four approximately equal segments, the boxplot suggests the values follow a uniform distribution. This is what one would expect if the digits were selected with a random procedure.

3.   Consider the 16 calorie amounts.
      For $Q_1 = P_{25}$, $L = (25/100)\cdot 16 = 4$ – a whole number, use 4.5.
      For $\tilde{x} = Q_2 = P_{50}$, $L = (50/100)\cdot 16 = 8$ – a whole number, use 8.5.
      For $Q_3 = P_{75}$, $L = (75/100)\cdot 16 = 12$ – a whole number, use 12.5.

   $\min = x_1 = 3.3$
   $Q_1 = (x_4+x_5)2 = (3.6+3.6)/2 = 3.6$
   $Q_2 = (x_8+x_9)2 = (3.7+3.8)/2 = 3.75$
   $Q_3 = (x_{12}+x_{13})2 = (3.9+4.0)/2 = 3.95$
   $\max = x_{16} = 4.1$

   These values are not necessarily representative of the cereals consumed by the general population. Although cereals like "Corn Flakes" and "Bran Flakes" are not identified by manufacturer, it is likely that the values given are for cereals produced by the major companies (General Mills, Kellogg's, etc) and not the less expensive generic imitations used by many consumers. In addition, each cereals is given equal weight in the data set – regardless of how popular (or unpopular) it is with consumers.

5.   Consider the 21 weights.
       For $Q_1 = P_{25}$, L = (25/100)·21 = 5.25 rounded up to 6.
       For $\tilde{x} = Q_2 = P_{50}$, L = (50/100)·21 = 10.5 rounded up to 11.
       For $Q_3 = P_{75}$, L = (75/100)·21 = 15.75 rounded up to 16.
   min = x_1 = .870
   Q_1 = x_6 = .891
   Q_2 = x_{11} = .908
   Q_3 = x_{16} = .924
   max = x_{21} = .983

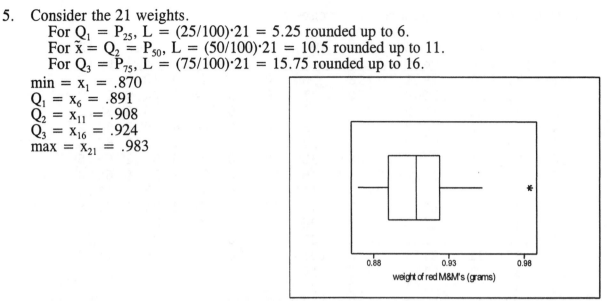

weight of red M&M's (grams)

Yes; assuming there is no significant difference in the weights of the coloring agents, the
weights of the red M&M's are likely to be representative of the weights of all the colors.

7.   Consider the 50 times.
       For $Q_1 = P_{25}$, L = (25/100)·50 = 12.5 rounded up to 13.
       For $\tilde{x} = Q_2 = P_{50}$, L = (50/100)·50 = 25 -- an integer, use 25.5.
       For $Q_3 = P_{75}$, L = (75/100)·50 = 37.5 rounded up to 38.
   min = x_1 = 0
   Q_1 = x_{13} = 0
   Q_2 = $x_{25.5}$ = (0+3)/2 = 1.5
   Q_3 = x_{38} = 39
   max = x_{50} = 414

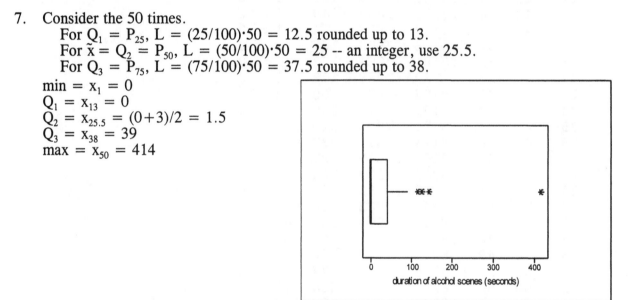

duration of alcohol scenes (seconds)

Based on the boxplot, the distribution appears to be extremely positively skewed.

The following ordered lists are used for exercises #9-12.

	#9 mal	#9 fem	#10 requ	#10 diet	#11 smo	#11 ets	#11 no	#12 clan	#12 rowl	#12 tols
01	31	21	.7901	.7758	0	0	0	43.9	70.9	51.9
02	32	24	.8044	.7760	1	0	0	58.2	74.0	52.8
03	32	25	.8062	.7771	1	0	0	64.4	78.6	58.4
04	32	26	.8073	.7773	3	0	0	69.4	79.2	64.2
05	33	26	.8079	.7802	17	0	0	72.7	79.5	65.4
06	35	26	.8110	.7806	32	0	0	72.9	80.2	68.5
07	36	26	.8126	.7811	35	0	0	73.1	82.5	69.4
08	36	27	.8128	.7813	44	0	0	73.4	83.7	71.4
09	37	28	.8143	.7822	48	0	0	76.3	84.3	71.6
10	37	30	.8150	.7822	86	1	0	76.4	84.6	72.2
11	38	30	.8150	.7822	87	1	0	79.8	85.3	73.6
12	39	31	.8152	.7826	103	1	0	89.2	86.2	74.4
13	39	31	.8152	.7830	112	1	0			
14	40	33	.8161	.7833	121	1	0			
15	40	33	.8161	.7837	123	1	0			
16	40	33	.8163	.7839	130	1	0			
17	41	34	.8165	.7844	131	1	0			
18	42	34	.8170	.7852	149	1	0			
19	42	34	.8172	.7852	164	1	0			
20	43	34	.8176	.7859	167	1	0			
21	43	35	.8181	.7861	173	2	0			
22	44	35	.8189	.7868	173	2	0			
23	45	35	.8192	.7870	198	3	0			
24	45	37	.8192	.7872	208	3	0			
25	46	37	.8194	.7874	210	3	0			
26	46	38	.8194	.7874	222	4	0			
27	47	39	.8207	.7879	227	13	0			
28	48	41	.8211	.7879	234	13	0			
29	48	41	.8299	.7881	245	17	0			
30	51	41	.8244	.7885	250	19	0			
31	53	42	.8244	.7892	253	45	0			
32	55	44	.8247	.7896	265	51	0			
33	56	49	.8251	.7907	266	69	0			
34	56	50	.8264	.7910	277	74	0			
35	60	60	.8284	.7923	284	178	1			
36	60	61	.8295	.7923	289	197	1			
37	61	61			290	241	9			
38	62	74			313	384	90			
39	76	80			477	543	244			
40					491	551	309			

9. Consider the 39 actor and 39 actress values.
    For $Q_1 = P_{25}$, L = (25/100)·39 = 9.75 rounded up to 10.
    For $\bar{x} = Q_2 = P_{50}$, L = (50/100)·39 = 19.5 rounded up to 20.
    For $Q_3 = P_{75}$, L = (75/100)·39 = 29.25 rounded up tp 30.

    For the actors,
    min = x_1 = 31
    $Q_1 = x_{10}$ = 37
    $Q_2 = x_{20}$ = 43
    $Q_3 = x_{30}$ = 51
    max = x_{39} = 76

    For the actresses,
    min = x_1 = 21
    $Q_1 = x_{10}$ = 30
    $Q_2 = x_{20}$ = 34
    $Q_3 = x_{30}$ = 41
    max = x_{39} = 80

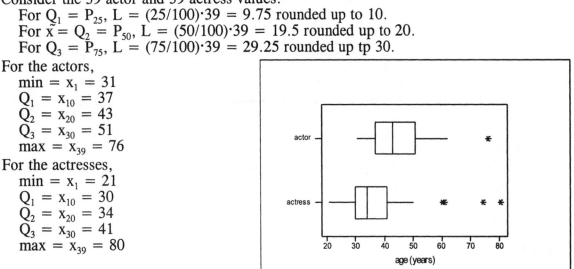

The ages for the actresses cover a wider range and cluster around a lower value than do the ages of the actors.

11. Consider the 40 values for each group.
    For $Q_1 = P_{25}$, L = (25/100)·40 = 10 – an integer, use 10.5.
    For $\tilde{x} = Q_2 = P_{50}$, L = (50/100)·40 = 20 – an integer, use 20.5.
    For $Q_3 = P_{75}$, L = (75/100)·40 = 30 – an integer, use 30.5.

    For the smokers,
    min = x_1 = 0
    $Q_1 = x_{10.5}$ = (86+87)/2 = 86.5
    $Q_2 = x_{20.5}$ = (167+173)/2 = 170
    $Q_3 = x_{30.5}$ = (250+253)/2 = 251.5
    max = x_{40} = 491

    For the ETS group,
    min = x_1 = 0
    $Q_1 = x_{10.5}$ = (1+1)/2 = 1
    $Q_2 = x_{20.5}$ = (1+2)/2 = 1.5
    $Q_3 = x_{30.5}$ = (19+45)/2 = 32
    max = x_{40} = 551

    For the no ETS group,
    min = x_1 = 0
    $Q_1 = x_{10.5}$ = (0+0)/2 = 0
    $Q_2 = x_{20.5}$ = (0+0)/2 = 0
    $Q_3 = x_{30.5}$ = (0+0)/2 = 0
    max = x_{40} = 309

Even though all three groups cover approximately the same range, the boxplot makes it clear that the typical values are highest for the smokers and essentially zero (except for outliers) for the no ETS group.

13. The following given values are also calculated in detail in exercise #11.
    $Q_1 = 86.5$ $Q_2 = 170$ $Q_3 = 251.5$
    a. IQR $= Q_3 - Q_1 = 251.5 - 86.5 = 165$
    b. Since $1.5(IQR) = 1.5(165) = 247.5$, the modified boxplot line extends to
       the smallest value above $Q_1 - 247.5 = 86.5 - 247.5 = -161$, which is 0
       the largest value below $Q_3 + 30 = 251.5 + 247.5 = 499$, which is 491
    c. Since $3.0(IQR) = 3.0(165) = 495$, mild outliers are x values for which
       $Q_1 - 495 \le x < Q_1 - 247.5$ or $Q_3 + 247.5 < x \le Q_3 + 495$
       $-408.5 \le x < -161$ $499 < x \le 746.5$
       There are no such values.
    d. Extreme outliers in this exercise are x values for which $x < -408.5$ or $x > 746.5$.
       There are no such values.

Review Exercises

1. The scores arranged in order are:
   42 43 46 46 47 48 49 49 50 51 51 51 51 51 52 52 54 54 54 54 54
   55 55 55 55 56 56 56 57 57 57 57 58 60 61 61 61 62 64 64 65 68 69
   preliminary values: $n = 43$, $\Sigma x = 2358$, $\Sigma x^2 = 130{,}930$
   a. $\bar{x} = (\Sigma x)/n = (2358)/43 = 54.8$
   b. $\tilde{x} = 55$
   c. M = 51,54 [bi-modal]
   d. m.r. $= (42 + 69)/2 = 55.5$
   e. R $= 69 - 42 = 27$
   f. s $= 6.2$ (from part g)
   g. $s^2 = [n(\Sigma x^2) - (\Sigma x)^2]/[n(n-1)]$
      $= [43(130{,}930) - (2358)^2]/[43(42)]$
      $= (69{,}826)/1806 = 38.7$
   h. For $Q_1 = P_{25}$, L $= (25/100) \cdot 43 = 10.75$ rounded up to 11.
      And so $Q_1 = x_{11} = 51$.
   i. For $Q_3 = P_{75}$, L $= (75/100) \cdot 43 = 32.25$ rounded up to 33.
      And so $Q_3 = x_{33} = 58$.
   j. For P_{10}, L $= (10/100) \cdot 43 = 4.3$ rounded up to 5.
      And so $P_{10} = x_5 = 47$.

2. a. $z = (x - \bar{x})/s$
      $z_{43} = (43 - 54.837)/6.218$
      $= -1.90$
   b. No; Kennedy's inaugural age is not unusual since $-2.00 < -1.90 < -2.00$.
   c. According to the Range Rule of Thumb, the usual values are within 2s of $\bar{x}$
      usual minimum: $\bar{x} - 2s = 54.8 - 2(6.2)$
                     $= 42.4$
      usual maximum: $\bar{x} + 2s = 54.8 + 2(6.2)$
                     $= 67.2$
      The ages 42, 68 and 69 are unusual.
   d. Yes; an age of 35 would be unusual – even if $\bar{x}$ and s were recalculated with the value
      35 included. It is likely that a presidential candidate of age 35 would find that age
      would be a campaign issue – especially during the primaries and other preliminary
      stages of the campaign, but probably not so much so if the candidate wins the
      nomination of a major party. Presumably a candidate who wins the nomination of a
      major party will have enough positive qualities to overcome an unusual age - after all,
      Ronald Reagan (the 69 on the list) was elected despite his "unusual" age.

3.

age			frequency
40	–	44	2
45	–	49	6
50	–	54	13
55	–	59	12
60	–	64	7
65	–	69	3
			43

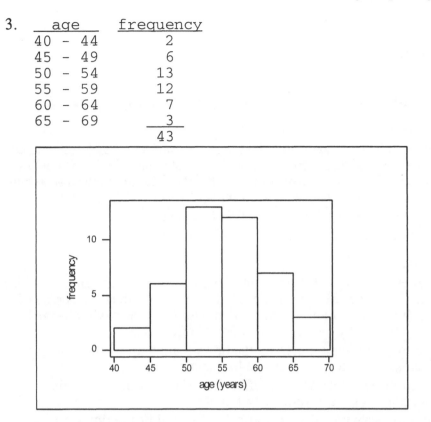

4. See the figure above.
    NOTE: Unlike other data, age is not reported to the nearest unit. The bars in a histogram extend from class boundary to class boundary. Because of the way that ages are reported, the boundaries here are 40, 45, 50, etc -- i.e., someone 44.9 years old still reports an age of 44 and crosses into the next class only upon turning 45, not upon turning 44.5.

5. From results in exercise #1,
    min $= x_1 = 42$
    $Q_1 = x_{11} = 51$
    $Q_2 = x_{22} = 55$
    $Q_3 = x_{33} = 58$
    max $= x_{43} = 69$

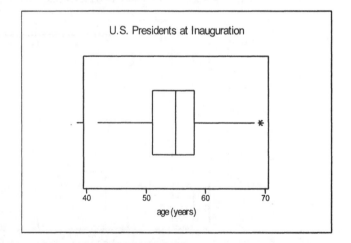

6. a. Since scores 48.6 to 61.0 are within 1·s of $\bar{x}$, the Empirical Rule for Data with a Bell-Shaped Distribution states that about 68% of such persons fall within those limits.
    b. Since scores 42.4 to 67.2 are within 2·s of $\bar{x}$, the Empirical Rule for Data with a Bell-Shaped Distribution states that about 95% of such persons fall within those limits.

7. In general, $z = (x - \bar{x})/s$.
   management, $z_{72} = (72 - 80)/12$
   $$= -0.67$$
   production, $z_{19} = (19 - 20)/5$
   $$= -0.20$$
   The score on the test for production employees has the better relative position since $-0.20 > -0.67$.

8. In general, the Range Rule of Thumb states that the range typically covers about 4 standard deviations -- with the lowest and highest scores being about 2 standard deviations below and above the mean respectively. While answers will vary, the following answers assume the cars driven by students have ages that range from $x_1 = 0$ years to $x_n = 12$ years.
   a. The estimated mean age is $(x_1 + x_n)/2 = (0 + 12)/2 = 6$ years.
   b. The estimated standard deviation of the ages is $R/4 = (x_n - x_1)/4 = (12 - 0)/4 = 3$ years.

9. Adding the same value to every time will increase all the times by the same amount. This changes the location of the times on the number line, but it does not affect the spread of the times or the shape of their distribution -- i.e., each measure of center will change by the amount added, but the measures of variation will not be affected.
   a. Since adding 5 minutes to every time moves all the times 5 units up on the number line, the mean will increase by 5 to $135 + 5 = 140$ minutes.
   b. Since adding 5 minutes to every time does not affect the spread of the times, the standard deviation will not change and remain 15 minutes.
   c. Since adding 5 minutes to every time does not affect the spread of the times, the variance will not change and will remain $15^2 = 225$.

10. Arranging the categories in decreasing order by frequency produces the following figure.

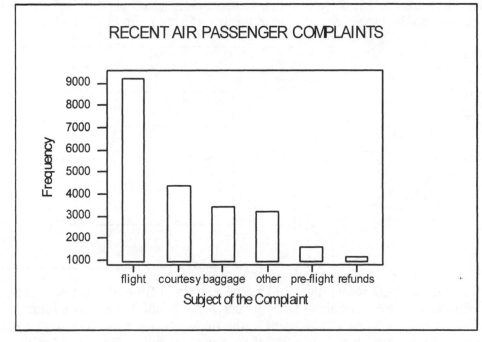

Cumulative Review Exercises

1. The scores arranged in order are:
    -241 -151 -125 -85 -65 20 20 27 30 41 80 105 140 186 325
    preliminary values: $n = 15$, $\Sigma x = 307$, $\Sigma x^2 = 289313$
    a. $\bar{x} = (\Sigma x)/n = (307)/15 = 20.5$
    $\tilde{x} = 27$
    $M = 20$
    m.r. $= (-241 + 325)/2 = 42$
    b. $R = 325 - (-241) = 566$
    $s^2 = [n(\Sigma x^2) - (\Sigma x)^2]/[n(n-1)]$
    $= [15(289313) - (307)^2]/[15(14)]$
    $= (4245446)/210$
    $= 20216.4$
    $s = 142.2$
    c. continuous. Even though the values were reported rounded to whole seconds, time can be any value on a continuum.
    d. ratio. Differences are consistent and there is a meaningful zero; 8 seconds is twice as much time as 4 seconds.

2. a. mode. The median requires at least ordinal level data, and the mean and the midrange require at least interval level data.
    b. convenience. The group was not selected by any process other than the fact that they happened to be the first names on the list.
    c. cluster. The population was divided into units (election precincts), some of which were selected at random to be examined in their entirety.
    d. The standard deviation, since it is the only measure of variation in the list. The standard deviation should be lowered to improve the consistency of the product.

3. No; using the mean of the state values counts each state equally, while states with more people will have a greater affect on the per capita consumption for the population in all 50 states combined. Use the state populations as weights to find the weighted mean.

Chapter 3

Probability

3-2 Fundamentals

1. A probability value is a number between 0 and 1 inclusive.
   a. 50-50 = 50% = .50
   b. 20% = .20
   c no chance = 0% = 0

3. Since $0 \leq P(A) \leq 1$ is always true, the following values less than 0 or greater than 1 cannot be probabilities.
   values less than 0: -1
   values greater than 1: 2 5/3 $\sqrt{2}$

5. Assuming boys (B) and girls (G) are equally likely, the 8 equally likely possibilities for a family of 3 children are given at the right. Rule #2 can be used to find the desired probabilities as follows.
   a. P(1G) = 3/8
   b. P(2G's) = 3/8
   c. P(3G's) = 1/8

   BBB
   GBB
   BGB
   BBG
   GGB
   GBG
   BGG

7. Let H = getting a home run.
   P(H) = 73/476 = .1534
   Yes; this is very different from his lifetime probability of 567/7932 = .0715. It is about twice as high.

9. Consider only the 85 women who were pregnant.
   a. Let W = the test wrongly concludes a woman is not pregnant.
      P(W) = 5/85 = .059
   b. No; since .059 > .05, it is not unusual for the test to be wrong for pregnant women..

10. Consider only the 14 women who were not pregnant.
    a. Let W = the test wrongly concludes a woman is pregnant.
       P(W) = 3/14 = .214
    b. No; since .214 > .05, it is not unusual for the test to be wrong for women who are not pregnant.

11. Let N = selecting someone who feels that secondhand smoke is not at all harmful.
    a. P(N) = 52/1038 = .0501
    b. By Rule #1, .0501 is an appropriate approximation for the proportion of people who believe that secondhand smoke is not at all harmful. No; since .0501 > .05, it is not unusual for someone to believe that secondhand smoke is not at all harmful.

13. Consider only the 2624 + 168,262 = 170,886 American Airlines passenger who were bumped.
    Let I = a passenger is bumped involuntarily.
    a. By Rule #1, an appropriate approximation is P(I) = 2624/168,262 = .0154
    b. Yes; since .0164 < .05, such involuntary bumpings are considered unusual.

15. a. Let C = selecting the correct birthdate.
    P(C) = 1/365
    b. Yes; since 1/365 = .003 < .05.
    c. Since P(C) = .003 is so small, C is unlikely to occur by chance. It appears that Mike had inside information and was not operating on chance.
    d. Depends. Most likely Mike was on the spot and gave an obviously incorrect response in an attempt to handle the situation humorously. If Kelly appreciates his spontaneity and sense of humor, the probability may be quite high.

17. a. Let B = a person's birthday is October 18.
    B is one of 365 (assumed) equally likely outcomes; use Rule #2.
    P(B) = 1/365 = .00274
    b. Let O = a person's birthday is in October.
    O includes 31 of 365 (assumed) equally likely outcomes; use Rule #2.
    P(N) = 31/365 = .0849
    c. Let D = a person's birthday was on a day that ends in the letter y.
    D includes all 7 of the (assumed) equally likely outcomes; use Rule #2.
    P(D) = 7/7 = 1
    NOTE: Since event D is a certainty, P(D) = 1 without reference to Rule #2 – i.e., even if all 7 outcomes are not equally likely

19. There were 132 + 880 = 1012 respondents.
    Let D = getting a respondent who gave the doorstop answer.
    D includes 132 of 1012 equally likely outcomes; use Rule #2.
    P(D) = 132/1012 = .130

21. The following table summarizes the relevant information from Data Set 27, where a win is defined as a positive result.

column	1	2	3	4	5	6	7	8	9	10	total
# of values	50	50	50	50	50	50	50	50	50	50	500
# of wins	6	6	10	8	7	6	8	9	6	11	77
# of 208's	0	1	1	3	1	1	3	2	0	1	13

    a. Let W = winning; use Rule #1.
    P(W) = 77/500 = .154
    b. Let R = running the deck; use Rule #1.
    P(R) = 13/500 = .026

23. a. Assuming boys (B) and girls (G) are equally likely, the 4 equally likely possibilities for a family of 2 children are given at the right. Rule #2 can be used to find the desired probabilities as follows.
    b. P(2G's) = 1/4
    c. P(1 of each)= 2/4 = 1/2

    BB
    BG
    GB
    GG

25. a. net profit = (winnings received) – (amount bet)
                  = $23 – $2
                  = $21
    b. payoff odds = (net profit):(amount bet)
                   = 21:2
    c. actual odds = P(lose)/P(win)
                   = (14/15)/(1/15)
                   = 14/1 or 14:1
    d. Payoff odds of 14:1 mean a profit of $14 for every $1 bet.
    A winning bet of $2 means a profit of $28, which implies winnings received of $30.

27. Chance fluctuations in testing situations typically prevent subjects from scoring exactly the same each time the test is given. When treated subjects show an improvement, there are two possibilities: (1) the treatment was effective and the improvement was due to the treatment or (2) the treatment was not effective and the improvement was due to chance. The accepted standard is that an event whose probability of occurrence is less than .05 is an unusual event. In this scenario, the probability that the treatment group shows an improvement even if the drug has no effect is calculated to be .04. If the treatment group does show improvement, then, either the treatment truly is effective or else an unusual event has occurred. Operating according to the accepted standard, one should conclude that the treatment was effective.

29. If the odds against A are a:b, then P(A) = b/(a+b).
Let M = Millennium wins the next race.
If the odds against M are 3:5, then a=3 and b=5 in the statement above.
P(M) = 5/(3+5)
      = 5/8 = .625

31. Let O = a person is born on October 18.
a. Every 4 years there are 1(366) + 3(365) = 1461 days.
   P(O) = 4/1461 [= .00273785, compare to 1/365 = .00273973]
b. Every 400 years there are 97(366) + 303(365) =146097 days.
   P(O) = 400/146097 [= .00273791, compared to .00273785 and .00273973]

33. This difficult problem will be broken into two events and a conclusion. Let L denote the length of the stick. Label the midpoint of the stick M, and label the first and second breaking points X and Y.
   ■ Event A: X and Y must be on opposite sides of M. If X and Y are on the same side of M, the side without X and Y will be longer than .5L and no triangle will be possible. Once X is set, P(Y is on same side as X) = P(Y is on opposite side from X) = .5.
   ■ Event B: The distance |XY| must be less than .5L. Assuming X and Y are on opposite sides of M, let Q denote the end closest to X and R denote the end closest to Y so that the following sketch applies.

```
|-----|-----------|---|------------|
Q     X           M  Y             R
```

In order to form a triangle, it must be true that |XY| = |XM| + |MY| < .5L. This happens only when |MY| < |QX|.
With all choices random, there is no reason for |QX| to be larger than |MY|, or vice-versa. This means that P(|MY| < |QX|) = .5.
   ■ Conclusion: For a triangle, we need both events A and B. Since P(A) = .5 and P(B occurs assuming A has occurred) = .5, the probability of a triangle is (.5)(.5) = .25. In the notation of section 3-4, P(A and B) = P(A)·P(B|A) =(.5)(.5)=.25.

NOTE: In situations where the laws of science and mathematics are difficult to apply (e.g., finding the likelihood that a particular thumbtack will land point up when it is dropped), probabilities are typically estimated using Rule #1. Suppose an instructor gives 100 sticks to a class of 100 students and says, "Break you stick at random 2 times so that you have 3 pieces." Does this mean that only about 25% of the students would be able to form a triangle with their 3 pieces? In theory, yes. In practice, no. It is likely the students would break the sticks "conveniently" and not "at random" – e.g., there would likely be fewer pieces less than one inch long than you would expect by chance, because such pieces would be difficult to break off by hand.

3-3 Addition Rule

1. a. No, it's possible for a cardiac surgeon to be a female.
   b. No, it's possible for a female college student to drive a motorcycle.
   c. Yes, a person treated with Lipitor would not be in the group that was given no treatment.

3. a. $P(\overline{A}) = 1 - P(A) = 1 - .05 = .95$
   b. Let B = a randomly selected woman over the age of 25 has a bachelor's degree.
   c. $P(\overline{B}) = 1 - P(B) = 1 - .218 = .782$

5. Make a chart like the one on the right.
   Let G = getting a green pod.
   Let W = getting a white flower.

		POD Green	Yellow	
FLOWER	Purple	5	4	9
	White	3	2	5
		8	6	14

   There are two approaches.
   * Use broad categories and allow for double-counting (i.e., the "formal addition rule").
   $P(G \text{ or } W) = P(G) + P(W) - P(G \text{ and } W)$
   $\quad\quad = 8/14 + 5/14 - 3/14$
   $\quad\quad = 10/14 = 5/7 \text{ or } .714$

   * Use individual mutually exclusive categories that involve no double-counting (i.e., the "intuitive addition rule"). For simplicity in this problem, we use GP for "G and P" and GW for "G and W" and so on, as indicated in the table.
   $P(G \text{ or } W) = P(GP \text{ or } GW \text{ or } YW)$
   $\quad\quad = P(GP) + P(GW) + P(YW)$
   $\quad\quad = 5/14 + 3/14 + 2/14$
   $\quad\quad = 10/14 = 5/7 \text{ or } .714$

   NOTE: In general, using broad categories and allowing for double-counting is a "more powerful" technique that "lets the formula do the work" and requires less analysis by the solver. Except when such detailed analysis is instructive, this manual uses the first approach.

NOTE: Throughout the manual we follow the pattern in exercise #5 of using the first letter [or other natural designation] of each category to represent that category. And so in exercises #9-13 P(S) = P(selecting a person who survived) and P(M) = P(selecting a man) and so on. If there is ever cause for ambiguity, the notation will be clearly defined. Since mathematics and statistics use considerable notation and formulas, it is important to clearly define what various letters stand for.

7. Let O = a person's birthday falls on October 18; then P(O) = 1/365.
   $P(\overline{O}) = 1 - P(O)$
   $\quad\quad = 1 - 1/365$
   $\quad\quad = 364/365 \text{ or } .997$

9. Make a chart like the one on the right.
   $P(W \text{ or } C) = P(W) + P(C) - P(W \text{ and } C)$
   $\quad\quad = 422/2223 + 109/2223 - 0/2223$
   $\quad\quad = 531/2223 = .239$
   Note: Let C stand for child – i.e., boys and girls together. And so P(C) = 109/2223, P(C and S) =56/2223, etc.

		FATE Survived	Died	
	Men	332	1360	1692
GROUP	Women	318	104	422
	Boys	29	35	64
	Girls	27	18	45
		706	1517	2223

11. Refer to exercise #9
    $P(C \text{ or } S) = P(C) + P(S) - P(C \text{ and } S)$
    $\quad\quad = 109/2223 + 706/2223 - 56/2223$
    $\quad\quad = 759/2223 = .341$

13. Make a chart like the one on the right.

$P(\overline{A}) = 1 - P(A)$

$= 1 - 40/100$

$= 60/100 = .60$

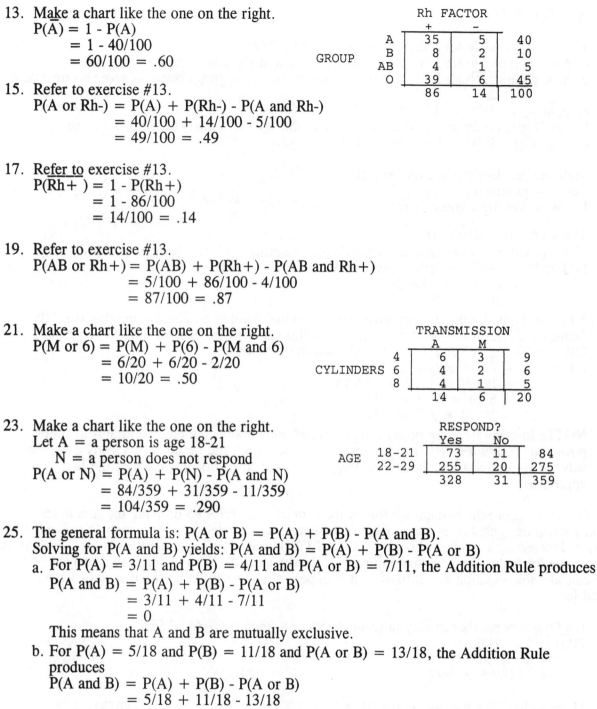

GROUP	Rh FACTOR		
	+	−	
A	35	5	40
B	8	2	10
AB	4	1	5
O	39	6	45
	86	14	100

15. Refer to exercise #13.

P(A or Rh-) = P(A) + P(Rh-) - P(A and Rh-)

$= 40/100 + 14/100 - 5/100$

$= 49/100 = .49$

17. Refer to exercise #13.

$P(Rh+) = 1 - P(\overline{Rh+})$

$= 1 - 86/100$

$= 14/100 = .14$

19. Refer to exercise #13.

P(AB or Rh+) = P(AB) + P(Rh+) - P(AB and Rh+)

$= 5/100 + 86/100 - 4/100$

$= 87/100 = .87$

21. Make a chart like the one on the right.

P(M or 6) = P(M) + P(6) - P(M and 6)

$= 6/20 + 6/20 - 2/20$

$= 10/20 = .50$

CYLINDERS	TRANSMISSION		
	A	M	
4	6	3	9
6	4	2	6
8	4	1	5
	14	6	20

23. Make a chart like the one on the right.

Let A = a person is age 18-21

N = a person does not respond

P(A or N) = P(A) + P(N) - P(A and N)

$= 84/359 + 31/359 - 11/359$

$= 104/359 = .290$

AGE	RESPOND?		
	Yes	No	
18-21	73	11	84
22-29	255	20	275
	328	31	359

25. The general formula is: P(A or B) = P(A) + P(B) - P(A and B).

Solving for P(A and B) yields: P(A and B) = P(A) + P(B) - P(A or B)

a. For P(A) = 3/11 and P(B) = 4/11 and P(A or B) = 7/11, the Addition Rule produces

P(A and B) = P(A) + P(B) - P(A or B)

$= 3/11 + 4/11 - 7/11$

$= 0$

This means that A and B are mutually exclusive.

b. For P(A) = 5/18 and P(B) = 11/18 and P(A or B) = 13/18, the Addition Rule produces

P(A and B) = P(A) + P(B) - P(A or B)

$= 5/18 + 11/18 - 13/18$

$= 3/18$

This means that A and B are not mutually exclusive.

27. If the *exclusive or* is used instead of the *inclusive or*, then the double-counted probability must be completely removed (i.e., must be subtracted twice) and the formula becomes

P(A or B) = P(A) + P(B) - 2·P(A and B)

3-4 Multiplication Rule: Basics

1. a. Independent, since getting a 5 when rolling a die does not affect the likelihood of getting heads when flipping a coin.
   b. Independent, since random selection means that the viewing preferences of the first person chosen in no way affect the viewing preferences of the second person.
   c. Dependent, since getting a positive response to a request for a date generally depends on the personal opinions of the one being asked – and dress is one of the factors that people consider when forming opinions of each other.
   NOTE: This manual views such subjective situations "in general" and without reading too much into the problem. The problem doesn't say, for example, how the invitation was extended – if it were extended over the phone, then dress would be irrelevant. And the problem doesn't say the nature of the relationship prior to the invitation – if the people already know each other fairly well, then the dress one happens to be wearing at the time of the invitation would probably have little if any affect on the response.

3. Let T = getting tails when tossing a coin.
   Let 3 = getting a three then rolling a die.
   $$P(T \text{ and } 3) = P(T) \cdot P(3 \mid T)$$
   $$= (1/2) \cdot (1/6)$$
   $$= 1/12$$

5. There are 1 blue + 3 green + 2 red + 1 yellow = 7 items.
   Let A = the first item selected is colored green.
   Let B = the second item selected is colored green.
   In general, $P(A \text{ and } B) = P(A) \cdot P(B \mid A)$
   a. $P(A \text{ and } B) = P(A) \cdot P(B \mid A)$
   $$= (3/7) \cdot (3/7)$$
   $$= 9/49 = .184$$
   b. $P(A \text{ and } B) = P(A) \cdot P(B \mid A)$
   $$= (3/7) \cdot (2/6)$$
   $$= 6/42 = .143$$

7. Let D = selecting a defective gas mask
   $P(D) = 10322/19218 = .5371$, for the first selection1from the population
   a. $P(D_1 \text{ and } D_2) = P(D_1) \cdot P(D_2 \mid D_1)$
   $$= (10322/19218) \cdot (10322/19218)$$
   $$= .288477$$
   b. $P(D_1 \text{ and } D_2) = P(D_1) \cdot P(D_2 \mid D_1)$
   $$= (10322/19218) \cdot (10321/19217)$$
   $$= .288464$$
   c. The results differ only slightly and are identical to the usual accuracy with which probabilities are reported.
   d. Selecting without replacement. Replacement could lead to re-testing the same unit, which would not yield any new information.

9. Let C = a student guesses the correct response.
   $P(C) = \frac{1}{2}$ for each question.
   a. $P(C_1 \text{ and } C_2 \text{ and } C_3 \text{ and } C_4 \text{ and } C_5 \text{ and } C_6 \text{ and } C_7 \text{ and } \overline{C}_8 \text{ and } \overline{C}_9 \text{ and } \overline{C}_{10})$
   $$= P(C_1) \cdot P(C_2) \cdot P(C_3) \cdot P(C_4) \cdot P(C_5) \cdot P(C_6) \cdot P(C_7) \cdot P(\overline{C}_8) \cdot P(\overline{C}_9) \cdot P(\overline{C}_{10})$$
   $$= (\frac{1}{2}) \cdot (\frac{1}{2}) \cdot (\frac{1}{2}) \cdot (\frac{1}{2}) \cdot (\frac{1}{2}) \cdot (\frac{1}{2}) \cdot (\frac{1}{2}) \cdot (\frac{1}{2}) \cdot (\frac{1}{2}) \cdot (\frac{1}{2})$$
   $$= 1/1032$$
   b. No; getting the first 7 correct and the last 3 incorrect is only one possible way of passing the test.

11. a. Let N = a person is born on November 27.
    $P(N)$ = 1/365, for each selection
    $P(N_1$ and $N_2) = P(N_1) \cdot P(N_2)$
    $= (1/365) \cdot (1/365)$
    $= 1/133225 = .00000751$

   b. Let S = a person's birthday is favorable to being the same.
    $P(S)$ = 365/365 for the first person, but then the second person must match
    P(the birthdays are the same) = $P(S_1$ and $S_2) = P(S_1) \cdot P(S_2 | S_1)$
    $= (365/365) \cdot (1/365)$
    $= 1/365 = .00274$

13. Since the sample size is no more than 5% of the size of the population, treat the selections as being independent even though the selections are made without replacement.
    Let G = a selected CD is good.
    $P(G)$ = .97, for each selection
    P(batch accepted) = P(all good)
    $= P(G_1$ and G_2 and...and $G_{12})$
    $= P(G_1) \cdot P(G_2) \cdot ... \cdot P(G_{12})$
    $= (.97)^{12}$
    $= .694$

15. Let G = a girl is born.
    $P(G)$ = .50, for each couple
    $P(G_1$ and G_2 and...and $G_{10}) = P(G_1) \cdot P(G_2) \cdot ... \cdot P(G_{10}) = (.50)^{10} = .000977$
    Yes, the gender selection method appears to be effective. The choice is between two possibilities: (1) the gender selection method has no effect and a very unusual event occurred or (2) the gender selection method is effective.

17. Let T = the tire named is consistent with all naming the same tire.
    $P(T)$ = 4/4 for the first person, but then the others must match him
    P(all name the same tire) = $P(T_1$ and T_2 and T_3 and $T_4)$
    $= P(T_1) \cdot P(T_2 | T_1) \cdot P(T_3 | T_1$ and $T_2) \cdot P(T_4 | T_1$ and T_2 and $T_3)$
    $= (4/4)(1/4)(1/4)(1/4)$
    $= 1/64 = .0156$

19. Since the sample size is no more than 5% of the size of the population, treat the selections as being independent even though the selections are made without replacement.
    Let N = a CD is not defective
    $P(N)$ = 1 – .02 = .98, for each selection
    $P(N_1$ and N_2 and...and $N_{15}) = P(N_1) \cdot P(N_2) \cdot ... \cdot P(N_{15})$
    $= (.98)^{15}$
    $= .739$

    No, there is not strong evidence to conclude that the new process is better. Even if the new process made no difference, the observed result is not considered unusual and could be expected to occur about 74% of the time.

21. Refer to the table at the right.
    $P(Pos_1$ and $Pos_2) = P(Pos_1) \cdot P(Pos_2)$
    $= (83/99) \cdot (.82/98)$
    $= .702$

		TEST RESULT		
		Pos	Neg	
PREGNANT?	Yes	80	5	85
	No	3	11	14
		83	16	99

23. Refer to the table for exercise #21.
    $P(Yes_1$ and $Yes_2) = P(Yes_1) \cdot P(Yes_2)$
    $= (85/99) \cdot (84/98)$
    $= .736$

25. let D = a birthday is different from any yet selected

$P(D_1) = 366/366$ NOTE: With nothing to match, it <u>must</u> be different.

$P(D_2 | D_1) = 365/366$

$P(D_3 | D_1 \text{ and } D_2) = 364/366$

...

$P(D_5 | D_1 \text{ and } D_2 \text{ and}...\text{and } D_4) = 362/366$

...

$P(D_{25} | D_1 \text{ and } D_2 \text{ and}...\text{and } D_{24}) = 342/366$

a. P(all different) = $P(D_1 \text{ and } D_2 \text{ and } D_3)$

   = $P(D_1) \cdot P(D_2) \cdot P(D_3)$

   = $(366/366) \cdot (365/366) \cdot (364/366)$

   = .992

b. P(all different) = $P(D_1 \text{ and } D_2 \text{ and}...\text{and } D_5)$

   = $P(D_1) \cdot P(D_2) \cdot ... \cdot P(D_5)$

   = $(366/366) \cdot (365/366) \cdot ... \cdot (362/366)$

   = .973

c. P(all different) = $P(D_1 \text{ and } D_2 \text{ and}...\text{and } D_{25})$

   = $P(D_1) \cdot P(D_2) \cdot ... \cdot P(D_{25})$

   = $(366/366) \cdot (365/366) \cdot ... \cdot (342/366)$

   = .432

NOTE: A program to perform this calculation can be constructed using a programming language, or using most spreadsheet or statistical software packages. In BASIC, for example, use

```
10 LET P=1
15 PRINT "How many birthdays?"
20 INPUT D
30  FOR K=1 TO D-1
40  LET P=P*(366-K)/366
50  NEXT K
55 PRINT "The probability they all are different is"
60 PRINT P
70 END
```

27. This is problem can be done by two different methods. In either case,

   let A = getting an ace

   S = getting a spade

* consider the sample space

   The first card could be any of 52 cards; for each first card, there are 51 possible second cards. This makes a total of $52 \cdot 51 = 2652$ equally likely outcomes in the sample space. How many of them are $A_1 S_2$?

   The aces of hearts, diamonds and clubs can be paired with any of the 13 clubs for a total of $3 \cdot 13 = 39$ favorable possibilities. The ace of spades can only be paired with any of the remaining 12 members of that suit for a total of 12 favorable possibilities. Since there are $39 + 12 = 51$ favorable possibilities among the equally likely outcomes,

   $P(A_1 S_2) = 51/2652$

   = .0192

* use the formulas

   Let As and Ao represent the ace of spades and the ace of any other suit respectively. Break $A_1 S_2$ into mutually exclusive parts so the probability can be found by adding and without having to consider double-counting.

   $P(A_1 S_2) = P[(As_1 \text{ and } S_2) \text{ or } (Ao_1 \text{ and } S_2)]$

   = $P(As_1 \text{ and } S_2) + P(Ao_1 \text{ and } S_2)$

   = $P(As_1) \cdot P(S_2 | As_1) + P(Ao_1) \cdot P(S_2 | Ao_1)$

   = $(1/52)(12/51) + (3/52)(13/51)$

   = $12/2652 + 39/2652$

   = $51/2652$

   = .0192

3-5 Multiplication Rule: Complements and Conditional Probability

1. If it is not true that "at least one of them has Group A blood," then "none of them has Group A blood."

3. If it is not true that "none of them is correct," then "at least one of them is correct."

5. Let L = selecting someone with long hair.
   Let W = selecting a woman.
   $P(W|L) = .90$ seems like a reasonable estimate. No; while someone with long hair is likely to be a woman, one cannot say such will "almost surely" be the case.

7. Let B = a child is a boy
   $P(B) = .5$, for each birth
$$
\begin{aligned}
P(\text{at least one girl}) &= 1 - P(\text{all boys}) \\
&= 1 - P(B_1 \text{ and } B_2 \text{ and } B_3 \text{ and } B_4 \text{ and } B_5) \\
&= 1 - P(B_1) \cdot P(B_2) \cdot P(B_3) \cdot P(B_4) \cdot P(B_5) \\
&= 1 - (.5) \cdot (.5) \cdot (.5) \cdot (.5) \cdot (.5) \\
&= 1 - .03125 \\
&= .96875
\end{aligned}
$$
   Yes; that probability is high enough to be "very confident" of getting at least one girl – since .03125 is less than .05, the accepted probability of an unusual event.

9. Let N = not getting cited for a traffic violation.
   $P(N) = .9$ for each such infraction
$$
\begin{aligned}
P(\text{at least violation citation}) &= 1 - P(\text{no violation citations}) \\
&= 1 - P(N_1 \text{ and } N_2 \text{ and } N_3 \text{ and } N_4 \text{ and } N_5) \\
&= 1 - P(N_1) \cdot P(N_2) \cdot P(N_3) \cdot P(N_4) \cdot P(N_5) \\
&= 1 - (.9) \cdot (.9) \cdot (.9) \cdot (.9) \cdot (.9) \\
&= 1 - .590 \\
&= .410
\end{aligned}
$$

11. Let G = getting a girl.
    $P(G) = .5$, for each birth
    $P(G_3 | \overline{G}_1 \text{ and } \overline{G}_2) = P(G_3) = .5$
    No; this is not the same as
$$
\begin{aligned}
P(G_1 \text{ and } G_2 \text{ and } G_3) &= P(G_1) \cdot P(G_2) \cdot P(G_3) \\
&= (.5) \cdot (.5) \cdot (.5) \\
&= .125
\end{aligned}
$$

13. Refer to the table at the right.
    $P(\text{Neg}|\text{No}) = 11/14 = .786$
    Since $P(\text{Neg}|\text{Yes}) = 5/85 = .059 > .05$, it would not be considered unusual for

		TEST RESULT		
		Pos	Neg	
PREGNANT?	Yes	80	5	85
	No	3	11	14
		83	16	99

    the test to read negative even if she were pregnant. If a woman wanted to be more sure that she were not pregnant, she could take another test. Assuming independence of tests (i.e., that errors occur at random and there is no biological reason why the test should err in her particular case), P(2 false negatives) = $(.059)^2 = .003$.

15. Let F = the alarm clock fails.
    $P(F) = .01$, for each clock
$$
\begin{aligned}
P(\text{at least one works}) &= 1 - P(\text{all fail}) = 1 - P(F_1 \text{ and } F_2 \text{ and } F_3) \\
&= 1 - P(F_1) \cdot P(F_2) \cdot P(F_3) \\
&= 1 - (.01)(.01)(.01) \\
&= 1 - .000001 \\
&= .999999
\end{aligned}
$$

Yes; the student appears to gain, because now the probability of a timely alarm is virtually a certainty. There may be something else going on, however, if the student "misses many classes because of malfunctioning alarm clocks" when his clock works correctly 99% of the time.

NOTE: Rounded to 3 significant digits as usual, the answer is 1.00. In cases when rounding to 3 significant digits produces a probability of 1.00, this manual gives the answer with sufficient significant digits to distinguish the answer from a certainty.

17. Let N = a person is HIV negative
    P(N) = .9, for each person in the at-risk population
    P(HIV positive result) = P(at least person is HIV positive)
    $$= 1 - P(\text{all persons HIV negative})$$
    $$= 1 - P(N_1 \text{ and } N_2 \text{ and } N_3)$$
    $$= 1 - P(N_1) \cdot P(N_2) \cdot P(N_3)$$
    $$= 1 - (.9)(.9)(.9)$$
    $$= 1 - .729$$
    $$= .271$$

NOTE: This plan is very efficient. Suppose, for example, there were 3,000 people to be tested. Only in .271 = 27.1% of the groups would a retest need to be done for each of the 3 individuals. Those $(.271) \cdot (1,000) = 271$ groups would generate $271 \cdot 3 = 813$ retests. The total number of tests required is then 1813 (1000 original tests + 813 retests), only 60% of the 3,000 tests that would have been required to test everyone individually.

19. Refer to the table at the right.
    This problem may be done two ways.
    * reading directly from the table
      P(M|D) = 1360/1517 = .897
    * using the formula
      P(M|D) = P(M and D)/P(D)
      $$= (1360/2223)/(1517/2223)$$
      $$= .6118/.6824$$
      $$= .897$$

		FATE		
		Survived	Died	
	Men	332	1360	1692
GROUP	Women	318	104	422
	Boys	29	35	64
	Girls	27	18	45
		706	1517	2223

NOTE: In general, the manual will use the most obvious approach -- which most often is the first one of applying the basic definition by reading directly from the table.

21. Refer to the table and comments for exercise #19.
    P([B or G]|S) = [29 + 27]/706
    $$= 56/706 = .0793$$

23. Let F = getting a seat in the front row.
    P(F) = 2/24 = 1/12 = .083
    P(at least one F in n rides) = 1 - P(no F in n rides) = $1 - (11/12)^n$.
    P(at least one F in n rides) > .95 requires that $(11/12)^n < .05$.
    According to the table at the right, this first occurs for n = 35.

    $(11/12)^{31} = .0674$
    $(11/12)^{32} = .0618$
    $(11/12)^{33} = .0566$
    $(11/12)^{34} = .0519$
    $(11/12)^{35} = .0476$

25. a. The table is given below: 300 = 0.3% of 100,000; 285 = 95% of 300; etc.

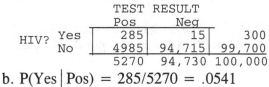

		TEST RESULT		
		Pos	Neg	
HIV?	Yes	285	15	300
	No	4985	94,715	99,700
		5270	94,730	100,000

   b. P(Yes|Pos) = 285/5270 = .0541

27. When 2 coins are tossed, there are 4 equally likely possibilities: HH HT TH TT.
    There are two possible approaches.
    * counting the sample space directly
        If there is at least one H, that leaves only 3 equally likely possibilities: HH HT TH
        Since HH is one of those 3, P(HH|at least one H) = 1/3
    * applying an appropriate formula
        Use P(B|A) = P(A and B)/P(A), where   B = HH
                                                                        A = at least one H
    P(HH|at least one H) = P(at least one H and HH)/P(at least one H)
                        = (1/4)/(3/4)
                        = 1/3

3-6 Probabilities Through Simulations

1. The digits 4,6,1,9,6 correspond to T,T,F,F,T.

3. Letting G and D represent good and defective fuses respectively, the digits 4,6,1,9,6
   correspond to G,G,D,G,G.

5. Using the 20 5-digit numbers to represent families of 5 children and letting odd digits
   represent girls produces the following list of 20 numbers of girls per family:
   2 3 4 0 2 3 2 3 5 1 2 3 2 3 3 1 2 3 4 2
   Letting x be the number of girls per simulated family gives the following distribution.

x	f	r.f.	true P(x)
0	1	.05	.03125
1	2	.10	.15625
2	7	.35	.31250
3	7	.35	.31250
4	2	.10	.15625
5	1	.05	.03125
	20	1.00	1.00000

   From the simulation we estimate $P(x \geq 2) \approx 17/20 = .85$. This compares very favorably
   with the true value of .8125, and in this instance the simulation produced a good estimate.

7. Using the 20 5-digit rows to represent groups of 5 people and letting 0 represent a left-
   hander, produces the following list of 20 numbers of left-handers per group:
   0 0 0 1 0 2 1 1 0 1 0 1 0 0 0 0 1 0 0 0
   Letting x be the number of left-handers per simulated group gives the following distribution.

x	f	r.f.	true P(x)
0	13	.65	.59049
1	6	.30	.32805
2	1	.05	.07290
3	0	.00	.00810
4	0	.00	.00045
5	0	.00	.00001
	20	1.00	1.00000

   From the simulation we estimate $P(x \geq 1) \approx 7/20 = .35$. This compares very favorably to
   the true value of .40951, and in this instance the simulation produced a good estimate.

9. The following Minitab commands produce the desired simulation, where 0 represents a boy and 1 represents a girl.
```
MTB > RANDOM 100 C1-C5;
SUBC> INTEGER 0 1.
MTB > LET C6 = C1+C2+C3+C4+C5
MTB > TABLE C6
```
Letting x be the number of girls per simulated family of 5 gives the following distribution.

x	f	r.f.	true P(x)
0	2	.02	.03125
1	16	.16	.15625
2	27	.27	.31250
3	31	.31	.31250
4	19	.19	.15625
5	5	.05	.03125
	100	1.00	1.00000

From the simulation we estimate $P(x \geq 2) \approx 82/100 = .82$. This compares very favorably to the true value of .8125, and in this instance the simulation produced a good estimate.

11. The following Minitab commands produce the desired simulation, where 0 represents a left-hander.
```
MTB > RANDOM 100 C1-C5;
SUBC> INTEGER 0 9.
MTB > CODE (1:9) TO 1 C1-C5 C1-C5
MTB > LET C6 = C1+C2+C3+C4+C5
MTB > LET C6 = 5-C6
MTB > TABLE C6
```

Letting x be the number of left-handers per simulated group gives the following distribution.

x	f	r.f.	true P(x)
0	60	.60	.59049
1	33	.33	.32805
2	5	.05	.07290
3	2	.02	.00810
4	0	.00	.00045
5	0	.00	.00001
	20	1.00	1.00000

From the simulation we estimate $P(x \geq 1) \approx .40$. This compares very favorably to the true value of .40951, and in this instance the simulation produced a good estimate.

13. The probability of originally selecting the door with the car is 1/3. You can either not switch or switch.

If you make a selection and do not switch, you expect to get the car 1/3 of the time.

If you make a selection and switch under the conditions given, there are 2 possibilities.

   A = you originally had the winning door (and switched to a losing door).

   B = you originally had a losing door (and switched to a winning door). [You cannot switch from a losing door to another losing door. If you have a losing door and the door Monty Hall opens is a losing door, then the only door to switch to is the winning door.]

   If you switch, P(A) = 1/3 and P(B) = 2/3.

Conclusion: the better strategy is to switch since

   P(winning without switching) = P(picking winning door originally) = 1/3

   P(winning with switching) = P(B above) = 2/3

NOTE: While the above analysis makes the simulation unnecessary, it can be done as follows.

   1. For n trials, randomly select door 1,2 or 3 to hold the prize.

   2. For n trials, randomly select door 1,2 or 3 as your original selection.

3. Determine m = the number of times the doors match.
4. If you don't switch, you win m/n of the time. [m/n should be about 1/3]
5. If you switch, you lose m/n of the time and you win (n-m)/n of the time. [(n-m)/n should be about 2/3]

The following Minitab commands produce the desired simulation, where column 1 indicates the location of the winning door and column 2 indicates your choice.

```
MTB > RANDOM 100 C1-C2;
SUBC> INTEGER 1 3.
MTB > LET C3 = C1-C2
MTB > TABLE C3
```

A zero in C3 indicates that C1 and C2 match and that the contestant's original choice was correct. For this simulation, the n=100 trials produced 24 0's in C3. The simulated contestant originally picked the winning door 24/100 = 24% of the time and a losing door 76% of the time. If he doesn't switch, he wins 24% of the time – and loses 76% of the time. If he switches, he moves from a winner to a loser 24% of the time – and from a loser to a winner 76% of the time.

15. No; his reasoning is not correct. No; the proportion of girls will not increase. Each family would consist of zero to several consecutive girls, followed by a boy. The types of possible families and their associated probabilities would be as follows:

P(B) = ½ = 1/2   =16/32
P(GB) = (½)·(½) = 1/4   =8/32
P(GGB) = (½)·(½)·(½) = 1/8   =4/32
P(GGGB) = (½)·(½)·(½)·(½) = 1/16   =2/32
P(GGGGB) = (½)·(½)·(½)·(½)·(½)   =1/32
etc.

Each collection of 32 families would be expected to look like this, where * represents one family with 5 or more girls and a boy:

B	B	B	B	B	B	B	B
B	B	B	B	B	B	B	B
GB	GB	GB	GB	GB	GB	GB	GB
GGB	GGB	GGB	GGB	GGGB	GGGB	GGGGB	*

A gender count reveals 32 boys and 31 or more girls, an approximately equally distribution of genders. In practice, however, familes would likely stop after having a certain number of children – whether they had a boy or not. If that number was 5 for each family, then * = GGGGG and the expected gender count for the 32 families would be an exactly equal distribution of 31 boys and 31 girls.

3-7 Counting

1. 6! = 6·5·4·3·2·1 = 720

3. $_{25}P_2$ = 25!/23! = (25·24·23!)/23! = 25·24 = 600
NOTE: This technique of "cancelling out" or "reducing" the problem by removing the factors 23! = 23·22·...·1 from both the numerator and the denominator is preferred over actually evaluating 25!, actually evaluating 23!, and then dividing those two very large numbers. In general, a smaller factorial in the denominator can be completely divided into a larger factorial in the numerator to leave only the "excess" factors not appearing the in the denominator. This is the technique employed in this manual -- e.g., see #5 below, where the 23! is cancelled from both the numerator and the denominator. In addition, $_nP_r$ and $_nC_r$ will always be integers; calculating a non-integer value for either expression indicates an error has been made. More generally, the answer to any <u>counting</u> problem (but not a <u>probability</u> problem) must always be a whole number; a fractional number indicates that an error has been made.

5   $_{25}C_2 = 25!/(2!23!) = (25·24)/2! = 300$

7. $_{52}C_5 = 52!/(5!47!) = (52·51·50·49·48)/5! = 2,598,960$

9.   Let W = winning the described lottery with a single selection.
     The total number of possible combinations is
     $$_{49}C_6 = 49!/(43!6!) = (49·48·47·46·45·44)/6!$$
     $$= 13,983,816.$$
     Since only one combination wins, P(W) = 1/13,983,816.

11. Let W = winning the described lottery with a single selection.
     The total number of possible combinations is
     $$_{59}C_6 = 59!/(53!6!) = (59·58·57·56·55·54)/6!$$
     $$= 45,057,474$$
     Since only one combination wins, P(W) = 1/45,047,474.

13. Let C = choosing the 4 oldest at random.
     The total number of possible groups of 4 is
     $$_{32}C_4 = 32!/4!28! = 32·31·30·29/4!$$
     $$= 35,960.$$
     Since only one group of 4 is the 4 oldest, P(C) = 1/35,960 = .0000278.
     Yes; P(C) is low enough to suggest the event did not happen randomly.

15. Let W = winning the described lottery with a single selection.
     The total number of possible permutations is
     $$_{42}C_6 = 42!/36! = 42·41·40·39·38·37 = 3,776,965,920.$$
     Since only one permutation wins, P(W) = 1/3,776,965,920.

17. Let Y = the 6 youngest are chosen.
     The total number of possible groups of 6 is $_{15}C_6 = 15!/(6!9!)$
     $$= 5005.$$
     Since only of the possibilities consists of the 6 youngest, P(Y) = 1/5005 = .000200.
     Yes; that probability is low enough to suggest the event did not occur by chance alone.

19. For 3 cities there are 3! = 6 possible sequences; and so there are 4 more possible routes.
     For 8 cities there are 8! = 40,320 possible sequences.

21. $_5C_2 = 5!/(2!3!) = (5·4)/2! = 10$

23. For 6 letters there are 6! =720 possible sequences.
     The only English word possible from those letters is "satire."
     Since there is only one correct sequence, P(correct sequence by chance) = 1/720.

25. Let O = opening the lock on the first try.
     The total number of possible "combinations" is 50·50·50 = 125,000.
     Since only of the possibilities is correct, P(O) = 1/125,000.

27. a. There are 2 possibilities (B,G) for each baby. The number of possible sequences is
        $$2·2·2·2·2·2·2·2 = 2^8 = 256$$
     b. The number of ways to arrange 8 items consisting of 4 B's and 4 G's is 8!/(4!4!) = 70.
     c. Let E = getting equal numbers of boys and girls in 8 births.
        Based on parts (a) and (b), P(E) 70/256 = .273

29. The 1st digit may be any of 8 (any digit except 0 or 1).
    The 2nd digit may be any of 2 (0 or 1).
    The 3rd digit may be any of 9 (any digit except whatever the 2nd digit is).
    There are $8 \cdot 2 \cdot 9 = 144$ possibilities.

31. Winning the game as described requires two events A and B as follows.
    A = selecting the correct 5 numbers between 1 and 47
    There are $_{47}C_5 = 47!/(42!5!) = 1,533,939$ possible selections.
    Since there is only 1 winning selection, $P(A) = 1/1,533,939$
    B = selecting the correct number between 1 and 27 in a separate drawing
    There are $_{27}C_1 = 27!/(26!1!) = 27$ possible selections.
    Since there is only 1 winning selection, $P(B) = 1/27$.
    P(winning the game) = P(A and B)
    $$= P(A) \cdot P(B|A)$$
    $$= (1/1,533,939) \cdot (1/27)$$
    $$= 1/41,416,353$$

33. There are 26 possible first characters, and 36 possible characters for the other positions.
    Find the number of possible names using 1,2,3,…,8 characters and then add to get the
    total.

characters	possible names		
1	26	=	26
2	26·36	=	936
3	26·36·36	=	33,696
4	26·36·36·36	=	1,213,056
5	26·36·36·36·36	=	43,670,016
6	26·36·36·36·36·36	=	1,572,120,576
7	26·36·36·36·36·36·36	=	56,596,340,736
8	26·36·36·36·36·36·36·36	=	2,037,468,266,496
		total =	2,095,681,645,538

35. a. The calculator factorial key gives $50! = 3.04140932 \times 10^{64}$.
    Using the approximation, $K = (50.5) \cdot \log(50) + .39908993 - .43429448(50)$
    $$= 85.79798522 + .39908933 - 21.71472400$$
    $$= 64.48235115$$
    and then $50! = 10^k$
    $$= 10^{64.48235115}$$
    $$= 3.036345215 \times 10^{64}$$
    NOTE: The two answers differ by 5.1×10^{61} (i.e., by 51 followed by 60 zeros – which is
    "zillions and zillions"). While the error may seem large, the numbers being dealt with
    are so large that the error is only $(5.1 \times 10^{61})/(3.04 \times 10^{64}) = 1.7\%$
    b. The number of possible routes is 300!.
    Using the approximation, $K = (300.5) \cdot \log(300) + .39908993 - .43429448(300)$
    $$= 614.4$$
    and then $300! = 10^{614.5}$
    Since the number of digits in 10^x is the next whole number above x, 300! has 615 digits.

Review Exercises

1. Refer to the table at the right.
   Let PT = the polygraph indicates truth
       PL = the polygraph indicates lie
       AT = the subject is actually telling the truth
       AL = the subject is actually telling a lie
   $P(AL) = 20/100 = .200$

		POLYGRAPH		
		Truth	Lie	
SUBJECT	Truth	65	15	80
	Lie	3	17	20
		68	32	100

2. Refer to the table and notation of exercise #1.
   $P(PL) = 32/100 = .320$

3. Refer to the table and notation of exercise #1.
   $$P(AL \text{ or } PL) = P(AL) + P(PL) - P(AL \text{ and } PL)$$
   $$= 20/100 + 32/100 - 17/100$$
   $$= 35/100 = .350$$

4. Refer to the table and notation of exercise #1.
   $$P(AT \text{ or } PT) = P(AT) + P(PT) - P(AT \text{ and } PT)$$
   $$= 80/100 + 68/100 - 65/100$$
   $$= 83/100 = .830$$

5. Refer to the table and notation of exercise #1.
   $$P(AT_1 \text{ and } AT_2) = P(AT_1) \cdot P(AT_2 \mid AT_1)$$
   $$= (80/100) \cdot (79/99) = .638$$

6. Refer to the table and notation of exercise #1.
   $$P(PL_1 \text{ and } PL_2) = P(PL_1) \cdot P(PL_2 \mid PL_1)$$
   $$= (32/100) \cdot (31/99) = .100$$

7. Refer to the table and notation of exercise #1.
   $P(AT \mid PL) = 15/32 = .469$

8. Refer to the table and notation of exercise #1.
   $P(PL \mid AT) = 15/80 = .1875$

9. Let B = a computer breaks down during the first two years
   a. $P(B) = 992/4000 = .248$
   b. $P(B_1 \text{ and } B_2) = P(B_1) \cdot P(B_2) = (.248) \cdot (.248)$
      $$= .0615$$
   c. $P(\text{at least 1 breaks down}) = 1 - P(\text{all good})$
      $$= 1 - P(\bar{B}_1 \text{ and } \bar{B}_2 \text{ and } \bar{B}_3)$$
      $$= 1 - P(\bar{B}_1) \cdot P(\bar{B}_2) \cdot P(\bar{B}_3)$$
      $$= 1 - (.752)^3 = .575$$

10. Since the sample size is no more than 5% of the size of the population, treat the selections as being independent even though the selections are made without replacement.
    Let G = a CD is good.
    $P(G) = .98$, for each selection
    $P(\text{reject batch}) = P(\text{at least one is defective})$
    $$= 1 - P(\text{all are good})$$
    $$= 1 - P(G_1 \text{ and } G_2 \text{ and } G_3 \text{ and } G_4)$$
    $$= 1 - P(G_1) \cdot P(G_2) \cdot P(G_3) \cdot P(G_4)$$
    $$= 1 - (.98)^4 = .0776$$
    NOTE: $P(\text{reject batch}) = 1 - (2450/2500) \cdot (2449/2499) \cdot (2448/2498) \cdot (2447/2497) = .0777$
    if the problem is done considering the effect of the non-replacement.

11. Let G = getting a girl
     P(G) = ½, for each child
     P(all girls) = P(G$_1$ and G$_2$ and G$_3$ and...and G$_{12}$)
     $\qquad\qquad$ = P(G$_1$)·P(G$_2$)·P(G$_3$)·...·P(G$_{12}$) = (½)^{12}
     $\qquad\qquad$ = 1/4096
     Yes; since the probability of the getting all girls by chance alone is so small, it appears the company's claim is valid.

12. a. Let W = the committee of 3 is the 3 wealthiest members.
        The number of possible committees of 3 is $_{10}$C$_3$ = 10!/(7!3!) = 120.
        Since only one of those committees is the 3 wealthiest members,
        $\quad$ P(W) = 1/120 = .00833
     b. Since order is important, the number of possible slates of officers is
        $_{10}$P$_3$ = 10!/7! = 720

13. Let E = a winning even number occurs.
     a. P(E) = 18/38 [=9/19]
     b. odds against E = P(E)/P(E)
        $\qquad\qquad\qquad$ = (20/38)/(18/38)
        $\qquad\qquad\qquad$ = 20/18 = 10/9, usually expressed as 10:9
     c. payoff odds = (net profit):(amount bet)
        If the payoff odds are 1:1, the net profit is $5 for a winning $5 bet.

14. Since the sample size is no more than 5% of the size of the population, treat the selections as being independent even though the selections are made without replacement.
     Let R = getting a Republican.
     P(R) = .30, for each selection
     P(R$_1$ and R$_2$ and...and R$_{12}$) = P(R$_1$)·P(R$_2$)·...·P(R$_{12}$)
     $\qquad\qquad\qquad\qquad\qquad\quad$ = (.30)^{12}
     $\qquad\qquad\qquad\qquad\qquad\quad$ = .000000531
     No, the pollster's claim is not correct.

15. Let L = a selected 27 year old male lives for one more year.
     P(L) = .9982, for each such person
     P(L$_1$ and L$_2$ and...and L$_{12}$) = P(L$_1$)·P(L$_2$)·...·P(L$_{12}$)
     $\qquad\qquad\qquad\qquad\qquad\quad$ = (.9982)^{12}
     $\qquad\qquad\qquad\qquad\qquad\quad$ = .979

16. a. Let W = winning the described lottery with a single selection.
        The total number of possible combinations is
        $_{52}$C$_6$ = 52!/(46!6!)
        $\qquad$ = (52·51·50·49·48·47)/6!
        $\qquad$ = 20,358,520.
        Since only one combination wins, P(W) = 1/20,358,520.
     b. Let W = winning the described lottery with a single selection.
        The total number of possible combinations is
        $_{30}$C$_5$ = 30!/(25!5!)
        $\qquad$ = (30·29·28·27·26)/5!
        $\qquad$ = 142,506.
        Since only one combination wins, P(W) = 1/142,506.

c. Two events need to occur to win the Big Game.
  *Let A = selecting the winning 5 from among the 50.
    The total number of possible combinations is
      $_{50}C_5 = 50!/(45!5!)$
        $= (50 \cdot 49 \cdot 48 \cdot 47 \cdot 46)/5!$
        $= 2,118,760.$
    Since only one combination wins, P(A) = 1/2,118,760.
  *Let B = selecting the winning 1 from among the 36, and P(B) = 1/36.
  P(winning the Big Game) = P(A and B)
                 $= P(A) \cdot P(B|A)$
                 $= (1/2,118,760) \cdot (1/36)$
                 $= 1/76,275,360$

Cumulative Review Exercises

1. scores arranged in order: 0 0 0 2 3 3 3 3 4 4 4 5 5 5 5 6 6 6 6 7 7
   summary statistics: n = 21, $\Sigma x = 84$, $\Sigma x^2 = 430$
   a. $\bar{x} = (\Sigma x)/n$
        $= 84/21 = 4.0$
   b. $\tilde{x} = 4$, the 11th score in the ordered list
   c. $s^2 = [n \cdot (\Sigma x^2) - (\Sigma x)^2]/[n(n-1)]$
        $= [21(430) - (84)^2]/[(21)(20)]$
        $= 4.7$
      $s = 2.2$
   d. $s^2 = 4.7$ [from part (c)]
   e. Yes; all of the scores are either 0 or positive.
   f. Let G = randomly selecting a positive value
      P(G) = 18/21 = 6/7.
   g. Let G = randomly selecting a positive value
      $P(G_1 \text{ and } G_2) = P(G_1) \cdot P(G_2|G_1)$
                  $= (18/21) \cdot (17/20)$
                  $= .729$
   h. Let E = the treatment is effective.
      P(E) = ½, under the given assumptions, for each person
      $P(E_1 \text{ and } E_2 \text{ and} \ldots \text{and } E_{18}) = P(E_1) \cdot P(E_2) \cdot \ldots \cdot P(E_{18})$
                       $= (½)^{18}$
                       $= 1/262,144$
      Yes; this probability is low enough to reject the idea that the treatment is ineffective.
      Yes; the treatment appears to be effective.

2. Let x = the height selected.
   The values identified on the boxplot are: x_1 [minimum] = 56.1
                        $P_{25} = 62.2$
                        $P_{50} = 63.6$
                        $P_{75} = 65.0$
                        x_n [maximum] = 71.1
   NOTE: Assume that the number of heights is so large that (1) repeated sampling does not affect the probabilities for subsequent selections and (2) it can be said that P_a has a% of the heights below it and (100-a)% of the heights above it.

a. The exact values of $\bar{x}$ and s cannot be determined from the information given. The Range Rule of Thumb [the highest and lowest values are approximately 2 standard deviations above and below the mean], however, indicates that $s \approx (x_n - x_1)/4$ and $\bar{x} \approx (x_1 + x_n)/2$. In this case we estimate $\bar{x} \approx (56.1 + 71.1)/2 = 63.6$

NOTE: The boxplot, which gives $\tilde{x} = P_{50} = 63.6$, is symmetric. This suggests that the original distribution is approximately so. In a symmetric distribution the mean and median are equal. This also suggests $\bar{x} \approx 63.6$.

b. $P(56.1 < x < 62.2) = P(x_1 < x < P_{25})$
$$= .25 - 0$$
$$= .25$$

c. $P(x < 62.2 \text{ or } x > 63.6) = P(x < 62.2) + P(x > 63.6)$
$$= P(x < P_{25}) + P(x > P_{50})$$
$$= .25 + .50$$
$$= .75$$

NOTE: No height is less than 62.2 <u>and</u> greater than 63.6 – i.e., the addition rule for mutually exclusive events can be used.

d. Let B = selecting a height between 62.2 and 63.6.

$P(B) = P(P_{25} < x < P_{50})$
$$= .50 - .25$$
$$= .25$$
$P(B_1 \text{ and } B_2) = P(B_1) \cdot P(B_2 \mid B_1)$
$$= (.25) \cdot (.25)$$
$$= .0625$$

e. Let S = a selected woman is shorter than the mean.

T = a selected woman is taller than the mean.

E = the event that a group of 5 women consists of 3 T's and 2 S's.

There are $2 \cdot 2 \cdot 2 \cdot 2 \cdot 2 = 32$ equally likely [see the NOTE below] sequences of T's and S's. The number of arrangements of 3 T's and 2 S's is $5!/(3!2!) = 10$.

$P(E) = 10/32 = .3125$

NOTE: The exact value of the mean cannot be determined from the information given.

From part (a) we estimate $\bar{x} \approx 63.6$, which is also the median. If this is true then *in this particular problem* $P(x < \bar{x}) = P(S) = .50$ and $P(x > \bar{x}) = P(T) = .50$. This means that all arrangements of T's and S's are equally likely and

$P(E) = $ (# of ways E can occur)/(total number of possible outcomes)

Chapter 4

Probability Distributions

4-2 Random Variables

1. a. Continuous, since height can be any value on a continuum.
   b. Discrete, since the number of points must be an integer.
   c. Continuous, since time can be any value on a continuum.
   d. Discrete, since the number of players must be an integer.
   e. Discrete, since salary is typically stated in whole dollars – and it cannot be stated in units smaller than whole number of cents.

NOTE: When working with probability distributions and formulas in the exercises that follow, always keep these important facts in mind.
   * If one of the conditions for a probability distribution does not hold, the formulas do not apply and produce numbers that have no meaning.
   * $\Sigma x \cdot P(x)$ gives the mean of the x values and must be a number between the highest and lowest x values.
   * $\Sigma x^2 \cdot P(x)$ gives the mean of the x^2 values and must be a number between the highest and lowest x^2 values.
   * $\Sigma P(x)$ must always equal 1.000.
   * Σx and Σx^2 have no meaning and should not be calculated.
   * The quantity $[\Sigma x^2 \cdot P(x) - \mu^2]$ cannot possibly be negative; if it is, then there is a mistake.
   * Always be careful to use the <u>unrounded</u> mean in the calculation of the variance and to take the square root of the <u>unrounded</u> variance to find the standard deviation.

3. This is a probability distribution since $\Sigma P(x)=1$ is true and $0 \le P(x) \le 1$ is true for each x.

x	P(x)	x·P(x)	x²	x²P(x)	
0	.125	0	0	0	$\mu = \Sigma x \cdot P(x)$
1	.375	.375	1	.375	= 1.500, rounded to 1.5
2	.375	.750	4	1.500	$\sigma^2 = \Sigma x^2 \cdot P(x) - \mu^2$
3	.125	.375	9	1.125	= 3.000 - (1.500)²
	1.000	1.500		3.000	= .750
					σ = .866, rounded to 0.9

5. This is not a probability distribution since $\Sigma P(x)=.94 \ne 1$.

7. This is a probability distribution since $\Sigma P(x)=1$ is true and $0 \le P(x) \le 1$ is true for each x.

x	P(x)	x·P(x)	x²	x²P(x)	
0	.512	0	0	0	$\mu = \Sigma x \cdot P(x)$
1	.301	.301	1	.301	= .730, rounded to 0.7
2	.132	.264	4	.528	$\sigma^2 = \Sigma x^2 \cdot P(x) - \mu^2$
3	.055	.165	9	.495	= 1.324 - (.73)²
	1.000	.730		1.324	= .791
					σ = .889, rounded to 0.9

9. This is a probability distribution since $\Sigma P(x)=1$ is true and $0 \le P(x) \le 1$ is true for each x.

x	P(x)	x·P(x)	x²	x²P(x)	
4	.1809	.7236	16	2.8944	$\mu = \Sigma x \cdot P(x)$
5	.2234	1.1170	25	5.5850	= 5.7871, rounded to 5.8
6	.2234	1.3404	36	8.0424	$\sigma^2 = \Sigma x^2 \cdot P(x) - \mu^2$
7	.3723	2.6061	49	18.2427	= 34.7645 - (5.7871)²
	1.0000	5.7871		34.7645	= 1.2740
					σ = 1.129, rounded to 1.1

No; since .1809 > .05, a four game sweep is not considered an unusual event.

11.

x	P(x)	x·P(x)
5	244/495	2.4646
-5	251/495	-2.5353
	495/495	-0.0707

E = Σx·P(x)
  = $-0.0707 [i.e., a loss of 7.07¢]

Since the expected loss is 7.07¢ for a $5 bet, the expected (or long run) loss is 7.07/5 = 1.41¢ for each $1 a person bets.

13. a. Mike "wins" $100,000 - $250 = $99,750 if he dies.
    Mike "loses" $250 if he lives.

b.

x	P(x)	x·P(x)
99750	.0015	149.625
-250	.9985	-249.625
	1.0000	-100.000

E = Σx·P(x)
  = -100.000   [i.e., a loss of $100]

c. Since the company is making a $100 profit at this price, it would break even if it sold the policy for $100 less – i.e. for $150. NOTE: This oversimplified analysis ignores the cost of doing business. If the costs (printing, salaries, etc.) associated with offering the policy is $25, for example, then the company's profit is only $75 and selling the policy for $75 less (i.e., for $175) would represent the break even point for the company.

d. Buying life insurance is similar to purchasing other services. You let someone make a profit off you by shoveling the snow in your driveway because the cost is worth the effort and frustration you save. You let someone make a profit off you by selling you life insurance because the cost is worth the peace you have from providing for the financial security of your heirs.

15. a. 10·10·10·10 = 10,000
    b. 1/10,000 = .0001
    c.

x	P(x)	x·P(x)
2787.50	.0001	.27875
-.50	.9999	-.49995
	1.0000	-.22120

E = Σx·P(x)
  = -.2212   [i.e., a loss of 22.12¢]

d. The Pick 4 game is the better of the two. Even though they both are losing propositions, the expected value is higher for the Pick 4 game – i.e., -22.12 > -22.5.

17. a. P(x=9) = .122
    b. P(x≥9) = P(x=9 or x=10 or x=11 or x=12 or x=13 or x=14)
             = P(x=9) + P(x=10) + P(x=11) + P(x=12) + P(x=13) + P(x=14)
             = .122 + .061 + .022 + .006 + .001 + .000
             = .212
    c. Part (b). In determining whether a particular event in some hierarchy of events is unusual, one should consider both that event and the events that are more extreme. Even though the probability of a particular event may be very low, that event might not be unusual in a practical sense. P(35 girls in 60 births) = .045 < .05, for example, but getting 35 girls in 60 births is one of the more common outcomes – it's just that there are so many possible outcomes that the probability of any one particular outcome is small.
    d. No; this is not evidence that the technique is effective. Getting 9 or more girls by chance alone is not unusual, since P(x≥9) =.212 > .05.

19. a. P(x≥11) = P(x=11 or x=12 or x=13 or x=14)
             = P(x=11) + P(x=12) + P(x=13) + P(x=14)
             = .022 + .006 + .001 + .000
             = .029
    b. Yes; this is evidence that the technique is effective. Getting 11 or more girls by chance alone would be unusual, since P(x≥11) = .029 < .05.

21. $P(x \geq 8)$ = P(x=8 or x=9 or x=10 or x=11 or x=12 or x=13 or x=14)
$\qquad$ = P(x=8) + P(x=9) + P(x=10) + P(x=11) + P(x=12) + P(x=13) + P(x=14)
$\qquad$ = .183 + .122 + .061 + .022 + .006 + .001 + .000
$\qquad$ = .395

No; Bob's claim is not valid. Getting 8 or more answers correct by chance alone is not unusual, since $P(x \geq 8) = .395 > .05$.

23. For every $1000, Bond A gives a profit of (.06)($1000) = $60 with probability .99.

```
    x        P(x)      x·P(x)        E = Σx·P(x)
    60        .99       59.40          = 49.40   [i.e., a profit of $49.40]
 -1000        .01      -10.00
             1.00       49.40
```

For every $1000, Bond B gives a profit of (.08)($1000) = $80 with probability .95.

```
    x        P(x)      x·P(x)        E = Σx·P(x)
    80        .95       76.00          = 26.00   [i.e., a profit of $26.00]
 -1000        .05      -50.00
             1.00       26.00
```

Bond A is the better bond since it was the higher expected value – i.e., 49.40 > 26.00. Since both bonds have positive expectations, either one would be a reasonable selection. Although bond A has a higher expectation, a person willing to assume more risk in hope of a higher payoff might opt for Bond B.

25. Let C and N represent correctly and not correctly calibrated altimeters respectively. For 8 C's and 2 N's, there are 7 possible samples of size n=3 – given with their probabilities at the right. Letting x be the number of N's produced the distribution and calculations given below.

$P(GGG) = (8/10) \cdot (7/9) \cdot (6/8) = 336/720$
$P(GGN) = (8/10) \cdot (7/9) \cdot (2/8) = 112/720$
$P(GNG) = (8/10) \cdot (2/9) \cdot (7/8) = 112/720$
$P(NGG) = (2/10) \cdot (8/9) \cdot (7/8) = 112/720$
$P(GNN) = (8/10) \cdot (2/9) \cdot (1/8) = 16/720$
$P(NGN) = (2/10) \cdot (8/9) \cdot (1/8) = 16/720$
$P(NNG) = (2/10) \cdot (1/9) \cdot (8/8) = \underline{16/720}$
$\qquad\qquad\qquad\qquad\qquad\qquad\qquad 720/720$

```
 x    P(x)      x·P(x)    x²    x²P(x)    μ = Σx·P(x)
 0  336/720         0     0        0        = 432/720 = .600, rounded to 0.6
 1  336/720   336/720     1  336/720      σ² = Σx²·P(x) - μ²
 2   48/720    96/720     4  192/720        = (528/720) - (432/720)² = .3733
    720/720   432/720        528/720       σ = .611, rounded to 0.6
```

27. The distribution and calculations are given below.

```
 x    P(x)    x·P(x)    x²    x²P(x)    μ = Σx·P(x)
 1     .2       .2       1      .2        = 3.0
 2     .2       .4       2      .8      σ² = Σx²·P(x) - μ²
 3     .2       .6       9     1.8        = 11.0 - (3.0)²
 4     .2       .8      16     3.2        = 2.0
 5     .2      1.0      25     5.0      σ = 1.414
      1.0      3.0            11.0
```

a. For n=5, $\mu = (n+1)/2$
$\qquad\qquad = 6/2 = 3.0$. This agrees with the calculations above.

b. For n=5, $\sigma^2 = (n^2-1)/12$
$\qquad\qquad = 24/12 = 2$.
$\qquad\qquad \sigma = \sqrt{2} = 1.414$. This agrees with the calculations above.

c. For n=20,
$\qquad \mu = (n+1)/2$
$\qquad\quad = 21/2 = 10.5$.
$\qquad \sigma^2 = (n^2-1)/12$
$\qquad\quad = 399/12 = 33.25$
$\qquad \sigma = \sqrt{33.25}$
$\qquad\quad = 5.766$, rounded to 5.8

4-3 Binomial Experiments

NOTE: The four requirements for a binomial experiment are
 #1 There are a fixed number of trials.
 #2 The trials are independent.
 #3 Each trial has two possible named outcomes.
 #4 The probabilities remain constant for each trial.

1. No; requirement #3 is not met. There are more than 2 possible outcomes.

3. No; requirement #3 is not met. There are more than 2 possible outcomes.

5. Yes; all four requirements are met.

7. No; requirement #3 is not met. There are more than 2 possible outcomes.

9. let W = guessing the wrong answer
        C = guessing the correct answer
   $P(W) = 4/5$, for each question
   $P(C) = 1/5$, for each question
   a. $P(WWC) = P(W_1 \text{ and } W_2 \text{ and } C_3)$
              $= P(W_1) \cdot P(W_2) \cdot P(C_3)$
              $= (4/5) \cdot (4/5) \cdot (1/5)$
              $= 16/125 = .128$
   b. There are 3 possible arrangements: WWC, WCW, CWW
      Following the pattern in part (a)
      $P(WWC) = P(W_1 \text{ and } W_2 \text{ and } C_3) = P(W_1) \cdot P(W_2) \cdot P(C_3) = (4/5) \cdot (4/5) \cdot (1/5) = 16/125$
      $P(WCW) = P(W_1 \text{ and } C_2 \text{ and } W_3) = P(W_1) \cdot P(C_2) \cdot P(W_3) = (4/5) \cdot (1/5) \cdot (4/5) = 16/125$
      $P(CWW) = P(C_1 \text{ and } W_2 \text{ and } W_3) = P(C_1) \cdot P(W_2) \cdot P(W_3) = (1/5) \cdot (4/5) \cdot (4/5) = 16/125$
   c. P(exactly one correct answer) = P(WWC or WCW or CWW)
                                    = P(WWC) + P(WCW) + P(CWW)
                                    $= 16/125 + 16/125 + 16/125$
                                    $= 48/125 = .384$

11. From Table A-1 in the .01 column and the 2-0 row, .980.

13. From Table A-1 in the .95 column and the 4-3 row, .171.

15. From Table A-1 in the .95 column and the 10-4 row, $.000^+$.

NOTE: To use the binomial formula, one must identify 3 quantities: n,x,p. Table A-1, for example, requires only these 3 values to supply a probability. Since what the text calls "q" always equals 1-p, it can be so designated without introducing unnecessary notation [just as no special notation is utilized for the quantity n-x, even though it appears twice in the binomial formula]. This has the additional advantage of ensuring that the probabilities p and 1-p sum to 1.00 and protecting against an error in the separate calculation and/or identification of "q." In addition, reversing the order of (n-x)! and x! in the denominator of the $_nC_x$ coefficient term seems appropriate. That agrees with the $n!/(n_1!n_2!)$ logic of the "permutation rule when some objects are alike" for n_1 objects of one type and n_2 objects of another type, and that places the "x" and "n-x" in the same order in both the denominator of the coefficient term and the exponents. Such a natural ordering also leads to fewer errors. Accordingly, this manual expresses the binomial formula as
    $P(x) = [n!/x!(n-x)!] \cdot p^x \cdot (1-p)^{n-x}$

17. $P(x) = [n!/x!(n-x)!] \cdot p^x \cdot (1-p)^{n-x}$
    $P(x=4) = [6!/4!2!] \cdot (.55)^4 \cdot (.45)^2$
    $= [15] \cdot (.55)^4 (.45)^2$
    $= [15] \cdot (.0915) \cdot (.2025) = .2779$

IMPORTANT NOTE: The intermediate values of 15, .0915 and .2025 are given to help those with incorrect answers to identify the portion of the problem in which the mistake was made. This practice will be followed in most problems (i.e., not just binomial problems) throughout the manual. In practice, all calculations can be done in one step on the calculator. You may choose to (or be asked to) write down such intermediate values for your own (or the instructor's) benefit, but <u>never round off in the middle of a problem</u>. <u>Do not write the values down on paper and then re-enter them in the calculator – use the memory to let the calculator remember with complete accuracy any intermediate values that will be used in subsequent calculations.</u> In addition, always make certain that the quantity [n!/x!(n-x)!] is a whole number and that the final answer is between 0 and 1.

19. $P(x) = [n!/x!(n-x)!] \cdot p^x \cdot (1-p)^{n-x}$
    $P(x=3) = [8!/3!5!] \cdot (1/4)^4 \cdot (3/4)^6$
    $= [56] \cdot (.25)^3 \cdot (.75)^5$
    $= .208$

21. $P(x \geq 5) = P(x=5 \text{ or } x=6)$
    $= P(x=5) + P(x=6)$
    $= .3283 + .1428$
    $= .4711$
    No; since .4711 > .05, this is not an unusual occurrence.

23. $P(x > 1) = 1 - P(x \leq 1)$
    $= 1 - P(x=0 \text{ or } x=1)$
    $= 1 - [P(x=0) + P(x=1)]$
    $= 1 - [.0005 + .0071]$
    $= 1 - [.0076]$
    $= .9924$
    NOTE: This also could be found as $P(x>1) = P(x=2 \text{ or } x=3 \text{ or } x=4 \text{ or } x=5 \text{ or } x=6)$. In general the manual will choose the most efficient technique for solving problems that may be approached in more than one way.
    Yes; since P(not having more than one on time) = 1 - .9924 = .0076 < .05, not having more than one on time would be an unusual occurrence.

25. Let x = the number of men that are color blind.
    binomial problem: n = 6 and p = .09, use the binomial formula
    $P(x) = [n!/x!(n-x)!] \cdot p^x \cdot (1-p)^{n-x}$
    $P(x=2) = [6!/2!4!] \cdot (.09)^2 \cdot (.91)^4$
    $= [15] \cdot (.0081) \cdot (.6857)$
    $= .0833$

27. Let x = the number of taxpayers that are audited.
    binomial problem: n = 5 and p = .01, use Table A-1
    a. $P(x=3) = .000^+$
    b. $P(x \geq 3) = P(x=3) + P(x=4) + P(x=5)$
    $= .000^+ + .000^+ + .000^+$
    $= .000^+$
    c. Since $.000^+ < .05$, "three or more are audited" is an unusual result and we conclude that factors other than chance are at work. It appears that either customers of the Hemingway Company are being targeted by the IRS for audits or that customers with the types of returns that are often audited are selecting the Hemingway Company to do

their returns.

NOTE: A classic scenario of this type is the hospital that has higher than average patient mortality. This can be due either to the care being below average or to the accepting of a higher than average proportion of more serious cases.

29. Let x = the number of booked passengers that actually arrive.
    binomial problem: n = 15 and p = .85, use the binomial formula
    $P(x) = [n!/x!(n-x)!] \cdot p^x \cdot (1-p)^{n-x}$
    $P(x=15) = [15!/15!0!] \cdot (.85)^{15} \cdot (.15)^0$
    $= [1] \cdot (.0874) \cdot (1)$
    $= .0874$
    No; the probability is not low enough to be of no concern – since .0874 > .05, not having enough seats is not an unusual event.

31. Let x = the number of households tuned to *60 Minutes*.
    binomial problem: n = 10 and p = .10, use Table A-1
    a. $P(x=0) = .107$
    b. $P(x \geq 1) = 1 - P(x=0)$
    $= 1 - .107$
    $= .893$
    c. $P(x \leq 1) = P(x=0) + P(x=1)$
    $= .107 + .268$
    $= .375$
    d. No; since .375 > .05, it would not be unusual to find at most one household watching *60 Minutes* when that show had a 20% share of the market.

33. Let x = the number of women hired.
    binomial problem: n = 20 and p = .50, use the binomial formula
    $P(x) = [n!/x!(n-x)!] \cdot p^x \cdot (1-p)^{n-x}$
    $P(x \leq 2) = P(x=0) + P(x=1) + P(x=2)$
    $= [20!/0!20!] \cdot (.5)^0 \cdot (.5)^{20} + [20!/1!19!] \cdot (.5)^1 \cdot (.5)^{19} + [20!/2!18!] \cdot (.5)^2 \cdot (.5)^{18}$
    $= [1] \cdot (1) \cdot (.00000095) + [20] \cdot (.5) \cdot (.00000191) + [190] \cdot (.25) \cdot (.00000381)$
    $= .00000095 + .00001907 + .00018120$
    $= .000201$
    Yes, the small probability of this result occurring by chance when all applicants are otherwise equally qualified does support a charge of gender discrimination.

35. The requested table is given in the .50 column of the second page of Table A-1 in the back of the book.
    Let x = the number of girls in 12 births.
    $P(x \geq 9) = P(x=9) + P(x=10) + P(x=11) + P(x=12)$
    $= .054 + .016 + .003 + .000^+$
    $= .073$
    No; this is not evidence that the gender selection technique is effective. Since .073 > .05, getting 9 or more girls is not an unusual event and could occur by chance alone.

37. Let x = the number of components tested to find 1st defect
    geometric problem: p = .2, use the geometric formula
    $P(x) = p \cdot (1-p)^{x-1}$
    $P(x=7) = (.2) \cdot (.8)^6$
    $= (.2) \cdot (.2621)$
    $= .0524$

39. Extending the pattern to cover 6 types of outcomes, where $\Sigma x = n$ and $\Sigma p = 1$,

$P(x_1, x_2, x_3, x_4, x_5, x_6) = [n!/(x_1! x_2! x_3! x_4! x_5! x_6!)] \cdot p_1^{x_1} \cdot p_2^{x_2} \cdot p_3^{x_3} \cdot p_4^{x_4} \cdot p_5^{x_5} \cdot p_6^{x_6}$

$n = 20$

$p_1 = p_2 = p_3 = p_4 = p_5 = p_6 = 1/6$

$x_1 = 5, x_2 = 4, x_3 = 3, x_4 = 2, x_5 = 3, x_6 = 3$

Use the multinomial formula.

$$
\begin{aligned}
P(x_1, x_2, x_3, x_4, x_5, x_6) &= [n!/(x_1! x_2! x_3! x_4! x_5! x_6!)] \cdot p_1^{x_1} \cdot p_2^{x_2} \cdot p_3^{x_3} \cdot p_4^{x_4} \cdot p_5^{x_5} \cdot p_6^{x_6} \\
&= [20!/(5!4!3!2!3!3!)] \cdot (1/6)^5 \cdot (1/6)^4 \cdot (1/6)^3 \cdot (1/6)^2 \cdot (1/6)^3 \cdot (1/6)^3 \\
&= [20!/(5!4!3!2!3!3!)] \cdot (1/6)^{20} \\
&= [1.955 \cdot 10^{12}] \cdot (2.735 \cdot 10^{-16}) \\
&= .000535
\end{aligned}
$$

4.4 Mean, Variance and Standard Deviation for the Binomial Distribution

1. $\mu = n \cdot p = (400) \cdot (.2) = 80.0$
   $\sigma^2 = n \cdot p \cdot (1-p) = (400) \cdot (.2) \cdot (.8) = 64.0$
   $\sigma = 8.0$
   minimum usual value $= \mu - 2\sigma$
   $= 80.0 - 2 \cdot (8.0)$
   $= 64.0$
   maximum usual value $= \mu + 2\sigma$
   $= 80.0 + 2 \cdot (8.0)$
   $= 96.0$

3. $\mu = n \cdot p = (1984) \cdot (3/4) = 1488.0$
   $\sigma^2 = n \cdot p \cdot (1-p) = (1984) \cdot (3/4) \cdot (1/4) = 372.0$
   $\sigma = 19.3$
   minimum usual value $= \mu - 2\sigma$
   $= 1488.0 - 2 \cdot (19.3)$
   $= 1449.4$
   maximum usual value $= \mu + 2\sigma$
   $= 1488.0 + 2 \cdot (19.3)$
   $= 1526.6$

5. Let x = the number of correct answers.
   binomial problem: $n = 10$, $p = .5$
   a. $\mu = n \cdot p = (10) \cdot (.5) = 5.0$
   $\sigma^2 = n \cdot p \cdot (1-p) = (10) \cdot (.5) \cdot (.5) = 2.50$
   $\sigma = 1.6$
   b. maximum usual value $= \mu + 2\sigma$
   $= 5.0 + 2 \cdot (1.6) = 8.2$
   No; since 7 is less than or equal to the maximum usual value, it would not be unusual for a student to pass by getting at least 7 correct answers.

7. Let x = the number of wins.
   binomial problem: $n = 100$, $p = 1/38$
   a. $\mu = n \cdot p = (100) \cdot (1/38) = 2.6$
   $\sigma^2 = n \cdot p \cdot (1-p) = (100) \cdot (1/38) \cdot (37/38) = 2.56$
   $\sigma = 1.6$
   b. minimum usual value $= \mu - 2\sigma$
   $= 2.6 - 2 \cdot (1.6) = -0.6$ [use 0, by practical constraints]
   No; since 0 is greater than or equal to the minimum usual value, it would not be unusual not to win once in the 100 trials.

9. Let x = the number of girls among the 15 births.
   binomial problem: n=15, p = .82

x	P(x)
0	.000+
1	.000+
2	.003
3	.014
4	.042
5	.092
6	.153
7	.196
8	.196
9	.153
10	.092
11	.042
12	.014
13	.003
14	.000+
15	.000+
	1.000

   a. The table at the right was constructed using the .50 column and the final group of values (for n=15) in Table A-1.
   b. $\mu = n{\cdot}p = (15){\cdot}(.50) = 7.5$
      $\sigma^2 = n{\cdot}p{\cdot}(1-p) = (15){\cdot}(.50){\cdot}(.50) = 3.75$
      $\sigma = 1.9$
   c. minimum usual value = $\mu - 2\sigma$
      $= 7.5 - 2{\cdot}(1.9)$
      $= 3.7$
      maximum usual value = $\mu + 2\sigma$
      $= 7.5 + 2{\cdot}(1.9)$
      $= 11.3$
   No; since 10 is between the minimum and maximum usual values, it would not be unusual to get that many girls. NOTE: Since the problem doesn't specify whether the couples were trying to conceive boys or girls, both extremes need to be considered.

11. Let x = the number of customers filing complaints.
    binomial problem: n = 850, p = .032
    a. $\mu = n{\cdot}p = (850){\cdot}(.032) = 27.2$
       $\sigma^2 = n{\cdot}p{\cdot}(1-p) = (850){\cdot}(.032){\cdot}(.968) = 26.33$
       $\sigma = 5.1$
    b. minimum usual value = $\mu - 2\sigma$
       $= 27.2 - 2{\cdot}(5.1) = 17.0$
    Yes; since 7 is less than the minimum usual value, it would be unusual to get that many complaints if the program had no effect and the rate were still 3.2%. This is sufficient evidence to conclude that the program is effective in lowering the rate of complaints.

13. Let x = the number that develop cancer.
    binomial problem: n = 420,000, p = .000340
    a. $\mu = n{\cdot}p = (420,000){\cdot}(.000340) = 142.8$
       $\sigma^2 = n{\cdot}p{\cdot}(1-p) = (420,000){\cdot}(.000340){\cdot}(.999660) = 142.75$
       $\sigma = 11.9$
    b. Unusual values are those outside $\mu \pm 2{\cdot}\sigma$
       $142.8 \pm 2{\cdot}(11.9)$
       142.8 ± 23.8
       119.0 to 166.6
    No; since 135 is within the above limits, it would not be considered an unusual result.
    c. These results do not support the publicized concern that cell phones are such a risk.

15. Let x = the number of adults who believe human cloning should not be allowed.
    a. $(1012){\cdot}(.89) = 901$
       NOTE: Actually, x/1012 rounds to 89% for $896 \le x \le 905$.
    b. binomial problem: n = 1012, p = .50
       $\mu = n{\cdot}p = (1012){\cdot}(.50) = 506.0$
       $\sigma^2 = n{\cdot}p{\cdot}(1-p) = (1012){\cdot}(.50){\cdot}(.50) = 253.0$
       $\sigma = 15.9$
    c. maximum usual value = $\mu + 2\sigma$
       $= 506.0 + 2{\cdot}(15.9) = 537.8$
    Yes; the results (at least 896 believe that human cloning should not be allowed) of the Gallup poll are unusually high if the assumed rate of 50% is true. Yes; this is evidence that the majority of adults believe that human cloning should not be allowed.

17. Let x = the number of baby girls.
    binomial problem: n = 100, p = .5
    μ = n·p = (100)·(.5) = 50.0
    σ^2 = n·p·(1-p) = (100)·(.5)·(.5) = 25.0
    σ = 5.0

    a. Yes: a probability histogram would reveal an approximately bell-shaped distribution. The most likely value is x=50, with the likelihoods decreasing symmetrically for x values smaller or larger than 50.
    b. Since 40 and 60 correspond to $\mu \pm 2 \cdot \sigma$, the empirical rule indicates that approximately 95% of the time x will fall between 40 and 60.
    c. Since 35 and 65 correspond to $\mu \pm 3 \cdot \sigma$, the empirical rule indicates that approximately 99.7% of the time x will fall between 35 and 65.
    d. Since 40 and 60 correspond to $\mu \pm 2 \cdot \sigma$, use Chebyshev's theorem with k=2. The portion of cases within k·σ of μ is at least $1 - 1/k^2$ = 1- 1/4
    $$= 3/4 \text{ (i.e., at least 75\%)}.$$

NOTE: Parts (b) and (d) agree, since 95% is "at ;east 75%." Chebyshev's theorem applies to all distributions and must be more general; the empirical rule applies only to bell-shaped distributions and can be more specific.

4-5 The Poisson Distribution

1. $P(x) = \mu^x \cdot e^{-\mu}/x!$ with μ = 2 and x = 3
   P(x=3) = $(2)^3 \cdot e^{-2}/3!$
   = (8)·(.1353)/6
   = .180

3. $P(x) = \mu^x \cdot e^{-\mu}/x!$ with μ = 100 and x = 99
   P(x=99) = $(100)^{99} \cdot e^{-100}/99!$
   This cannot be evaluated directly on most calculators because the numbers are too large. Even many statistical software packages cannot evaluate this.
   Using Excel, however, yields Poisson(100,99,0) = .039661.
   An approximation is possible using logarithms and exercise #35 of section 3-7 as follows.
   $n! \approx 10^k$, where k = (n+.5)·(log n) + .39908993 - .43429448·n
   for 99, k = (99.5)·(log 99) + .39908993 - .43429446·(99) = 155.9696383
   P(x=99) $\approx (100)^{99} \cdot e^{-100}/10^{155.9696383}$
   log [P(x=99)] $\approx$ 99·(log 100) - 100·(log e) - 155.9696383·(log 10)
   $\approx$ 99·(2) - 100·(.434294482) - 155.9696383·(1)
   $\approx$ -1.399086463
   P(x=99) $\approx 10^{-1.399086463}$
   $\approx$ .03989
   NOTE: Statdisk gives P(x=99) = .03986.

NOTE: In the problems that follow, remember to store the unrounded value for μ for use in the Poisson calculations.

5. a. The number of atoms lost is 1,000,000 - 977,287 = 22,713.
      μ = 22,713/365 = 62.2
   b. $P(x) = \mu^x \cdot e^{-\mu}/x!$ with μ = 62.2 and x = 50
      P(x=50) = $(62.2)^{50} \cdot e^{-62.2}/50!$
      = .0155

7. Let x = the number of horse-kick deaths per corps per year
   $P(x) = \mu^x \cdot e^{-\mu}/x!$ with $\mu = 196/280 = .70$
   a. $P(x=0) = (.70)^0 \cdot e^{-.70}/0! = (1) \cdot (.4966)/1 = .4966$
   b. $P(x=1) = (.70)^1 \cdot e^{-.70}/1! = (.70) \cdot (.4966)/1 = .3476$
   c. $P(x=2) = (.70)^2 \cdot e^{-.70}/2! = (.4900) \cdot (.4966)/2 = .1217$
   d. $P(x=3) = (.70)^3 \cdot e^{-.70}/3! = (.3430) \cdot (.4966)/6 = .0284$
   e. $P(x=4) = (.70)^4 \cdot e^{-.70}/4! = (.2401) \cdot (.4966)/24 = .0050$

   The following table compares the actual relative frequencies to the Poisson probabilities.

x	f	r.f.	P(x)	
0	144	.5143	.4966	
1	91	.3250	.3476	note: r.f. = f/Σf
2	32	.1143	.1217	
3	11	.0393	.0284	
4	2	.0071	.0050	
5 or more	0	.0000	.0007	(by subtraction)
	280	1.0000	1.0000	

   The agreement between the observed relative frequencies and the probabilities predicted by the Poisson formula is very good.

NOTE: The observed/predicted comparison in the above exercise is made using relative frequencies. It could have been made using frequencies. The choice is arbitrary.

9. Let x = the number of wins in 200 trials.
   binomial problem: n = 200, p = 1/38
   Poisson approximation appropriate since n = 200 $\geq$ 100 and np = 200·(1/38) = 5.26 $\leq$ 10.
   $P(x) = \mu^x \cdot e^{-\mu}/x!$ with $\mu = np = 200 \cdot (1/38) = 5.263$
   a. $P(x=0) = (5.263)^0 \cdot e^{-5.263}/0! = (1) \cdot (.00518)/1 = .00518$
   b. $P(x \geq 1) = 1 - P(x=0) = 1 - .00518 = .99482$
   c. $P(x=1) = (5.263)^1 \cdot e^{-5.263}/1! = (5.263) \cdot (.00518)/1 = .02726$
      $P(x=2) = (5.263)^2 \cdot e^{-5.263}/2! = (27.701) \cdot (.00518)/2 = .07173$
      $P(x=3) = (5.263)^3 \cdot e^{-5.263}/3! = (145.794) \cdot (.00518)/6 = .12584$
      $P(x=4) = (5.263)^4 \cdot e^{-5.263}/4! = (767.336) \cdot (.00518)/24 = .16558$
      $P(x=5) = (5.263)^5 \cdot e^{-5.263}/5! = (4038.611) \cdot (.00518)/120 = .17430$
      P(lose money) = P(x=0) + P(x=1) + P(x=2) + P(x=3) + P(x=4) + P(x=5)
               = .00518 + .02726 + .07173 + .12584 + .16558 + .17430 = .5699
   d. P(win money) = 1 - .5699 = .4301

11. Let x = the number of successes in 100 trials.
    binomial problem: n = 100, p = .1
    Poisson approximation appropriate since n = 100 $\geq$ 100 and np = 100·(.1) = 10 $\leq$ 10.
    $P(x) = \mu^x \cdot e^{-\mu}/x!$ with $\mu = np = 100 \cdot (.1) = 10$
    $P(x=101) = (10)^{101} \cdot e^{-10}/101!$
       This cannot be evaluated directly on most calculators because the numbers are too large. Even many statistical software packages cannot evaluate this.
       Using Excel, however, yields Poisson(10,101,0) = 4.17×10^{-31}
       An approximation is possible using logarithms and exercise #35 of section 3-7 as follows.
       $n! \approx 10^k$, where $k = (n+.5) \cdot (\log n) + .39908993 - .43429448 \cdot n$
       for 101, $k = (101.5) \cdot (\log 101) + .39908993 - .43429446 \cdot (101) = 159.9739689$
    $P(x=101) \approx (10)^{101} \cdot e^{-10}/10^{159.9739689}$
    $\log [P(x=101)] \approx 101 \cdot (\log 10) - 10 \cdot (\log e) - 159.97396789 \cdot (\log 10)$
                  $\approx 101 \cdot (1) - 10 \cdot (.434294482) - 159.9739689 \cdot (1)$
                  ≈ -63.31691373
       $P(x=101) \approx 10^{-63.31691373}$
                  $\approx 4.82 \times 10^{-64}$
    It appears that the Poisson value is so small that it can be considered 0, agreeing with the fact that x=101 is impossible in such a binomial distribution.

Review Exercises

1. a. A random variable is a characteristic that assumes a single value (usually a numerical value), determined by chance, for each outcome of an experiment.

   b. A probability distribution is a statement of the possible values a random variable can assume and the probability associated with each of those values. To be valid it must be true for each value x in the distribution that $0 \leq P(x) \leq 1$ and that $\Sigma P(x) = 1$.

   c. Yes, the given table is a valid probability distribution because it meets the definition and conditions in part (b) above.

   NOTE: The following table summarizes the calculations for parts (d) and (e).

x	P(x)	x·P(x)	x²	x²P(x)
0	.08	0	0	0
1	.05	.05	1	.05
2	.10	.20	4	.40
3	.13	.39	9	1.17
4	.15	.60	16	2.40
5	.21	1.05	25	5.25
6	.09	.54	26	3.24
7	.19	1.33	49	9.31
	1.00	4.16		21.82

   d. $\mu = \Sigma x \cdot P(x) = 4.16$, rounded to 4.2

   e. $\sigma^2 = \Sigma x^2 \cdot P(x) - \mu^2$
      $= 21.82 - (4.16)^2$
      $= 4.5144$
      $\sigma = 2.1247$, rounded to 2.1

   f. No; since $P(x=0) = .08 > .05$.

2. Let x = the number of TV's tuned to *West Wing*.
   binomial problem: n = 20 and p = .15, use the binomial formula

   a. $E(x) = \mu = n \cdot p = (20) \cdot (.15) = 3.0$

   b. $\mu = n \cdot p = (20) \cdot (.15) = 3.0$

   c. $\sigma^2 = n \cdot p \cdot (1-p) = (20) \cdot (.15) \cdot (.85) = 2.55$
      $\sigma = 1.6$

   d. $P(x) = [n!/x!(n-x)!] \cdot p^x \cdot (1-p)^{n-x}$
      $P(x=5) = [20!/5!15!] \cdot (.15)^5 \cdot (.85)^{15}$
      $= [15504] \cdot (.0000759) \cdot (.0874)$
      $= .103$

   e. approach #1: minimum usual value $= \mu - 2\sigma = 3.0 - 2 \cdot (1.6) = -0.2$
      No; since 0 is not less than the minimum usual value, it would not be unusual to find that 0 sets are tuned to *West Wing*.
      approach #2: $P(x=5) = [20!/0!20!] \cdot (.15)^0 \cdot (.85)^{20}$
      $= (1) \cdot (1) \cdot (.0388)$
      $= .0388$
      Yes; since .0388 < .05, it would be unusual to find that 0 sets are tuned to *West Wing*.
      resolution: Both approaches are guidelines, and not hard and fast rules. Approach #1 best applies when the probability distribution is approximately bell-shaped and the underlying question is whether it is considered unusual to get a result as extreme as or more extreme than the one under consideration. Approach #2 best applies when the underlying question is about one particular result only, often because that result is the lower or upper limit of the probability distribution. In this instance, follow approach #2.

3. Let x = the number of companies that test for drug abuse.
   binomial problem: n = 10 and p = .80, use Table A-1
   a. $P(x=5) = .026$
   b. $P(x \geq 5) = P(x=5) + P(x=6) + P(x=7) + P(x=8) + P(x=9) + P(x=10)$
      $= .026 + .088 + .201 + .302 + .268 + .107$
      $= .992$
   c. $\mu = n \cdot p = (10) \cdot (.80) = 8.0$
      $\sigma^2 = n \cdot p \cdot (1-p) = (10) \cdot (.80) \cdot (.20) = 1.60$
      $\sigma = 1.3$
   d. Unusual values are those outside $\mu \pm 2 \cdot \sigma$
      $8.0 \pm 2 \cdot (1.3)$
      8.0 ± 2.6
      5.4 to 10.6
   No; since 6 is within these limits, it would not be unusual to find that 6 of the 10 companies test for substance abuse.

4. Let x = the number of workers fired for inability to get along with others.
   binomial problem: n = 5 and p = .17, use the binomial formula
   $P(x) = [n!/x!(n-x)!] \cdot p^x \cdot (1-p)^{n-x}$
   a. $P(x=4) = [5!/4!1!] \cdot (.17)^4 \cdot (.83)^1 = (5) \cdot (.000835) \cdot (.83) = .00347$
      $P(x=5) = [5!/5!0!] \cdot (.17)^5 \cdot (.83)^0 = (1) \cdot (.000142) \cdot (1) = .000142$
      $P(x \geq 4) = P(x=4) + P(x=5)$
      $= .00347 + .00014$
      $= .00361$
   b. Yes; since $.00361 < .05$, $x \geq 4$ would be an unusual result for a company with the standard rate of 17%.

5. Let x = the number of deaths per year.
   Poisson problem: $P(x) = \mu^x \cdot e^{-\mu}/x!$
   a. $\mu = 7/365 = .019$
   b. $P(x=0) = (.019)^0 \cdot e^{-.019}/0! = (1) \cdot (.9810)/1 = .9810$
   c. $P(x=1) = (.019)^1 \cdot e^{-.019}/1! = (.109) \cdot (.9810)/1 = .0188$
   d. $P(x \geq 1) = 1 - [P(x=0) + P(x=1)]$
      $= 1 - [.9810 + .0188]$
      $= 1 - .9998$
      $= .0002$
   e. No; since $.0002 < .05$ by such a wide margin, having more than one death per day is such a rare event that no such contingency plans are necessary.
   NOTE: This decision was based on the deaths of village residents, not on deaths occurring within the village limits. If there are other circumstances to consider (e.g., a nearby highway prone to fatal accidents), then having such contingency plans would be advisable.

Cumulative Review Exercises

1. a. The table below at the left was used to calculate the mean and standard deviation.
   $\overline{x} = [\Sigma(f \cdot x)]/[\Sigma f] = 128/73 = 1.8$
   $s^2 = \{[\Sigma f] \cdot [\Sigma(f \cdot x^2)] - [\Sigma(f \cdot x)]^2\}/\{[(\Sigma f)] \cdot [(\Sigma f) - 1]\}$
   $= [(73) \cdot (730) - (128)^2]/[(73) \cdot (72)]$
   $= 36906/5256$
   $= 7.022$
   $s = 2.6$

x	f	f·x	f·x²		x	r.f.
0	47	0	0		0	.644
1	3	3	3		1	.041
2	1	2	4		2	.014
3	0	0	0		3	.000
4	3	12	48		4	.041
5	11	55	275		5	.151
6	3	18	108		6	.041
7	3	21	147		7	.041
8	1	8	64		8	.014
9	1	9	81		9	.014
	73	128	730			1.000

b. The table is given above at the right and was constructed using r.f. $= f/(\Sigma f) = f/70$.

c. The first two columns of the table below constitute the requested probability distribution.
The other columns were added to calculate the mean and the standard deviation.

x	P(x)	x·P(x)	x²	x²P(x)
0	.1	0	0	0
1	.1	.1	1	.1
2	.1	.2	4	.4
3	.1	.3	9	.9
4	.1	.4	16	1.6
5	.1	.5	25	2.5
6	.1	.6	36	3.6
7	.1	.7	49	4.9
8	.1	.8	64	6.4
9	.1	.9	81	8.1
	1	4.5		28.5

$\mu = \Sigma x \cdot P(x)$
$= 4.5$
$\sigma^2 = \Sigma x^2 \cdot P(x) - \mu^2$
$= 28.5 - (4.5)^2$
$= 8.25$
$\sigma = 2.8723$, rounded to 2.9

d. The presence of so many 0's make it clear that the last digits do not represent a random sample. Factors other than accurate measurement (which, like social security numbers and other such data, would represent random final digits) determined the recorded distances.

2. Let x = the number of such cars that fail the test.
binomial problem: n = 20 and p = .01, use the binomial formula

a. $E(x) = \mu = n \cdot p = (20) \cdot (.01) = 0.2$

b. $\mu = n \cdot p = (20) \cdot (.01) = 0.2$
$\sigma^2 = n \cdot p \cdot (1-p) = (20) \cdot (.01) \cdot (.99) = .198$
$\sigma = 0.4$

c. $P(x) = [n!/x!(n-x)!] \cdot p^x \cdot (1-p)^{n-x}$
$P(x=0) = [20!/0!20!] \cdot (.01)^0 \cdot (.99)^{20}$
$= (1) \cdot (1) \cdot (.818)$
$= .818$
$P(x \geq 1) = 1 - P(x=0)$
$= 1 - .818$
$= .182$

d. Unusual values are those outside $\mu \pm 2 \cdot \sigma$
$0.2 \pm 2 \cdot (0.4)$
0.2 ± 0.8
-0.6 to 1.0
Yes; since 3 is outside these limits, it would be considered an unusual result.

e. Let F = a randomly selected car fails the test.
$P(F) = .01$, for each selection
$P(F_1 \text{ and } F_2) = P(F_1) \cdot P(F_2)$, for independent events
$= (.01) \cdot (.01)$
$= .0001$

Chapter 5

Normal Probability Distributions

5-2 The Standard Normal Distribution

1. The height of the rectangle is .5. Probability corresponds to area, and the area of a rectangle is (width)·(height).
   P(x < 50.3) = (width)·(height)
   = (50.3 - 50.0)·(.5)
   = (.3)·(.5)
   = .15

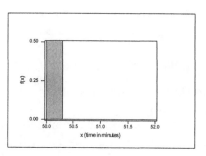

3. The height of the rectangle is .5. Probability corresponds to area, and the area of a rectangle is (width)·(height).
   P(50.5 < x < 50.8) = (width)·(height)
   = (50.8-50.5)·(.5)
   = (.3)·(.5)
   = .15

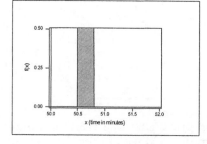

5. The height of the rectangle is 1/6. Probability corresponds to area, and the area of a rectangle is (width)·(height).
   P(x > 10) = (width)·(height)
   = (12-10)·(1/6)
   = (2)·(1/6)
   = .333

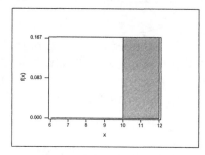

7. The height of the rectangle is 1/6. Probability corresponds to area, and the area of a rectangle is (width)·(height).
   P(7 < x < 10) = (width)·(height)
   = (10-7)·(1/6)
   = (3)·(1/6)
   = .5

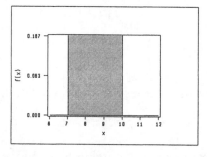

NOTE: The sketch is the key to exercises 9-28. It tells whether to subtract two Table A-2 probabilities, to subtract a Table A-2 probability from 1.0, etc. For the remainder of chapter 5, THE ACCOMPANYING SKETCHES ARE NOT TO SCALE and are intended only as aides to help the reader understand how to use the tabled values to answers the questions. In addition, the probability of any single point in a continuous distribution is zero – i.e., P(x=a) = 0 for any single point a. For normal distributions, therefore, this manual ignores P(x=a) and uses P(x<a) = 1 - P(x>a).

9.   P(z < -.25) = .4013

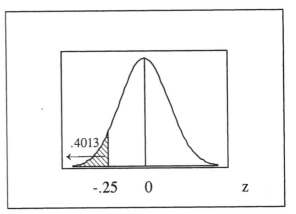

11.   P(z < .25) = .5987

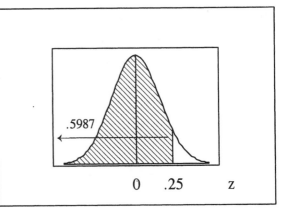

13.   P(z > 2.33)
        = 1 - P(z < 2.33)
        = 1 - .9901
        = .0099

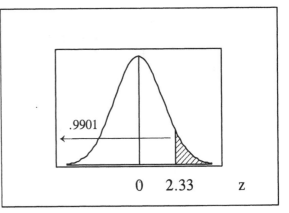

15.   P(z > -2.33)
        = 1 - P(z < -2.33)
        = 1 - .0099
        = .9901

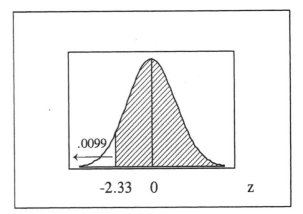

17.   P(.50 < z < 1.50)
          = P(z < 1.50) - P(z < .50)
          = .9332 - .6915
          = .2417

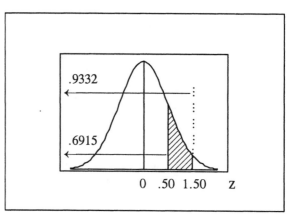

19.   P(-2.00 < z < -1.00)
          = P(z < -1.00) - P(z < -2.00)
          = .1587 - .0228
          = .1359

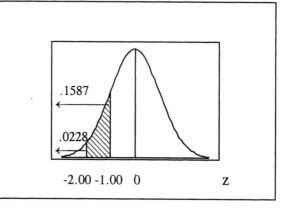

21.   P(-2.67 < z < 1.28)
          = P(z < 1.28) - P(z < -2.67)
          = .8997 - .0038
          = .8959

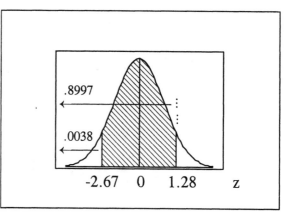

23.   P(-.52 < z < 3.75)
          = P(z < 3.75) - P(z < -.52)
          = .9999 - .3015
          = .6984

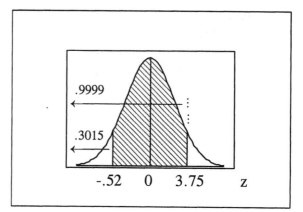

25.   P(z > 3.57)
       = 1 - P(z < 3.57)
       = 1 - .9999
       = .0001

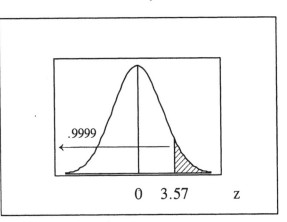

27.   P(z > 0)

       = 1 - P(z < 0)
       = 1 - .5000
       = .5000

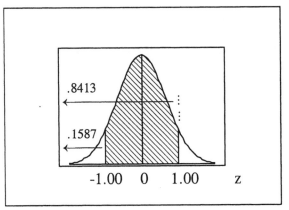

29.   P(-1.00 < z < 1.00)
       = P(z < 1.00) - P(z < -1.00)
       = .8413 - .1587
       = .6826 or 68.26%

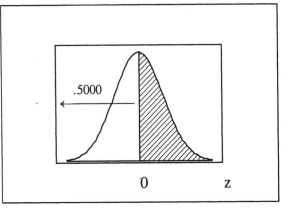

31.   P(-3.00 < z < 3.00)
       = P(z < 3.00) - P(z < -3.00)
       = .9987 - .0013
       = .9974 or 99.74%

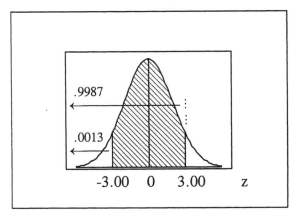

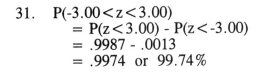

33. $P(-1.96 < z < 1.96)$
       $= P(z < 1.96) - P(z < -1.96)$
       $= .9750 - .0250$
       $= .9500$

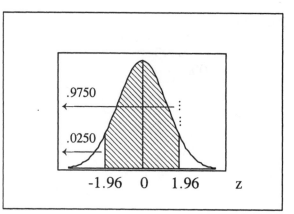

35. $P(z > -2.575)$
       $= 1 - P(z < -2.575)$
       $= 1 - .0050$
       $= .9950$

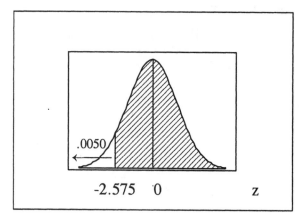

NOTE: The sketch is the key to exercises 37-40. It tells what probability (i.e., cumulative area) to look up when reading Table A-2 "backwards." It also provides a check against gross errors by indicating at a glance whether a z score is above or below 0.

37.   For P_{90}, the cumulative area is .9000.
       The closest entry is .8997, for which
       $z = 1.28$.

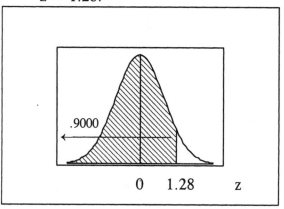

39.   For the lowest 5%, the cumulative
       area is .0500 – indicated by an
       asterisk, for which $z = -1.645$.

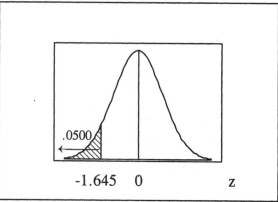

41. Rewrite each of the given statements in terms of z, recalling that z is the number of standard deviations a score is from the mean.

a. P(-1.00 < z < 1.00)
       = P(z < 1.00) - P(z < -1.00)
       = .8413 - .1587
       = .6826 or 68.26%

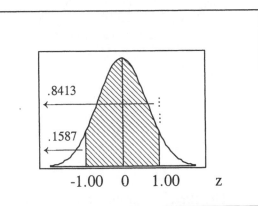

b. P(-1.96 < z < 1.96)
       = P(z < 1.96) - P(z < -1.96)
       = .9750 - .0250
       = .9500 or 95.00%

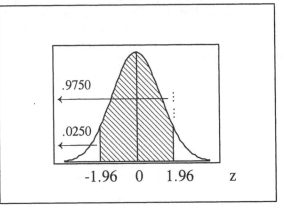

c. P(-3.00 < z < 3.00)
       = P(z < 3.00) - P(z < -3.00)
       = .9987 - .0013
       = .9974 or 99.74%

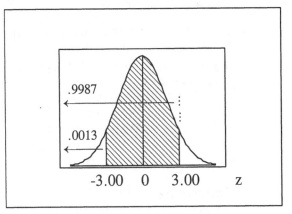

d. P(-1.00 < z < 2.00)
       = P(z < 2.00) - P(z < -1.00)
       = .9772 - .1587
       = .8185 or 81.85%

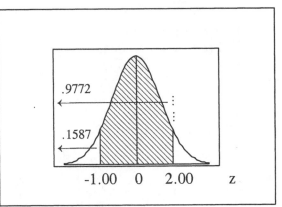

e. P(z < -2.00 or z > 2.00)
$$= P(z < -2.00) + P(z > 2.00)$$
$$= P(z < -2.00) + [1 - P(z < 2.00)]$$
$$= .0228 + [1 - .9772]$$
$$= .0228 + .0228$$
$$= .0456$$

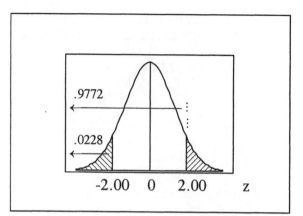

43.   The sketches are the key. They tell what probability (i.e., cumulative area) to look up when reading Table A-2 "backwards." They also provides a check against gross errors by indicating whether a score is above or below zero.

a. P(0 < z < a) = P(z < a) - P(z < 0)
   .3907 = P(z < a) - .5000
   .8907 = P(z < a)
       a = 1.23

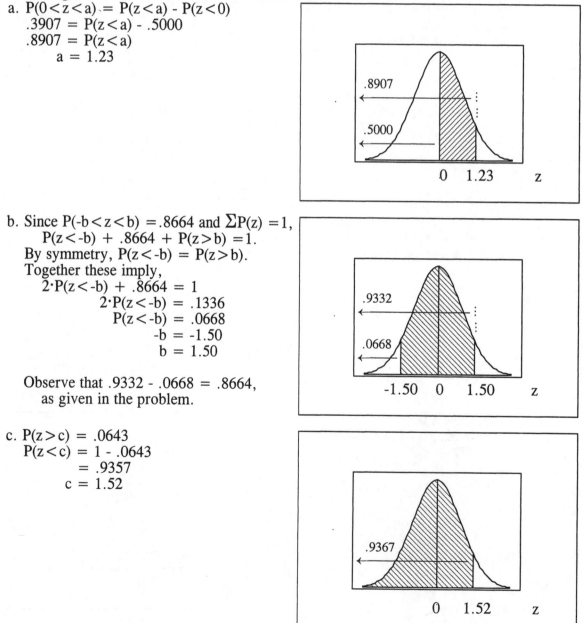

b. Since P(-b < z < b) = .8664 and ΣP(z) = 1,
   P(z < -b) + .8664 + P(z > b) = 1.
   By symmetry, P(z < -b) = P(z > b).
   Together these imply,
       2·P(z < -b) + .8664 = 1
             2·P(z < -b) = .1336
               P(z < -b) = .0668
                      -b = -1.50
                       b = 1.50

   Observe that .9332 - .0668 = .8664,
       as given in the problem.

c. P(z > c) = .0643
   P(z < c) = 1 - .0643
            = .9357
          c = 1.52

d. $P(z > d) = .9922$
   $P(z < d) = 1 - .9922$
   $\quad\quad\quad = .0078$
   $\quad\quad d = -2.42$

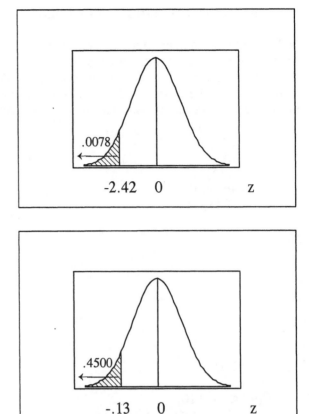

e. $P(z < e) = .4500$ [closest entry is .4483]
   $\quad\quad e = -.13$

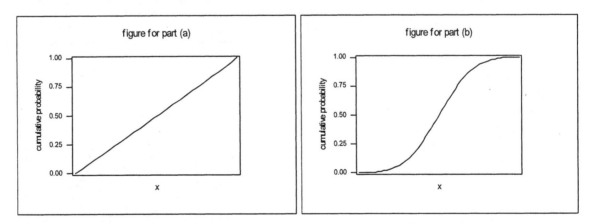

45. a. Moving across the x values from the minimum to the maximum will accumulate probability at a constant rate from 0 to 1. The result will be a straight line (i.e., with constant slope) from (minimum x, 0) to (maximum x, 1). The sketch is given below on the left.

b. Moving across the x values from the minimum to the maximum will accumulate probability slowly at first, and a larger rate near the middle, and more slowly again at the upper end. The result will be a curve that has a variable slope – starting near 0, gradually increasing until reaching a maximum at μ, and then gradually decreasing again toward 0. The sketch is given above on the right.

5-3 Applications of Normal Distributions

NOTE: In each nonstandard normal distribution, x scores are converted to z scores using the formula $z = (x-\mu)/\sigma$ and rounded to two decimal places. For "backwards" problems, solving the preceding formula for x yields $x = \mu + z\sigma$. The area (cumulative probability) given in Table A-2 will be designated by A. As in the previous section, drawing and labeling the sketch is the key to successful completion of the exercises.

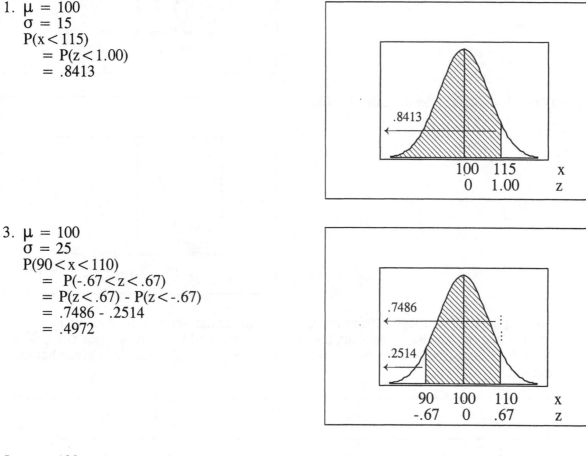

1. $\mu = 100$
   $\sigma = 15$
   $P(x < 115)$
   $\quad = P(z < 1.00)$
   $\quad = .8413$

3. $\mu = 100$
   $\sigma = 25$
   $P(90 < x < 110)$
   $\quad = P(-.67 < z < .67)$
   $\quad = P(z < .67) - P(z < -.67)$
   $\quad = .7486 - .2514$
   $\quad = .4972$

5. $\mu = 100$
   $\sigma = 15$
   For P_{20}, $A = .2000$ [.2005] and $z = -.84$.
   $x = \mu + z\sigma$
   $\quad = 100 + (-.84)(15)$
   $\quad = 100 - 12.6$
   $\quad = 87.4$

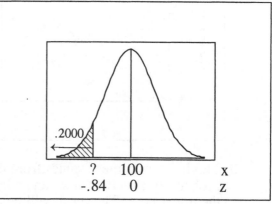

7. $\mu = 100$
   $\sigma = 15$
   For the top 15%,
      $A = .8500$ [.8508] and $z = 1.04$.
   $x = \mu + z\sigma$
      $= 100 + (1.04)(15)$
      $= 100 + 15.6$
      $= 115.6$

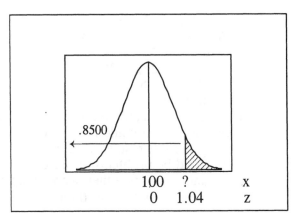

9. $\mu = 98.20$
   $\sigma = .62$
   a. $P(x > 100.6)$
      $= P(z > 3.87)$
      $= 1 - P(< 3.87)$
      $= 1 - .9999$
      $= .0001$
   Yes; the cutoff is appropriate in that there is a small probability of saying that a healthy person has a fever, but many people with low grade fevers may erroneously be labeled healthy.

   b. For the top 5%, $A = .9500$ and $z = 1.645$.
      $x = \mu + z\sigma$
         $= 98.20 + (1.645)(.62)$
         $= 98.20 + 1.02$
         $= 99.22$

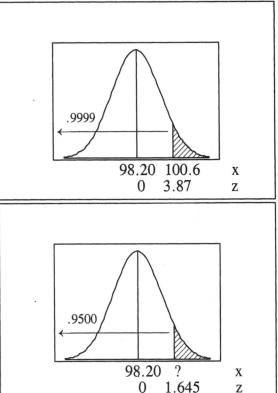

11. $\mu = 998$
    $\sigma = 202$
    a. $P(x < 1100)$
       $= P(z < .50)$
       $= .6915$ or 69.15%

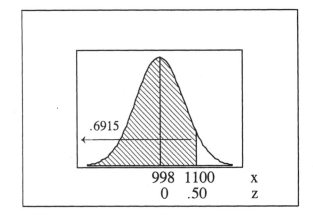

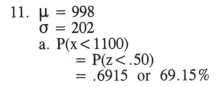

b. For the top 40%,
    A = .6000 [.5987] and z = .25.
    $x = \mu + z\sigma$
      $= 998 + (.25)(202)$
      $= 998 + 50.5$
      $= 1048.5$
The minimum required score would change each year, and it could not be determined until all the test results were in. In addition, there could be problems because the applicants might feel they were competing against each other rather than a standard.

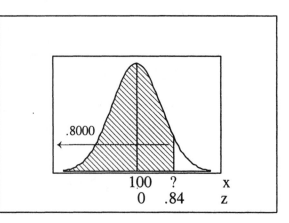

13. $\mu = 8.2$
    $\sigma = 1.1$
    a. $P(x < 5.0)$
        $= P(z < -2.91)$
        $= .0018$

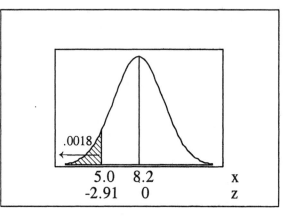

b. For the shortest 1%,
    A = .0100 [.0099] and z = -2.33.
    $x = \mu + z\sigma$
      $= 8.2 + (-2.33)(1.1)$
      $= 8.2 - 2.6$
      $= 5.6$ years
For a "nicer" figure, consider a warranty of 5.5 years = 66 months [for which $P(x < 5.5) < P(x < 5.6) = .01$].

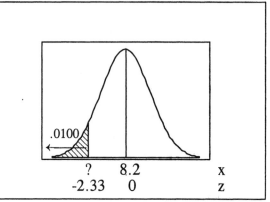

15. $\mu = .91470$
    $\sigma = .03691$
    a. $P(x < .88925)$
        $= P(z < -.69)$
        $= .2451$ or 24.51%, close to 25%

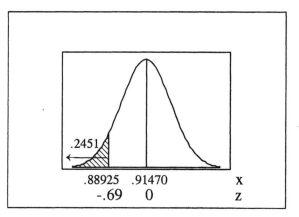

b. For the lowest 25%,
   A = .2500 [.2514] and z = -.67.
   $x = \mu + z\sigma$
   $= .91470 + (-.67)(.03691)$
   $= .91470 - .02473$
   $= .88997$, close to .88925

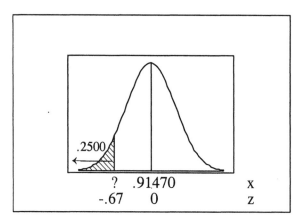

NOTE: Comparing the results using the normal distribution in parts (a) and (b) above to the actual values is not quite fair for two reasons.
1. There are differences in precision. While the weights in the problem were measured to the nearest .00001 gram, percentiles are stated to the whole percent (i.e., the nearest .01) and using the closest entry in Table A-2 limits the accuracy to 3 digits (since the z scores are accurate to the nearest 0.01). A more appropriate comparison uses 2 digits to see that 25% = 25% in part (a) and 3 digits to see that .890 ≈ .889 in part (b).
2. The normal distribution is continuous, but the actual data is discrete. Since the 25th percentile for n=100 scores is $(x_{25}+x_{26})/2$, it might be more appropriate to use A=.2550 instead of A=.2500 in any comparisons. Adjustments to handle discrepancies between continuous and discrete variables are covered in section 5-6.

17. $\mu = 63.6$
    $\sigma = 2.5$
    $P(x > 70)$
    $= P(z > 2.56)$
    $= 1 - P(z < 2.56)$
    $= 1 - .9948$
    $= .0052$ or .52%

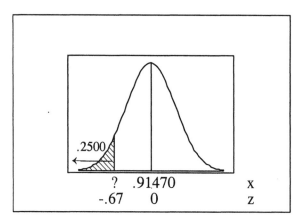

19. $\mu = 63.6$
    $\sigma = 2.5$
    $P(66.5 < x < 71.5)$
    $= P(1.16 < z < 3.16)$
    $= P(z < 3.16) - P(z < 1.16)$
    $= .9992 - .8770$
    $= .1222$
The percentage meeting the requirement is 12.22%. Yes; the Rockettes are taller than the general population of women.

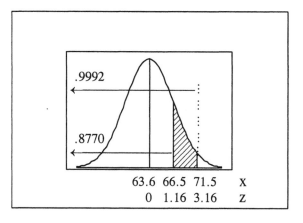

21. normal distribution with $\mu = 143$ lbs and $\sigma = 29$ lbs

   a. The z scores are always unit-free. Because the numerator and the denominator of the fraction $z = (x-\mu)/\sigma$ have the same units, the units will divide out.

   b. For a population of size N, $\mu = \Sigma x/N$ and $\sigma^2 = \Sigma(x-\mu)^2/N$.
   As shown below, $\mu_z = 0$ and $\sigma_z = 1$ will be true for *any* set of z scores.

   $$\begin{aligned}\Sigma z = \Sigma[(x-\mu)/\sigma] &= (1/\sigma)\cdot[\Sigma(x-\mu)] \\ &= (1/\sigma)\cdot[\Sigma x - \Sigma\mu] \\ &= (1/\sigma)\cdot[N\mu - N\mu] \\ &= (1/\sigma)\cdot 0 = 0 \end{aligned}$$

   $$\begin{aligned}\Sigma z^2 = \Sigma[(x-\mu)/\sigma]^2 &= (1/\sigma^2)\cdot[\Sigma(x-\mu)^2] \\ &= (1/\sigma^2)\cdot[N\sigma^2] \\ &= N\end{aligned}$$

   $$\mu_z = (\Sigma z)/N$$
   $$= 0/N = 0$$
   $$\sigma_z^2 = \Sigma(z-\mu_z)^2/N$$
   $$= \Sigma(z-0)^2/N = \Sigma z^2/N = N/N = 1$$
   $$\sigma_z = \sqrt{1} = 1$$

   Since rescaling the scores will not change the basic shape of the distribution, the z scores will still have a normal distribution.

   c. For a population of size N, $\mu = \Sigma x/N$ and $\sigma^2 = \Sigma(x-\mu)^2/N$.
   As shown below, multiplying each value by a constant multiplies the mean and the standard deviation by that constant.

   Let $y = k\cdot x$
   $$\begin{aligned}\mu_y = \Sigma y/N &= [\Sigma(k\cdot x)]/N \\ &= [k\cdot(\Sigma x)]/N \\ &= k\cdot[(\Sigma x)/N] \\ &= k\cdot\mu_x \end{aligned}$$
   $$\begin{aligned}\sigma_y^2 = \Sigma[y - \mu_y]^2/N &= \Sigma[k\cdot x - k\cdot\mu_x]^2/N \\ &= k^2\cdot\Sigma[x-\mu_x]^2/N \\ &= k^2\cdot\sigma_x^2; \; \sigma_y \\ &= k\cdot\sigma_x \end{aligned}$$

   Since changing from pounds to kilograms multiplies each weight by .4536,
   $$\mu = (.4536)(143) = 64.9 \text{ kg}$$
   $$\sigma = (.4536)(29) = 13.2 \text{ kg}$$

23. normal distribution with $\mu = 25$ and $\sigma = 5$

   a. For a population of size N, $\mu = \Sigma x/N$ and $\sigma^2 = \Sigma(x-\mu)^2/N$.
   As shown below, adding a constant to each score will increase the mean by that amount but not affect the standard deviation. In non-statistical terms, shifting everything by k units does not affect the spread of the scores

   Let $y = x+k$
   $$\begin{aligned}\mu_y = [\Sigma(x+k)]/N &= [\Sigma x + Nk]/N \\ &= (\Sigma x)/N + (Nk)/N \\ &= \mu_x+k \end{aligned}$$
   $$\begin{aligned}\sigma_y^2 = \Sigma[y - \mu_y]^2/N &= \Sigma[(x+k) - (\mu_x+k)]^2/N \\ &= \Sigma[x - \mu_x]^2/N \\ &= \sigma_x^2 \end{aligned}$$
   $$\sigma_y = \sigma_x$$

   If the teacher adds 50 to each grade
   new mean $= 25 + 50 = 75$
   new standard deviation $= 5$ (same as before)

b. No; curving the scores should consider the variation. Had the test been more appropriately constructed, it is not likely that every student would score exactly 50 points higher. If the typical student score increased by 50, we would expect the better students to increase by more than 50 and the poorer students to increase by less than 50. This would make the scores more spread out and increase the standard deviation.

c. For the top 10%, A=.9000 [.8997] and z=1.28.
$$x = \mu + z\sigma$$
$$= 25 + (1.28)(5) = 25 + 6.4 = 31.4$$
For the bottom 70%, A=.7000 [.6985] and z=.52.
$$x = \mu + z\sigma$$
$$= 25 + (.52)(5) = 25 + 2.6 = 27.6$$
For the bottom 30%, A=.3000 [.3015] and z=-.52.
$$x = \mu + z\sigma$$
$$= 25 + (-.52)(5) = 25 - 2.6 = 22.4$$
For the bottom 10%, A=.1000 [.1003] and z=-1.28.
$$x = \mu + z\sigma$$
$$= 25 + (-1.28)(5) = 25 - 6.4 = 18.6$$
This produces the grading scheme given at the right.

A: higher than 31.4
B: 27.6 to 31.4
C: 22.4 to 27.6
D: 18.6 to 22.4
E: less than 18.6

d. The curving scheme in part (c) is fairer because it takes into account the variation as discussed in part (b). Assuming the usual 90-80-70-60 letter grade cutoffs, the percentage of A's under the scheme in part (a) with $\mu=75$ and $\sigma=5$ is
$$P(x>90) = P(z>3.00)$$
$$= 1 - P(z<3.00)$$
$$= 1 - .9987$$
$$= .0013 \text{ or } .13\%.$$
This is considerably less than the 10% A's under the scheme in part (c) and reflects the fact that the variation in part (a) is unrealistically small.

25. a. For P_{67}, A=.6700 and z=.44
$$x_{SAT} = \mu + z\sigma$$
$$= 998 + (.44)(202)$$
$$= 998 + 89$$
$$= 1087$$
$$x_{ACT} = \mu + z\sigma$$
$$= 20.9 + (.44)(4.6)$$
$$= 20.9 + 2.0$$
$$= 22.9$$

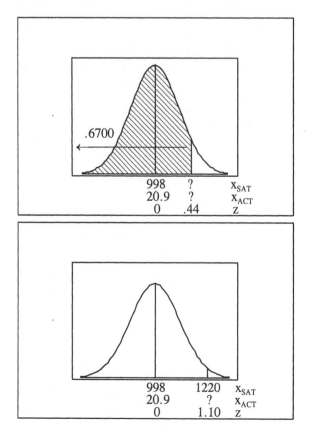

b. $z_{SAT} = (x - \mu)/\sigma$
$$= (1220 - 998)/202 = 1.10$$
$$x_{ACT} = \mu + z\sigma$$
$$= 20.9 + (1.10)(4.6) = 26.0$$

5-4 Sampling Distributions and Estimators

1. No; while the long run average of the values of the sample proportion equals the population proportion, sampling variability produces a distribution of sample proportions. The value of any one sample proportion is a member of that sampling distribution, but not necessarily equal to its long run average.

3. No; the histogram from one sample reflects the shape of the population distribution of individual scores. The shape of the sampling distribution of sample means is not necessarily the same as the shape of the population distribution of individual scores.

5. For the original population,
$\mu = \Sigma x/N = (10+6+5)/3 = 21/3 = 7$.
   a. The 9 equally likely samples and their means are given at the immediate right.
   b. Each of the 9 means has probability 1/9. The sampling distribution, a list of the means and the total probability of each, is given at the far right
   c. $\mu_{\bar{x}} = \Sigma\bar{x}\cdot P(\bar{x}) = 63/9 = 7$.
   d. Yes. Yes, the sample mean is an unbiased estimator of the population mean.

sample	$\bar{x}$
5,5	5.0
5,6	5.5
5,10	7.5
6,5	5.5
6,6	6.0
6,10	8.0
10,5	7.5
10,6	8.0
10,10	10.0
	63.0

$\bar{x}$	$P(\bar{x})$	$\bar{x}\cdot P(\bar{x})$
5.0	1/9	5/9
5.5	2/9	11/9
6.0	1/9	6/9
7.5	2/9	15/9
8.0	2/9	16/9
10.0	1/9	10/9
	9/9	63/9

7. For the original population,
$\mu = \Sigma x/N$
$= (85+79+82+73+78)/5$
$= 397/5 = 79.4$
   a. The 25 equally likely samples and their means are given at the immediate right.
   b. Each of the 25 means has probability 1/25. The sampling distribution, a list of the means and the total probability of each, is given at the far right
   c. $\mu_{\bar{x}} = \Sigma\bar{x}\cdot P(\bar{x})$
   $= 1985/25 = 79.4$
   d. Yes. Yes, the sample mean is an unbiased estimator of the population mean.

sample	$\bar{x}$
73,73	73.0
73,78	75.5
73,79	76.0
73,82	77.5
73,85	79.0
78,73	75.5
78,78	78.0
78,79	78.5
78,82	80.0
78,85	81.5
79,73	76.0
79,78	78.5
79,79	79.0
79,82	80.5
79,85	82.0
82,73	77.5
82,78	80.0
82,79	80.5
82,82	82.0
82,85	83.5
85,73	79.0
85,78	81.5
85,79	82.0
85,82	83.5
85,85	85.0
	1985.0

$\bar{x}$	$P(\bar{x})$	$\bar{x}\cdot P(\bar{x})$
73.0	1/25	73/25
75.5	2/25	151/25
76.0	2/25	152/25
77.5	2/25	155/25
78.0	1/25	78/25
78.5	2/25	157/25
79.0	3/25	237/25
80.0	2/25	160/25
80.5	2/25	161/25
81.5	2/25	163/25
82.0	3/25	246/25
83.5	2/25	167/25
85.0	1/25	85/25
	25/25	1985/25

9. For the original population, the proportion of females is p = 3/4.
   a. Selecting with replacement, there are 4^2 = 16 equally likely samples.

   Let x = the number of females per sample.
   binomial problem: n = 2 and p = 3/4, use the binomial formula
   $$P(x) = [n!/x!(n-x)!] \cdot p^x \cdot (1-p)^{n-x}$$
   $$P(x=0) = [2!/0!2!] \cdot (3/4)^0 \cdot (1/4)^2 = [1] \cdot (1) \cdot (1/16) = 1/16$$
   $$P(x=1) = [2!/1!1!] \cdot (3/4)^1 \cdot (1/4)^1 = [2] \cdot (3/4) \cdot (1/4) = 6/16$$
   $$P(x=2) = [2!/2!0!] \cdot (3/4)^2 \cdot (1/4)^0 = [1] \cdot (9/16) \cdot (1) = 9/16$$
   This means that of the 16 samples, there are
   1 with $\hat{p}$ = 0/2 = 0   [namely, MM]
   6 with $\hat{p}$ = 1/2 = .50   [namely, MA MB MC AM BM CM]
   9 with $\hat{p}$ = 2/2 = 1.00 [namely, AA AB AC BA BB BC CA CB CC]
   b. The sampling distribution, a list of the sample proportions and the probability
   of each, is as follows.

$\hat{p}$	$P(\hat{p})$	$\hat{p} \cdot P(\hat{p})$
0.00	1/16	0/16
0.50	6/16	3/16
1.00	9/16	9/16
	16/16	12/16

   c. $\mu_{\hat{p}} = \Sigma \hat{p} \cdot P(\hat{p}) = 12/16 = 3/4$
   d. Yes. Yes, the sample proportion is an unbiased estimator of the population proportion.

11. For the original population, the proportion of Democrats is 10/13 = .769.
    a. One procedure is as follows. Generate a list of random numbers, find their remainders
       when divided by 13, let the 10 remainders 0-9 correspond to the 10 Democrats and let the
       3 remainders 10-12 correspond to the 3 Republicans. So that each remainder has the
       same chance to occur, limit the usable random numbers to a whole multiple of 13. Using
       the 7x13 = 91 two-digit numbers from 00 to 90 and discarding all others, for example,
       allows each remainder 7 chances to appear among the usable random numbers. Applying
       this procedure to the first two digits of each 5-digit sequence in the list of random
       numbers for the exercises in section 3-6 yields the following.
       46 gives remainder 7, a Democrat
       99 is not usable
       72 gives remainder 7, a Democrat
       44 gives remainder 5, a Democrat
       86 gives remainder 8, a Democrat
       00 gives remainder 0, a Democrat
    b. The proportion of Democrats in the sample in part (a) is 5/5 = 1.00.
    c. The proportion in part (b) is a statistic, because it was calculated from a sample.
    d. No; the above sample proportion does not equal the true population proportion of .769.
       No; samples of size n=5 only produce sample proportions of 0,.20,.40,.60,.80 or 1.00.
    e. The sample proportion is an unbiased estimator of the population proportion. If all
       possible 13^5 = 371,293 samples were listed, the mean of those sample proportions would
       equal the true population proportion.

13. The population mean is $\mu = \Sigma x/N = (62+46+68+64+57) = 297/5 = 59.4$.
    a. All possible samples of size n=2 form a population of $N = {}_5C_2 = 10$ members as follows.

sample	$\overline{x}$	$\overline{x}-\mu_{\overline{x}}$	$(\overline{x}-\mu_{\overline{x}})^2$
46-57	51.5	-7.9	62.41
46-62	54.0	-5.4	29.16
46-64	55.0	-4.4	19.36
46-68	57.0	-2.4	5.76
57-62	59.5	.1	.01
57-64	60.5	1.1	1.21
57-68	62.5	3.1	9.61
62-64	63.0	3.6	12.96
62-68	65.0	5.6	31.36
64-68	66.0	6.6	43.56
	594.0	0	215.40

    $\mu_{\overline{x}} = \Sigma \overline{x}/N = 594.0/10 = 59.4$

    $\sigma_{\overline{x}}^2 = \Sigma(\overline{x}-\mu_{\overline{x}})^2/N = 215.40/10 = 21.540$
    $\sigma_{\overline{x}} = 4.64$

b. All possible samples of size n=3 form a population of $N={}_5C_3=10$ members as follows.

sample	$\overline{x}$	$\overline{x}-\mu_{\overline{x}}$	$(\overline{x}-\mu_{\overline{x}})^2$
46-57-62	55.00	-4.40	19.36
46-57-64	55.67	-3.73	13.94
46-57-68	57.00	-2.40	5.76
46-62-64	57.33	-2.07	4.27
46-62-68	58.67	-.73	.54
46-64-68	59.33	-.07	.00
57-62-64	61.00	1.60	2.56
57-62-68	62.33	2.93	8.60
57-64-68	63.00	3.60	12.96
62-64-68	64.67	5.27	27.74
	594.00	0	95.73

$\mu_{\overline{x}} = \Sigma\overline{x}/N = 594.00/10 = 59.4$

$\sigma_{\overline{x}}^2 = \Sigma(\overline{x}-\mu_{\overline{x}})^2/N = 95.73/10 = 9/573$

$\sigma_{\overline{x}} = 3.09$

c. All possible samples of size n=4 form a population of $N={}_5C_4=5$ members as follows.

sample	$\overline{x}$	$\overline{x}-\mu_{\overline{x}}$	$(\overline{x}-\mu_{\overline{x}})^2$
46-57-62-64	57.25	-2.15	4.6225
46-57-62-68	58.25	-1.15	1.3225
46-57-64-68	58.75	-.65	.4225
46-62-64-68	60.00	.60	.3600
57-62-64-68	62.75	3.35	11.2225
	297.00	0	17.9500

$\mu_{\overline{x}} = \Sigma\overline{x}/N = 297.00/5 = 59.4$

$\sigma_{\overline{x}}^2 = \Sigma(\overline{x}-\mu_{\overline{x}})^2/N = 17.9500/5 = 3.5900$

$\sigma_{\overline{x}} = 1.89$

d. Yes; each of the above sampling distributions has a mean equal to the population mean.

e. As the sample size increases, the variation of the sampling distribution decreases.

15. The 27 equally likely samples of size n=3 are listed below, followed by their medians. The sampling distribution of the medians is given at the right.

m	P(m)	m·P(m)
1	7/27	7/27
2	13/27	26/27
5	7/27	35/27
	27/27	68/27

```
111-1  121-1  151-1  211-1  221-2  251-2  511-1  521-2  551-5
112-1  122-2  152-2  212-2  222-2  252-2  512-2  522-2  552-5
115-1  125-2  155-5  215-2  225-2  255-5  515-5  525-5  555-5
```

The means for the above samples are as follows. The sampling distribution for the means is given at the right.

```
3/3   4/3   7/3    4/3   5/3   8/3    7/3   8/3   11/3
4/3   5/3   8/3    5/3   6/3   9/3    8/3   9/3   12/3
7/3   8/3   11/3   8/3   9/3   12/3   11/3  12/3  15/3
```

$\overline{x}$	P($\overline{x}$)	$\overline{x}$·P($\overline{x}$)
3/3	1/27	3/81
4/3	3/27	12/81
5/3	3/27	15/81
6/3	1/27	6/81
7/3	3/27	21/81
8/3	6/27	48/81
9/3	3/27	27/81
11/3	3/27	33/81
12/3	3/27	36/81
15/3	1/27	15/81
	27/27	216/81

$\mu_m = \Sigma m\cdot P(m) = 68/27 = 2.52$

$\mu_{\overline{x}} = \Sigma\overline{x}\cdot P(\overline{x}) = 216/81 = 2.67$

Since $\mu = 8/3 = 2.67$, the sample mean is a good estimator for the population mean, but the sample median is not. The sampling distributions indicate that

5-5 The Central Limit Theorem

NOTE: When using individual scores (i.e., making a statement about one x score from the original distribution), convert x to z using the mean and standard deviation of the x's and $z = (x-\mu)/\sigma$. When using a sample of n scores (i.e., making a statement about), convert $\overline{x}$ to z using the mean and standard deviation of the $\overline{x}$'s and $z = (\overline{x}-\mu_{\overline{x}})/\sigma_{\overline{x}}$.

IMPORTANT NOTE: After calculating $\sigma_{\overline{x}}$, STORE IT in the calculator to recall it with total accuracy whenever it is needed in subsequent calculations. DO NOT write it down on paper rounded off (even to several decimal places) and then re-enter it in the calculator whenever it is needed. This avoids both round-off errors and recopying errors

1. a. normal distribution
   $\mu = 172$
   $\sigma = 29$
   $P(x < 167)$
      $= P(z < -.17)$
      $= .4325$

   b. normal distribution,
       since the original distribution is so
   $\mu_{\bar{x}} = \mu = 172$
   $\sigma_{\bar{x}} = \sigma/\sqrt{n} = 29/\sqrt{36} = 4.833$
   $P(\bar{x} < 167)$
      $= P(z < -1.03)$
      $= .1515$

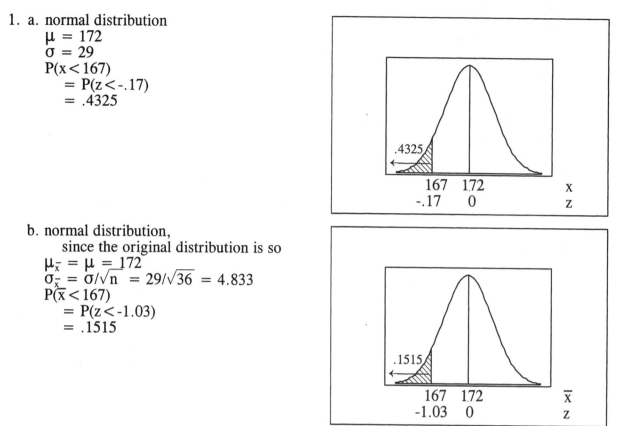

NOTE: Remember that the figures are for illustrative purposes, as an aid to solving the problem, and are not to scale. In scaled drawings, the shaded area would be very large (almost ½ the figure) in part 1(a) and very small in part 1(b).

3. a. normal distribution
   $\mu = 172$
   $\sigma = 29$
   $P(170 < x < 175)$
      $= P(-.07 < z < .10)$
      $= P(z < .10) - P(z < -.07)$
      $= .5398 - .4721$
      $= .0677$

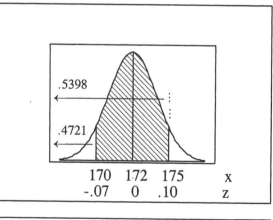

   b. normal distribution,
       since the original distribution is so
   $\mu_{\bar{x}} = \mu = 172$
   $\sigma_{\bar{x}} = \sigma/\sqrt{n} = 29/\sqrt{64} = 3.625$
   $P(170 < \bar{x} < 175)$
      $= P(-.55 < z < .83)$
      $= P(z < .83) - P(z < -.55)$
      $= .7967 - .2912$
      $= .5055$

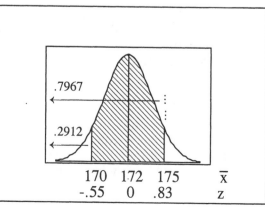

5. a. normal distribution,
       since the original distribution is so
   $\mu_{\bar{x}} = \mu = 172$
   $\sigma_{\bar{x}} = \sigma/\sqrt{n} = 29/\sqrt{25} = 5.8$
   $P(\bar{x} > 160)$
   $= P(z > -2.07)$
   $= 1 - P(z < -2.07)$
   $= 1 - .0192$
   $= .9808$
   b. When the original distribution is normal,
   the sampling distribution of the $\bar{x}$'s is
   normal regardless of the sample size.

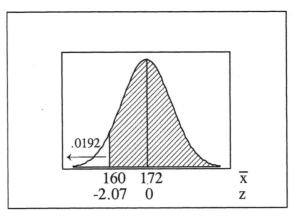

7. a. normal distribution
   $\mu = 143$
   $\sigma = 29$
   $P(140 < x < 211)$
   $= P(-.10 < z < 2.34)$
   $= P(z < 2.34) - P(z < -.10)$
   $= .9904 - .4602$
   $= .5302$

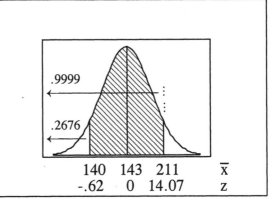

   b. normal distribution,
      since the original distribution is so
   $\mu_{\bar{x}} = \mu = 143$
   $\sigma_{\bar{x}} = \sigma/\sqrt{n} = 29/\sqrt{36} = 4.833$
   $P(140 < \bar{x} < 211)$
   $= P(-.62 < z < 14.07)$
   $= P(z < 14.07) - P(z < -.62)$
   $= .9999 - .2676$
   $= .7323$
   c. The information from part (a) is more
   relevant, since the seat will be occupied by
   one woman at a time.

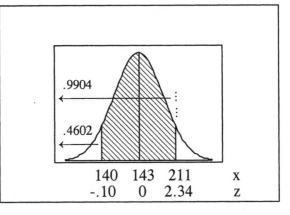

9. a. normal distribution,
      since the original distribution is so
   $\mu_{\bar{x}} = \mu = 14.4$
   $\sigma_{\bar{x}} = \sigma/\sqrt{n} = 1.0/\sqrt{2} = .707$
   $P(\bar{x} > 16.0)$
   $= P(z > 2.26)$
   $= 1 - P(z < 2.26)$
   $= 1 - .9881$
   $= .0119$
   b. No. Yes; even if all the seats were occupied
   by 2 men, there would be a problem only
   about 1% 0f the time – in practice, most
   men probably sit with a female or a child.

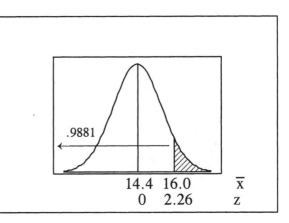

11. a. normal distribution,
   by the Central Limit Theorem
  $\mu_{\bar{x}} = \mu = 12.00$
  $\sigma_{\bar{x}} = \sigma/\sqrt{n} = .11/\sqrt{36} = .0183$
  $P(\bar{x} \geq 12.19)$
   $= P(z \geq 10.36)$
   $= 1 - P(z < 10.36)$
   $= 1 - .9999$
   $= .0001$

 b. No; if $\mu = 12$, then an extremely rare event
  has occurred. No; there is no cheating,
  since the actual amount is more than the
  advertised amount.

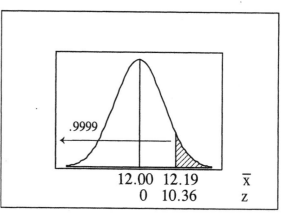

13. a. normal distribution,
   by the Central Limit Theorem
  $\mu_{\bar{x}} = \mu = 8.2$
  $\sigma_{\bar{x}} = \sigma/\sqrt{n} = 1.1/\sqrt{50} = .156$
  $P(\bar{x} \leq 7.8)$
   $= P(z \leq -2.57)$
   $= .0051$

 b. Yes; it appears that the TV sets sold by the
  Portland Electronics store are of less than
  average quality.

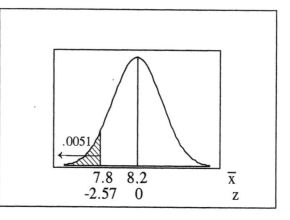

15. a. normal distribution,
   by the Central Limit Theorem
  $\mu_{\bar{x}} = \mu = .941$
  $\sigma_{\bar{x}} = \sigma/\sqrt{n} = .313/\sqrt{40} = .0495$
  $P(\bar{x} \leq .882)$
   $= P(z < -1.19)$
   $= .1170$

 b. No; since $.1170 > .05$, getting such a mean
  under the previous conditions would not be
  considered unusual.

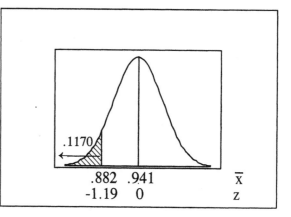

17. normal distribution,
  since the original distribution is so
  $\mu_{\bar{x}} = \mu = 27.44$
  $\sigma_{\bar{x}} = \sigma/\sqrt{n} = 12.46/\sqrt{4872} = .179$
  $P(\bar{x} > 27.88)$
   $= P(z > 2.46)$
   $= 1 - P(z < 2.46)$
   $= 1 - .9931$
   $= .0069$
 The system is currently acceptable and can
 expect to be overloaded only $(.0069)(52) = .36$
 weeks a year -- or about once every three years.

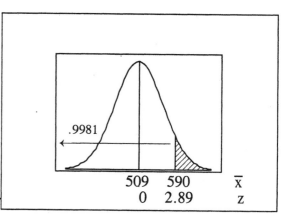

19. normal distribution,
    since the original distribution is so
    $\mu_{\bar{x}} = \mu = 172$
    $\sigma_{\bar{x}} = \sigma/\sqrt{n} = 29/\sqrt{16} = 7.25$
    This is a "backwards" normal problem,
    with A = .9760 and z = 1.96.
    $\bar{x} = \mu_{\bar{x}} + z\sigma_{\bar{x}}$
    $= 172 + (1.96)(7.25)$
    $= 172 + 14.21$
    $= 186.21$
    For 16 men, the total weight is
    $(16)(186.21) = 2979$ lbs.

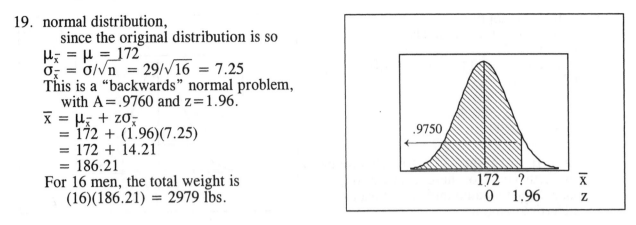

21. normal distribution, since the original distribution is so
    NOTE: Since 8/120 = .0667 > .05, use the finite population correction factor
    $\mu_{\bar{x}} = \mu = 143$
    $\sigma_{\bar{x}} = [\sigma/\sqrt{n}]\cdot\sqrt{(N-n)/(N-1)}$
    $= [29/\sqrt{8}]\cdot\sqrt{(120-8)/(120-1)}$
    $= [29/\sqrt{8}]\cdot\sqrt{(112)/(119)}$
    $= 9.947$
 a. A total weight of 1300 for 8 women
        corresponds to $\bar{x} = 1300/8 = 162.5$.
    $P(\bar{x} \le 162.5)$
        $= P(z \le 1.96)$
        $= .9750$

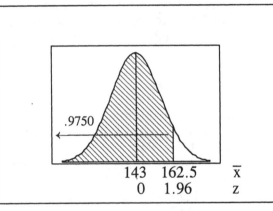

 b. normal distribution,
        since the original distribution is so
    $\mu_{\bar{x}} = \mu = 143$
    $\sigma_{\bar{x}} = 9.947$ (as calculated above)
    This is a "backwards" normal problem,
        with A = .9900 [.9901] and z = 2.33.
    $\bar{x} = \mu_{\bar{x}} + z\sigma_{\bar{x}}$
        $= 143 + (2.33)(9.947)$
        $= 143 + 23.18$
        $= 166.18$
    For 8 women, the total weight is
        $(8)(166.18) = 1329$ lbs.

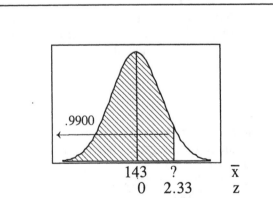

23. normal distribution,
by the Central Limit Theorem
$$\mu_{\bar{x}} = \mu = .500$$
$$\sigma_{\bar{x}} = \sigma/\sqrt{n} = .289/\sqrt{100} = .0289$$
$$P(.499 < \bar{x} < .501)$$
$$= P(-.03 < z < .03)$$
$$= P(z < .03) - P(z < -.03)$$
$$= .5120 - .4880$$
$$= .0240$$

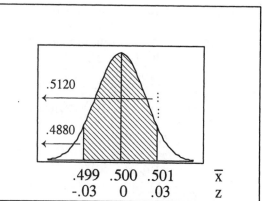

Since $.0240 < .05$, the event is "unusual" in that the sample mean will not fall within those narrow limits very often. Since the probability of getting a result "as extreme or more extreme" would be very close to ½ (whether the sample mean was less than or greater than $\mu = .5000$), the event is not of the "extreme" sort that leads one to question the assumptions associated with the procedure. The key idea is the difference between "unusual in the sense of rare" and "unusual in the sense of extreme." Since there are so many possible results, the probability of any one them (even the most "normal" ones) occurring is small.

5-6 Normal Distribution as Approximation to Binomial Distribution

NOTE: As in the previous sections, P(E) represents the probability of an event E; this manual uses $P_c(E)$ to represent the probability of an event E with the continuity correction applied.

1. the area to the right of 15.5; in symbols, $P(x > 15) = P_c(x > 15.5)$

3. the area to the left of 99.5; in symbols, $P(x < 100) = P_c(x < 99.5)$

5. the area to the left of 4.5; in symbols, $P(x \leq 4) = P_c(x < 4.5)$

7. the area from 7.5 to 10.5; in symbols, $P(8 \leq x \leq 10) = P_c(7.5 < x < 10.5)$

IMPORTANT NOTE: As in the previous sections, store σ in the calculator so that it may be recalled with complete accuracy whenever it is needed in subsequent calculations.

9. binomial: n = 14 and p = .50
a. from Table A-1, $P(x=9) = .122$
b. normal approximation appropriate since
$$np = 14(.50) = 7 \geq 5$$
$$n(1-p) = 14(.50) = 7 \geq 5$$
$$\mu = np = 14(.50) = 7$$
$$\sigma = \sqrt{np(1-p)} = \sqrt{14(.50)(.50)} = 1.871$$
$$P(x=9)$$
$$= P_c(8.5 < x < 9.5)$$
$$= P(.80 < z < 1.34)$$
$$= P(z < 1.34) - P(z < .80)$$
$$= .9099 - .7881$$
$$= .1218$$

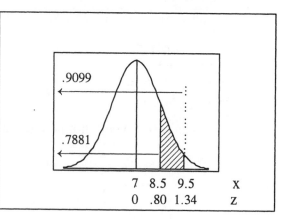

11. binomial: n = 15 and p = .90
a. from Table A-1, $P(x \geq 14) = P(x=14) + P(x=15)$
$$= .343 + .206$$
$$= .549$$
b. normal approximation <u>not</u> appropriate since $n(1-p) = 15(.10) = 1.5 < 5$

13. let x = the number of girls born
binomial: n = 100 and p = .50
normal approximation appropriate since
$\quad$ np = 100(.50) = 50 ≥ 5
$\quad$ n(1-p) = 100(.50) = 50 ≥ 5
μ = np = 100(.50) = 50
σ = $\sqrt{np(1-p)}$ = $\sqrt{100(.50)(.50)}$ = 5.000
P(x > 55)
$\quad$ = P_c(x > 55.5)
$\quad$ = P(z > 1.10)
$\quad$ = 1 - P(z < 1.10)
$\quad$ = 1 - .8643
$\quad$ = .1357

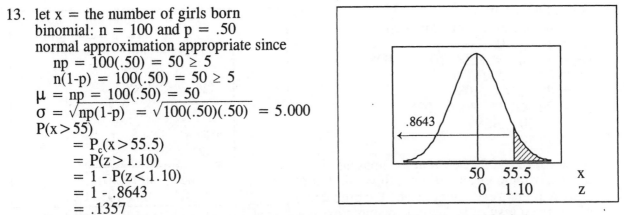

No; since .1357 > .05, it is not unusual to get more than 55 girls in 100 births.

15. let x = the number of correct responses
binomial: n = 100 and p = .50
normal approximation appropriate since
$\quad$ np = 100(.50) = 50 ≥ 5
$\quad$ n(1-p) = 100(.50) = 50 ≥ 5
μ = np = 100(.50) = 50
σ = $\sqrt{np(1-p)}$ = $\sqrt{100(.50)(.50)}$ = 5.000
P(x ≥ 60)
$\quad$ = P_c(x > 59.5)
$\quad$ = P(z > 1.90)
$\quad$ = 1 - P(z < 1.90)
$\quad$ = 1 - .9713
$\quad$ = .0287

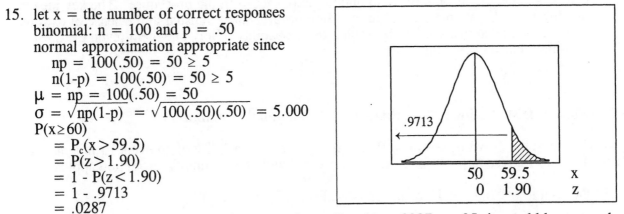

No; the probability of passing by chance is small – since .0287 < .05, it would be unusual.

17. let x = the number with yellow pods
binomial: n = 580 and p = .25
normal approximation appropriate since
$\quad$ np = 580(.25) = 145 ≥ 5
$\quad$ n(1-p) = 580(.75) = 435 ≥ 5
μ = np = 580(.25) = 145
σ = $\sqrt{np(1-p)}$ = $\sqrt{580(.25)(.75)}$ = 10.428
P(x ≥ 152)
$\quad$ = P_c(x > 151.5)
$\quad$ = P(z > .62)
$\quad$ = 1 - P(z < .62)
$\quad$ = 1 - .7324
$\quad$ = .2676

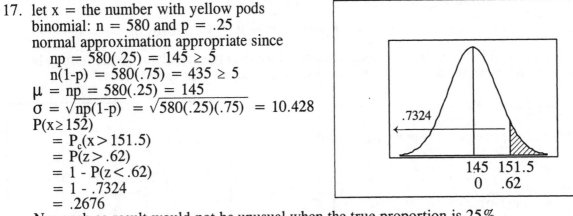

No; such as result would not be unusual when the true proportion is 25%.

19. let x = the number of color blind men
    binomial: n = 600 and p = .09
    normal approximation appropriate since
       np = 600(.09) = 54 ≥ 5
       n(1-p) = 600(.91) = 546 ≥ 5
    μ = np = 600(.09) = 54
    σ = $\sqrt{np(1-p)}$ = $\sqrt{600(.09)(.91)}$ = 7.010
    P(x≥50)
        = P_c(x>49.5)
        = P(z>-.64)
        = 1 - P(z<-.64)
        = 1 - .2611
        = .7389

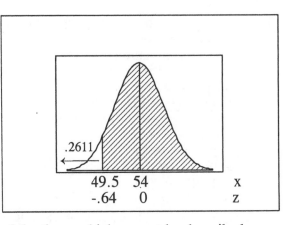

No; they will get the desired number about 3/4 of the time, which cannot be described as "very confident."

21. let x = the number of booked arrivals
    binomial: n = 400 and p = .85
    normal approximation appropriate since
       np = 400(.85) = 340 ≥ 5
       n(1-p) = 400(.15) = 60 ≥ 5
    μ = np = 400(.85) = 340
    σ = $\sqrt{np(1-p)}$ = $\sqrt{400(.85)(.15)}$ = 7.141
    P(x>350)
        = P_c(x>350.5)
        = P(z>1.47)
        = 1 - P(z<1.47)
        = 1 - .9292
        = .0708

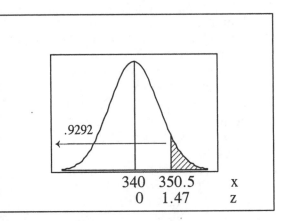

Yes; the airline would probably be willing to allow overbooking to occur 7% of the time (i.e., about 1 in every 14 flights) in order to fly with fewer empty seats.

23. let x = the number of women hired
    binomial: n = 62 and p = .50
    normal approximation appropriate since
       np = 62(.50) = 31 ≥ 5
       n(1-p) = 62(.50) = 31 ≥ 5
    μ = np = 62(.50) = 31
    σ = $\sqrt{np(1-p)}$ = $\sqrt{62(.50)(.50)}$ = 3.937
    P(x≤21)
        = P_c(x<21.5)
        = P(z<-2.41)
        = .0080

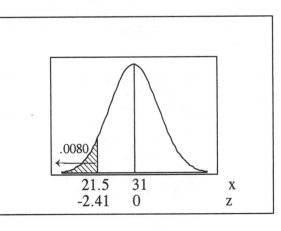

Yes; since the resulting probability is so small, it does support a charge of such discrimination.

25. let x = the number with group O blood
    binomial: n = 400 and p = .45
    normal approximation appropriate since
        np = 400(.45) = 180 ≥ 5
        n(1-p) = 400(.55) = 220 ≥ 5
    μ = np = 400(.45) = 180
    $\sigma = \sqrt{np(1-p)} = \sqrt{400(.45)(.55)}$ = 9.950
    P(x ≥ 177)
        = P_c(x > 176.5)
        = P(z > -.35)
        = 1 - P(z < -.35)
        = 1 - .3632 = .6368
    Yes; the pool is "likely" to be sufficient,
    but it wouldn't be unusual for it not to be sufficient either.

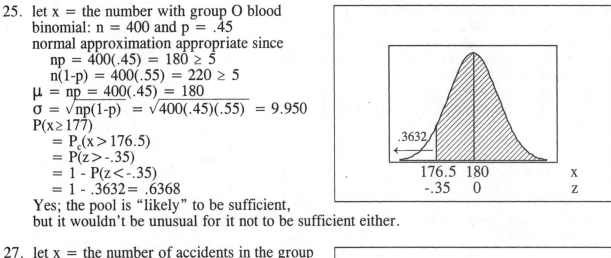

26.

27. let x = the number of accidents in the group
    binomial: n = 500 and p = .34
    normal approximation appropriate since
        np = 500(.34) = 170 ≥ 5
        n(1-p) = 500(.66) = 330 ≥ 5
    μ = np = 500(.34) = 170
    $\sigma = \sqrt{np(1-p)} = \sqrt{500(.34)(.66)}$ = 10.592
    A 40% accident rate implies
        x = 500(.40) = 200.
    P(x ≥ 200)
        = P_c(x > 199.5)
        = P(z > 2.79)
        = 1 - P(z < 2.79)
        = 1 - .9974
        = .0026
    Yes; either a very unusual sample occurred, or the true NYC rate us higher than 34%.
    NOTE: The number of in the survey that had accidents last year was not given. Any
    198 ≤ x ≤ 202 rounds to 40%. The conclusion will be the same for any value within those
    limits.

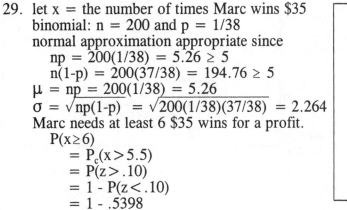
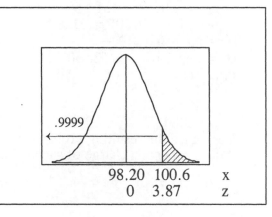

29. let x = the number of times Marc wins $35
    binomial: n = 200 and p = 1/38
    normal approximation appropriate since
        np = 200(1/38) = 5.26 ≥ 5
        n(1-p) = 200(37/38) = 194.76 ≥ 5
    μ = np = 200(1/38) = 5.26
    $\sigma = \sqrt{np(1-p)} = \sqrt{200(1/38)(37/38)}$ = 2.264
    Marc needs at least 6 $35 wins for a profit.
        P(x ≥ 6)
            = P_c(x > 5.5)
            = P(z > .10)
            = 1 - P(z < .10)
            = 1 - .5398
            = .4602

31. a. binomial: $n = 4$ and $p = .350$

$$P(x \geq 1) = 1 - P(x=0)$$
$$= 1 - [4!/0!4!] \cdot (.350)^0 (.650)^4$$
$$= 1 - .1785$$
$$= .8215$$

b. binomial: $n = 56 \cdot 4 = 224$ and $p = .350$
normal approximation appropriate since

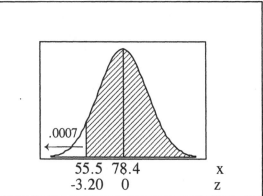

$np = 224(.350) = 78.4 \geq 5$
$n(1-p) = 224(.650) = 145.6 \geq 5$
$\mu = np = 224(.350) = 78.4$
$\sigma = \sqrt{np(1-p)} = \sqrt{224(.350)(.650)} = 7.139$
$P(x \geq 56)$
$= P_c(x > 55.5)$
$= P(z > -3.20)$
$= 1 - P(z < -3.20)$
$= 1 - .0007$
$= .9993$

c. let H = getting at least one hit in 4 times at bat
$P(H) = .8215$ [from part (a) above]
for 56 consecutive games, $[P(H)]^{56} = [.8125]^{56} = .0000165$

d. The solution below employs the methods and notation of parts (a) and (c) above.
for $[P(H)]^{56} > .10$, it is required that
$$P(H) > (.10)^{1/56}$$
$$P(H) > .9597$$
for $P(H) = P(x \geq 1) > .9597$, it is required that
$$1 - P(x=0) > .9597$$
$$.0403 > P(x=0)$$
$$.0403 > [4!/0!4!] \cdot p^0 (1-p)^4$$
$$.0403 > (1-p)^4$$
$$(.0403)^{1/4} > 1 - p$$
$$p > 1 - (.0403)^{1/4}$$
$$p > 1 - .448$$
$$p > .552$$

5-7 Determining Normality

1. Not normal. The points exhibit a systematic "snaking" pattern around a straight lin passing through the data.

3. Not normal. It would require a "broken" line with four segments of four different slopes to pass through the data.

NOTE: The tables on the next page give the information for working exercises #5-8 and #9-12 by hand. While these exercises can be answered using computer software, seeing the actual procedure and calculations involved promotes a better understanding of the concepts and processes of this section. For each problem,

x = the n original scores arranged in order
cp = the cumulative probability values 1/2m, 3/2m, 5/2n,..., (2n-1)/2n
z = the z scores have the cp value to their left.

#	5. rainfall			6. heads			7. M&M's			8. conductivity		
	x	cp	z	x	cp	z	x	cp	z	x	cp	z
1	.00	.009	-2.35	35.5	.01	-2.33	.856	.015	-2.17	28.1	.008	-2.40
2	.00	.028	-1.91	35.7	.03	-1.88	.858	.045	-1.69	29.3	.025	-1.97
3	.00	.047	-1.67	39.2	.05	-1.64	.860	.076	-1.43	30.2	.041	-1.74
4	.00	.066	-1.51	39.6	.07	-1.48	.866	.106	-1.25	30.3	.057	-1.58
5	.00	.085	-1.37	39.7	.09	-1.34	.867	.136	-1.10	30.5	.074	-1.45
6	.00	.104	-1.26	39.8	.11	-1.23	.871	.167	-0.97	32.1	.090	-1.34
7	.00	.123	-1.16	39.9	.13	-1.13	.875	.197	-0.85	32.7	.107	-1.25
8	.00	.142	-1.07	40.1	.15	-1.04	.876	.227	-0.75	32.8	.123	-1.16
9	.00	.160	-0.99	40.2	.17	-0.95	.889	.258	-0.65	33.5	.139	-1.08
10	.00	.179	-0.92	40.2	.19	-0.88	.897	.288	-0.56	33.8	.156	-1.01
11	.00	.198	-0.85	40.2	.21	-0.81	.898	.318	-0.47	35.9	.172	-0.95
12	.00	.217	-0.78	40.4	.23	-0.74	.900	.348	-0.39	40.5	.189	-0.88
13	.00	.236	-0.72	40.4	.25	-0.67	.902	.379	-0.31	40.6	.205	-0.82
14	.00	.255	-0.66	40.7	.27	-0.61	.902	.409	-0.23	41.5	.221	-0.77
15	.00	.274	-0.60	40.9	.29	-0.55	.904	.439	-0.15	42.4	.238	-0.71
16	.00	.292	-0.55	40.9	.31	-0.50	.909	.470	-0.08	43.2	.254	-0.66
17	.00	.311	-0.49	40.9	.33	-0.44	.909	.500	0.00	44.3	.270	-0.61
18	.00	.330	-0.44	40.9	.35	-0.39	.914	.530	0.08	45.6	.287	-0.56
19	.00	.349	-0.39	40.9	.37	-0.33	.914	.561	0.15	46.5	.303	-0.51
20	.00	.368	-0.34	41.0	.39	-0.28	.919	.591	0.23	46.7	.320	-0.47
21	.00	.387	-0.29	41.0	.41	-0.23	.920	.621	0.31	46.7	.336	-0.42
22	.00	.406	-0.24	41.0	.43	-0.18	.921	.652	0.39	47.1	.352	-0.38
23	.00	.425	-0.19	41.0	.45	-0.13	.923	.682	0.47	48.1	.369	-0.33
24	.00	.443	-0.14	41.0	.47	-0.08	.928	.712	0.56	48.3	.385	-0.29
25	.00	.462	-0.09	41.1	.49	-0.03	.930	.742	0.65	48.5	.402	-0.25
26	.00	.481	-0.05	41.1	.51	0.03	.930	.773	0.75	48.5	.418	-0.21
27	.00	.500	0.00	41.2	.53	0.08	.932	.803	0.85	48.5	.434	-0.17
28	.00	.519	0.05	41.3	.55	0.13	.936	.833	0.97	48.6	.451	-0.12
29	.00	.538	0.09	41.4	.57	0.18	.955	.864	1.10	49.0	.467	-0.08
30	.00	.557	0.14	41.7	.59	0.23	.965	.894	1.25	49.0	.484	-0.04
31	.00	.575	0.19	41.7	.61	0.28	.976	.924	1.43	49.7	.500	0.00
32	.00	.594	0.24	41.7	.63	0.33	.988	.955	1.69	49.8	.516	0.04
33	.00	.613	0.29	41.7	.65	0.39	1.033	.985	2.17	49.8	.533	0.08
34	.00	.632	0.34	41.8	.67	0.44				49.9	.549	0.12
35	.00	.651	0.39	41.9	.69	0.50				49.9	.566	0.17
36	.00	.670	0.44	41.9	.71	0.55				49.9	.582	0.21
37	.00	.689	0.49	42.0	.73	0.61				50.3	.598	0.25
38	.01	.708	0.55	42.0	.75	0.67				50.4	.615	0.29
39	.01	.726	0.60	42.2	.77	0.74				50.5	.631	0.33
40	.01	.745	0.66	42.2	.79	0.81				51.0	.648	0.38
41	.02	.764	0.72	42.3	.81	0.88				51.0	.664	0.42
42	.06	.783	0.78	42.3	.83	0.95				51.2	.680	0.47
43	.06	.802	0.85	42.4	.85	1.04				51.3	.697	0.51
44	.06	.821	0.92	42.5	.87	1.13				51.7	.713	0.56
45	.08	.840	0.99	42.6	.89	1.23				51.9	.730	0.61
46	.08	.858	1.07	42.8	.91	1.34				52.0	.746	0.66
47	.12	.877	1.16	42.8	.93	1.48				52.1	.762	0.71
48	.14	.896	1.26	42.8	.95	1.64				52.2	.779	0.77
49	.18	.915	1.37	43.2	.97	1.88				52.4	.795	0.82
50	.27	.934	1.51	43.2	.99	2.33				53.6	.811	0.88
51	.31	.953	1.67							55.2	.828	0.95
52	.64	.972	1.91							56.8	.844	1.01
53	.64	.991	2.35							56.8	.861	1.08
54										57.0	.877	1.16
55										57.1	.893	1.25
56										57.3	.910	1.34
57										57.7	.926	1.45
58										57.8	.943	1.58
59										57.8	.959	1.74
60										58.4	.975	1.97
61										59.2	.992	2.40

5. No. The frequency distribution and histogram for the Wednesday rain data are given below. The distribution is not normal because it is not symmetric – being positively skewed and having no values below the modal class.

rainfall	frequency
-.05 - .05	41
.05 - .15	7
.15 - .25	1
.25 - .35	2
.35 - .45	0
.45 - .55	0
.55 - .65	2
	53

7. Yes. The frequency distribution and histogram for the M&M weights are given below. The weights appear approximately normally distributed, with frequencies tapering off in both directions about the modal class.

weight (mg)	frequency
838 - 862	3
863 - 887	5
888 - 912	9
913 - 937	11
938 - 962	1
963 - 987	2
988 - 1012	1
1013 - 1037	1
	33

NOTE: The normal quantile plots for exercises #9-12 may be constructed using the appropriate columns from the table on the previous page. Exercise #10 most clearly illustrates the process: *each of the 50 scores is 2% of the data set and the first score covers .00-.02, the midpoint of which is .01.* Plot the x values on the horizontal axis and the corresponding z values (calculated as indicated from the cp values) on the vertical axis. Making a "vertically consistent-scaled" plot with the z scores determined from the normal probability distribution is equivalent to making a "vertically stretch-scaled" plot using the cumulative probabilities. As with most exercises in this section, the judgment as to whether the points can be approximated by a straight line is subjective.

9. No. Since the points do not lie close to a straight line, conclude that the population distribution is not approximately normal.

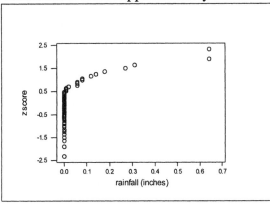

11. Yes. Since the points lie close to a straight line, conclude the population distribution is approximately normal.

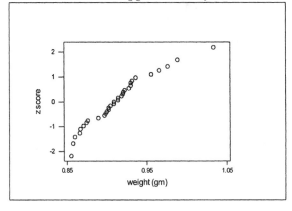

13. The two histograms are given below. The heights (on the left) appear to be approximately normally distributed, while the cholesterol levels (at the right) appear to be positively skewed. Many natural phenomena are normally distributed. Height is a natural phenomenon unaffected by human input; cholesterol levels are humanly influenced (by diet, exercise, medication, etc.) in ways that might alter any naturally occurring distribution.

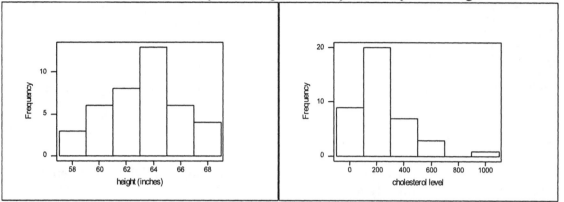

15. The corresponding z scores in the table below at the left were determined as follows.
   (1) Arrange the n scores in order and place them in the x column.
   (2) For each x_i, calculate the cumulative probability using $cp_i = (2i-1)/2n$ for $i = 1,2,\ldots,n$.
   (3) For each cp_i, find the z_i for which $P(z<z_i) = cp_i$ for $i = 1,2,\ldots,n$.

   The resulting normal quantile plot at the right indicates the data appear to come from a population with a normal distribution.

i	x	cp	z
1	73	.10	−1.28
2	78	.30	−.52
3	79	.50	.00
4	82	.70	.52
5	85	.90	1.28

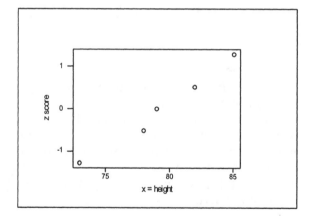

17. No. The z scores from the cumulative probability must be used, and not the z scores for the raw data. The z formula is a linear transformation.
   $$z = (x - \mu)/\sigma$$
   $$= (1/\sigma)\cdot x - (\mu/\sigma)$$
   $$= a\cdot x + b$$
   The (x,z) pairs will all lie on the straight line $z = (1/\sigma)\cdot x - (\mu/\sigma)$ regardless of the shape of the data.

Review Exercises

1. a. normal distribution
$\mu = 178.1$
$\sigma = 40.7$
$P(x > 260)$
$= P(z > 2.01)$
$= 1 - P(z < 2.01)$
$= 1 - .9778$
$= .0222$

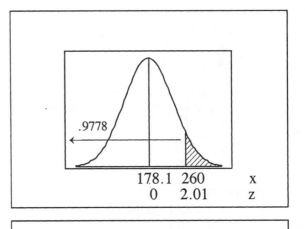

b. normal distribution
$\mu = 178.1$
$\sigma = 40.7$
$P(170 < x < 200)$
$= P(-.20 < z < .54)$
$= P(z < .54) - P(z < -.20)$
$= .7054 - .4207$
$= .2847$

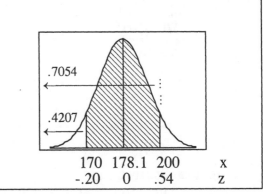

c. normal distribution,
since the original distribution is so
$\mu_{\bar{x}} = \mu = 178.1$
$\sigma_{\bar{x}} = \sigma/\sqrt{n} = 40.7/\sqrt{9} = 13.567$
$P(170 < \bar{x} < 200)$
$= P(-.60 < z < 1.61)$
$= P(z < 1.61) - P(z < -.60)$
$= .9463 - .2743$
$= .6720$

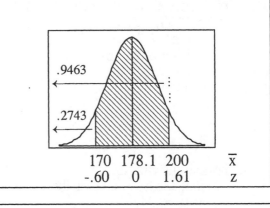

d. For the top 3%,
$A = .9700 \ [.9699]$ and $z = 1.88$
$x = \mu + z\sigma$
$= 178.1 + (1.88)(40.7)$
$= 178.1 + 76.5$
$= 254.6$

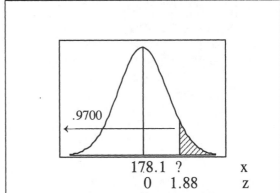

2. a. normal distribution

$\mu = 3420$

$\sigma = 495$

$P(x < 2200)$

$\quad = P(z < -2.46)$

$\quad = .0069$ or $.69\%$

For 900 births, we expect

$\quad (.0069)(900) \approx 6$ to be at risk.

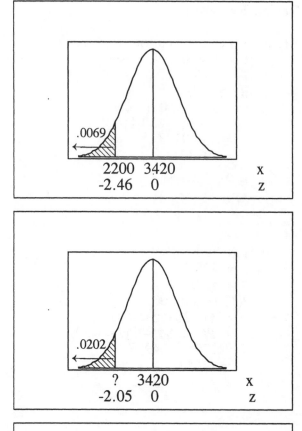

b. For the lowest 2%,

$\quad A = .0200 \ [.0202]$ and $z = -2.05$

$x = \mu + z\sigma$

$\quad = 3420 + (-2.05)(495)$

$\quad = 3420 - 1015$

$\quad = 2405$

c. normal distribution,

$\quad$ since the original distribution is so

$\mu_{\bar{x}} = \mu = 3420$

$\sigma_{\bar{x}} = \sigma/\sqrt{n} = 495/\sqrt{16} = 123.75$

$P(\bar{x} > 3700)$

$\quad = P(z > 2.26)$

$\quad = 1 - P(z < 2.26)$

$\quad = 1 - .9881$

$\quad = .0019$

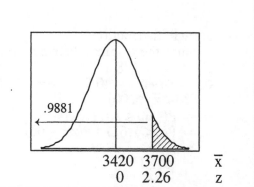

d. normal distribution,

$\quad$ since the original distribution is so

$\mu_{\bar{x}} = \mu = 3420$

$\sigma_{\bar{x}} = \sigma/\sqrt{n} = 495/\sqrt{49} = 70.714$

$P(3300 < \bar{x} < 3700)$

$\quad = P(-1.70 < z < 3.96)$

$\quad = P(z < 3.96) - P(z < -1.70)$

$\quad = .9999 - .0446$

$\quad = .9553$

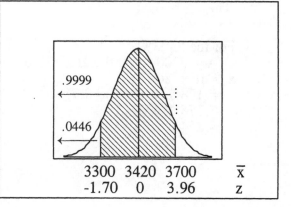

3. let x = the number of offspring with blue eyes
   binomial: n = 100 and p = .25
   normal approximation appropriate since
   $\quad$ np = 100(.25) = 25 $\geq$ 5
   $\quad$ n(1-p) = 100(.75) = 75 $\geq$ 5
   μ = np = 100(.25) = 25
   σ = $\sqrt{np(1-p)}$ = $\sqrt{100(.25)(.75)}$ = 4.330
   $P(x \leq 19)$
   $\quad$ = $P_c(x < 19.5)$
   $\quad$ = $P(z < -1.27)$
   $\quad$ = .1020
   No; since .1020 > .05, it is not very unusual
   to get 19 or fewer such offspring among 100.

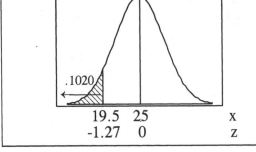

4. a. normal distribution
   μ = 69.0
   σ = 2.8
   $P(64 < x < 78)$
   $\quad$ = $P(-1.79 < z < 3.21)$
   $\quad$ = $P(z < 3.21) - P(z < -1.79)$
   $\quad$ = .9993 - .0367
   $\quad$ = .9626

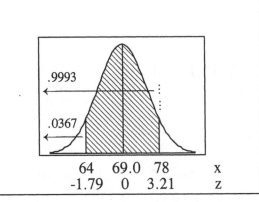

b. For the shortest 2%,
   $\quad$ A=.0200 [.0202] and z=-2.05.
   x = μ + zσ
   $\quad$ = 69.0 + (-2.05)(2.8)
   $\quad$ = 69.0 - 5.7
   $\quad$ = 63.3
   For the tallest 2%,
   $\quad$ A=.9800 [.9798] and z=2.05.
   x = μ + zσ
   $\quad$ = 69.0 + (2.05)(2.8)
   $\quad$ = 6.90 + 5.7
   $\quad$ = 74.7
   The new minimum and maximum
   requirements would be 63.3 inches and 74.7 inches respectively.

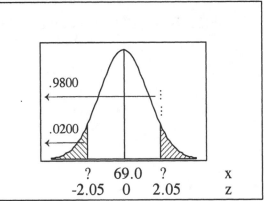

c. normal distribution,
   $\quad$ since the original distribution is so
   $\mu_{\bar{x}}$ = μ = 69.0
   $\sigma_{\bar{x}}$ = $\sigma/\sqrt{n}$ = 2.8/$\sqrt{64}$ = .350
   $P(\bar{x} > 68.0)$
   $\quad$ = $P(z > -2.86)$
   $\quad$ = 1 - $P(z < -2.86)$
   $\quad$ = 1 - .0021
   $\quad$ = .9979

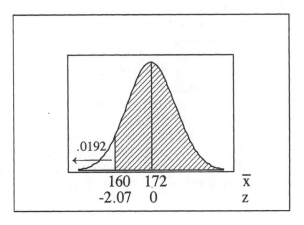

5. Probability corresponds to area. The calculations at the right verify that the height is 2.5 from 11.8 to 12.2. Since there is no probability outside those limits, the height outside those limits is 0.

$$\text{total area} = 1$$
$$(\text{width})(\text{height}) = 1$$
$$(12.2 - 11.8)(\text{height}) = 1$$
$$.4(\text{height}) = 1$$
$$\text{height} = 1/.4 = 2.5$$

a. P(x < 12.0) = (width)(height)
= (12.0-11.8)(2.5)
= (.2)(2.5)
= .50

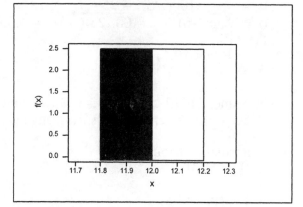

b. P(11.2 < x < 12.7) = P(11.2 < x < 11.8) + P(11.8 < x < 12.2) + P(12.2 < x < 12.7)
= (width)(height) + (width)(height) + (width)(height)
= (11.8 - 11.2)(0) + (12.2 - 11.8)(2.5) + (12.7 - 12.2)(0)
= (.6)(0) + (.4)(2.5) + (.5)(0)
= 0 + 1 + 0
= 1, a certainty

c. P(x > 12.2) = (width)(height)
= (width)·0
= 0, an impossibility

d. P(11.9 < x < 12.0) = (width)(height)
= 12.0 - 11.9)(2.5)
= (.1)(2.5)
= .25

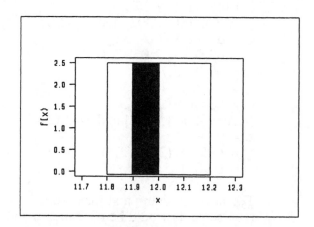

6. a. Since n=100, the Central Limit Theorem guarantees that the sampling distribution of the means of different samples will be normal – regardless of the shape of the original distribution.
b. $\sigma_{\bar{x}} = \sigma/\sqrt{n} = 512/\sqrt{100} = 51.2$ lb
c. Let x = the number who voted in the last election.
binomial: n = 1200 and p = ?
For each per person define the variable y as follows:
y = 1 if the person voted
y = 0 if the person did not vote.
The sample proportion x/n is really a sample mean, since x/n = Σy/n. Since n=1200, the Central Limit Theorem guarantees that the sampling distribution will be normal – regardless of the shape of the original distribution.

7. let x = the number of women selected
   binomial: n = 20 and p = .30
   normal approximation appropriate since
      np = 20(.30) = 6 ≥ 5
      n(1-p) = 20(.70) = 14 ≥ 5
   μ = np = 20(.25) = 6
   σ = $\sqrt{np(1-p)}$ = $\sqrt{20(.30)(.70)}$ = 2.049
   P(x≤2)
      = P_c(x<2.5)
      = P(z<-1.71)
      = .0436

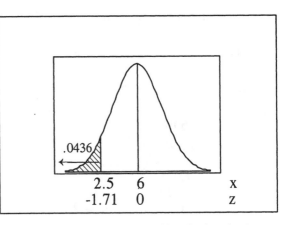

Yes; since .0436 < .05, it would be unusual
to get 2 or fewer women by chance alone.
Either an unusual event has occurred or there is some other factor (e.g., discrimination)
affecting the hiring process.

NOTE: Since the above probability calculated using the normal approximation is close to
.05, it would be good to calculate the more accurate exact binomial probability.

P(x≤2) = P(x=0) + P(x=1) + P(x=2)
   = [20!/(0!20!](.30)^0(.70)^{20} + [20!/(1!19!](.30)^1(.70)^{19} + [20!/(2!18!](.30)^2(.70)^{18}
   = [1](1)(.00080) + [20](.30)(.00114) + [190](.09)(.00162)
   = .00080 + .00684 + .02785
   = .0355

This probability is even smaller, confirming the previous conclusion.

8. Arrange the n=70 scores in order, calculate the cumulative probability for each score, and
   determine the z score for each cumulative probability. The frequency histogram (bell-
   shaped) and normal quantile plot (points approximately in a straight line) suggest that the
   data come from a population with a normal distribution.

i	x	cp	z		i	x	cp	z
01	3.407	.007	-2.45		36	3.588	.507	.02
02	3.450	.021	-2.03		37	3.590	.521	.05
03	3.464	.036	-1.80		38	3.598	.536	.09
04	3.468	.050	-1.64		39	3.600	.550	.13
05	3.475	.064	-1.52		40	3.601	.564	.16
06	3.482	.079	-1.41		41	3.604	.579	.20
07	3.491	.093	-1.32		42	3.604	.593	.23
08	3.492	.107	-1.24		43	3.611	.607	.27
09	3.494	.121	-1.17		44	3.617	.621	.31
10	3.494	.136	-1.10		45	3.621	.636	.35
11	3.506	.150	-1.04		46	3.622	.650	.39
12	3.507	.164	-.98		47	3.625	.664	.42
13	3.508	.179	-.92		48	3.632	.679	.46
14	3.511	.193	-.87		49	3.635	.693	.50
15	3.516	.207	-.82		50	3.635	.707	.55
16	3.521	.221	-.77		51	3.638	.721	.59
17	3.522	.236	-.72		52	3.639	.736	.63
18	3.522	.250	-.67		53	3.643	.750	.67
19	3.526	.264	-.63		54	3.643	.764	.72
20	3.531	.279	-.59		55	3.645	.779	.77
21	3.532	.293	-.55		56	3.647	.793	.82
22	3.535	.307	-.50		57	3.654	.807	.87
23	3.542	.321	-.46		58	3.660	.821	.92
24	3.545	.336	-.42		59	3.665	.836	.98
25	3.548	.350	-.39		60	3.666	.850	1.04
26	3.569	.364	-.35		61	3.667	.864	1.10
27	3.573	.379	-.31		62	3.671	.879	1.17
28	3.576	.393	-.27		63	3.673	.893	1.24
29	3.577	.407	-.23		64	3.678	.907	1.32
30	3.580	.421	-.20		65	3.687	.921	1.41
31	3.582	.436	-.16		66	3.688	.936	1.52
32	3.582	.450	-.13		67	3.718	.950	1.64
33	3.583	.464	-.09		68	3.723	.964	1.80
34	3.585	.479	-.05		69	3.725	.979	2.03
35	3.588	.493	-.02		70	3.726	.993	2.45

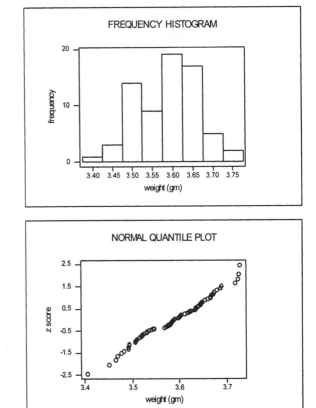

Cumulative Review Exercises

1. The $n=8$ ordered scores are: 55 59 62 63 66 66 66 67
    summary values: $\Sigma x = 504$, $\Sigma x^2 = 31{,}876$
    a. $\bar{x} = (\Sigma x)/n = (504)/8 = 63.0$ mm
    b. $\tilde{x} = (63 + 66)/2 = 64.5$ mm
    c. $M = 66$ mm
    d. $s^2 = [n(\Sigma x^2) - (\Sigma x)^2]/[n(n-1)]$
        $= [8(31{,}876) - (504)^2]/[8(7)]$
        $= 992/56$
        $= 17.714$
        $s = 4.209$ rounded to 4.2 mm
    e. $z = (x-\bar{x})/s$
        $z_{59} = (59 - 63.0)/4.209 = -.95$
    f. r.f. of scores greater than 59 is $6/8 = .75 = 75\%$
    g. normal: $\mu = 63.0$ and $\sigma = 4.299$
        $P(x > 59) = P(z > -.95)$
        $= 1 - P(z < .95)$
        $= 1 - .1711$
        $= .8289$ or 82.89%
    h. Ratio, since distances between scores are meaningful and there is a natural zero.
    i. Continuous, since length can be any value on a continuum.

2. a. Let L = a person is left-handed.
        $P(L) = .10$, for each random selection
        $P(L_1 \text{ and } L_2 \text{ and } L_3) = P(L_1) \cdot P(L_2) \cdot P(L_3) = (.10) \cdot (.10) \cdot (.10) = .001$
    b. Let N = a person is not left-handed
        $P(N) = .90$, for each random selection
        $P(\text{at least one left-hander}) = 1 - P(\text{no left-handers})$
        $= 1 - P(N_1 \text{ and } N_2 \text{ and } N_3)$
        $= 1 - P(N_1) \cdot P(N_2) \cdot P(N_3)$
        $= 1 - (.90) \cdot (.90) \cdot (.90)$
        $= 1 - .729$
        $= .271$
    c. binomial: $n = 3$ and $p = P(\text{left-hander}) = .10$
        normal approximation <u>not</u> appropriate since
        $np = 3(.10) = 0.3 < 5$
    d. binomial: $n = 50$ and $p = .10$
        $\mu = np = 50(.10) = 5$
    e. binomial: $n = 50$ and $p = .10$
        $\sigma = \sqrt{np(1-p)} = \sqrt{50(.10)(.90)} = 2.121$
    f. An unusual score is one that is more than 2 standard deviations from the mean.
        $z = (x-\mu)/\sigma$
        $z_8 = (8-5)/2.212$
        $= 1.41.$
        Since 8 is 1.41 standard deviations from the mean, it would <u>not</u> be an unusual score.

Chapter 6

Estimates and Sample Sizes

6-1 Estimating a Population Proportion

IMPORTANT NOTE: When calculating confidence intervals using the formula

$$\hat{p} \pm E$$
$$\hat{p} \pm z_{\alpha/2}\sqrt{\hat{p}\hat{q}/n}$$

do not round off in the middle of the problem. This may be accomplished conveniently on most calculators having a memory as follows.

(1) Calculate $\hat{p}$ = x/n and STORE the value
(2) Calculate E as 1 - RECALL = * RECALL = ÷ n = $\sqrt{}$ * $z_{\alpha/2}$ =
(3) With the value of E showing on the display, the upper confidence limit is calculated by + RECALL.
(4) With the value of the upper confidence limit showing on the display, the lower confidence limit is calculated by - RECALL ± + RECALL

You must become familiar with your own calculator. [Do your homework using the same type of calculator you will be using for the exams.] The above procedure works on most calculators; make certain you understand why it works and verify whether it works on your calculator. If it does not seem to work on your calculator, or if your calculator has more than one memory so that you can STORE both $\hat{p}$ and E at the same time, ask your instructor for assistance.

NOTE: It should be true that $0 \le \hat{p} \le 1$ and that E ≤ .5 [usually, <u>much</u> less than .5]. If such is not the case, an error has been made.

1. For .99 confidence, α = 1-.99 = .01 and $\alpha/2$ = .01/2 = .005.
   For the upper .005, A = .9950 and z = 2.575.
   $z_{\alpha/2} = z_{.005} = 2.575$

3. For .98 confidence, α = 1-.98 = .02 and $\alpha/2$ = .02/2 = .01.
   For the upper .01, A = .9900 [.9901] and z = 2.33.
   $z_{\alpha/2} = z_{.01} = 2.33$

5. Let L = the lower confidence limit; U = the upper confidence limit.
   $\hat{p}$ = (L + U)/2 = (.220 + .280)/2 = .500/2 = .250
   E = (U - L)/2 = (.280 - .220)/2 = .060/2 = .030
   The interval can be expressed as .250 ± .030.

7. Let L = the lower confidence limit; U = the upper confidence limit.
   $\hat{p}$ = (L + U)/2 = (.604 + .704)/2 = 1.308/2 = .654
   E = (U - L)/2 = (.704 - .604)/2 = .100/2 = .050
   The interval can be expressed as .654 ± .050.

9. Let L = the lower confidence limit; U = the upper confidence limit.
   $\hat{p}$ = (L + U)/2 = (.444 + .484)/2 = .928/2 = .464
   E = (U - L)/2 = (.484 - .444)/2 = .040/2 = .020

11. Let L = the lower confidence limit; U = the upper confidence limit.
    $\hat{p}$ = (L + U)/2 = (.632 + .678)/2 = 1.310/2 = .655
    E = (U - L)/2 = (.678 - .632)/2 = .046/2 = .023

13. $\alpha = .05$ and $\hat{p} = x/n = 200/800 = .25$
$E = z_{.025}\sqrt{\hat{p}\hat{q}/n} = 1.960\sqrt{(.25)(.75)/800} = .0300$

15. $\alpha = .01$ and $\hat{p} = x/n = [450/1000] = .45$
$E = z_{.005}\sqrt{\hat{p}\hat{q}/n} = 2.575\sqrt{(.45)(.55)/1000} = .0405$

17. $\alpha = .05$ and $\hat{p} = x/n = 300/400 = .750$
$\hat{p} \pm z_{.025}\sqrt{\hat{p}\hat{q}/n}$
$.750 \pm 1.96\sqrt{(.750)(.250)/400}$
$.750 \pm .042$
$.708 < p < .792$

19. $\alpha = .02$ and $\hat{p} = x/n = 176/1655 = .1063$
$\hat{p} \pm z_{.01}\sqrt{\hat{p}\hat{q}/n}$
$.1063 \pm 2.33\sqrt{(.1063)(.8937)/1655}$
$.1063 \pm .0177$
$.089 < p < .124$

21. $\hat{p}$ unknown, use $\hat{p} = .5$
$n = [(z_{.005})^2\hat{p}\hat{q}]/E^2$
$= [(2.575)^2(.5)(.5)]/(.060)^2 = 460.46$, rounded up to 461

23. $\hat{p} = .15$
$n = [(z_{.025})^2\hat{p}\hat{q}]/E^2$
$= [(1.96)^2(.185)(.815)]/(.05)^2 = 231.69$, rounded up to 232

25. a. We are 95% certain that the interval from 4.891% to 5.308% contains the true percentage of 18-20 year old males who drove within the last month while impaired from alcohol.
b. Yes, since any rate greater than 0% presents a risk to society and is preventable.
c. The upper confidence interval limit of 5.308% would be a responsible conservative estimate.

27. NOTE: This problem is limited to whole percent accuracy. Any x value from 293 ($293/1025 = .2859$) to 302 ($302/1025 = .2946$) gives a $\hat{p}$ that rounds to the given 29%. Since the exact value of x was not given, three decimal accuracy is not possible.
a. $\hat{p} = x/n = [297.25/1025] = .29 = 29\%$
b. $\hat{p} \pm z_{.005}\sqrt{\hat{p}\hat{q}/n}$
$.29 \pm 2.575\sqrt{(.29)(.71)/1025}$
$.29 \pm .04$
$.25 < p < .33$
$25\% < p < 33\%$
c. The upper confidence interval limit of 33% would be a responsible estimate for determining the maximum impact.

29. a. $\alpha = .05$ and $\hat{p} = x/n = 152/(152+428) = 152/580 = .262$
$\hat{p} \pm z_{.025}\sqrt{\hat{p}\hat{q}/n}$
$.262 \pm 1.96\sqrt{(.262)(.738)/580}$
$.262 \pm .036$
$.226 < p < .298$
$22.6\% < p < 29.8\%$
b. No; since the confidence interval includes 25%, there is no contradiction.

31. a. $\alpha = .01$ and $\hat{p} = x/n = 7/221 = .032$
    NOTE: Since $\hat{p} = x/n = 3.2\%$ was given, we may assume $x = \hat{p}\cdot n = .032\cdot221 = 7$.
    No other <u>integer</u> value x/221 rounds to .032.
    $\hat{p} \pm z_{.005}\sqrt{\hat{p}\hat{q}/n}$
    $.032 \pm 2.575\sqrt{(.032)(.968)/221}$
    $.032 \pm .030$
    $.00134 < p < .0620$
    $0.134\% < p < 6.20\%$
    b. Since 1.8% is within the interval in part (a), it cannot be ruled out as the correct value
    for the true proportion of Ziac users that experience dizziness. The test does not
    provide evidence that Ziac users experience any more dizziness than non-users – i.e.,
    the test does not provide evidence that dizziness is an adverse reaction to Ziac.

33. $\hat{p}$ unknown, use $\hat{p} = .50$
    $n = [(z_{.005})^2\hat{p}\hat{q}]/E^2 = [(2.575)^2(.50)(.50)]/(.02)^2 = 4144.14$, rounded up to 4145

35. a. $\hat{p} = .86$
    $n = [(z_{.03})^2\hat{p}\hat{q}]/E^2$
    $= [(1.88)^2(.86)(.14)]/(.03)^2 = 472.82$, rounded up to 473
    b. $\hat{p}$ unknown, use $\hat{p} = .5$
    $n = [(z_{.03})^2\hat{p}\hat{q}]/E^2$
    $= [(1.88)^2(.5)(.5)]/(.03)^2 = 981.78$, rounded up to 982
    c. In general, results from a self-selected sample are not valid. It is not appropriate to
    assume that those who choose to respond will be representative of the general population.

37. a. $\alpha = .10$ <u>and</u> $\hat{p} = x/n = 7/80 = .0875$
    $\hat{p} \pm z_{.01}\sqrt{\hat{p}\hat{q}/n}$
    $.0875 \pm 1.645\sqrt{(.0875)(.9125)/80}$
    $.0875 \pm .0520$
    $.036 < p < .139$
    b. $\hat{p} = .0875$
    $n = [(z_{.02})^2\hat{p}\hat{q}]/E^2 = [(2.05)^2(.0875)(.9125)]/(.03)^2 = 372.83$, rounded up to 373
    c. Yes; since 0.25% = .0025 is not in within the confidence interval in part (a), we may
    safely conclude that women have a lower rate of red/green color blindness than men do.

39. a. $\alpha = .05$ <u>and</u> $\hat{p} = x/n = 135/420000 = .000321$
    $\hat{p} \pm z_{.025}\sqrt{\hat{p}\hat{q}/n}$
    $.000321 \pm 1.96\sqrt{(.000321)(.999679)/420000}$
    $.000321 \pm .000054$
    $.000267 < p < .000376$
    $.0267\% < p < .0376\%$
    b. No; since the confidence interval includes .0340%, the results are consistent with those
    for the general population.

41. a. $\alpha = .05$ <u>and</u> $\hat{p} = x/n = 6/123 = .0488$
    $\hat{p} \pm z_{.025}\sqrt{\hat{p}\hat{q}/n}$
    $.0488 \pm 1.96\sqrt{(.0488)(.9512)/123}$
    $.0488 \pm .0381$
    $.0107 < p < .0868$
    $1.07\% < p < 8.68\%$
    b. $\alpha = .05$ <u>and</u> $\hat{p} = x/n = 811/1115 = .727$
    $\hat{p} \pm z_{.025}\sqrt{\hat{p}\hat{q}/n}$
    $.727 \pm 1.96\sqrt{(.727)(.273)/1115}$
    $.727 \pm .026$
    $.701 < p < .753$
    $70.1\% < p < 75.3\%$

c. Yes; the confidence intervals do not overlap. About 72.7% of the hunters in general wear orange. If wearing orange is unrelated to some other event involving hunters (e.g., eating eggs for breakfast), we expect about 72.7% of the hunters involved in that event to wear orange. If being mistaken for game is unrelated to wearing orange, we expect about 72.7% of those mistaken for game to be wearing orange. Since the proportion of hunters mistaken for game while wearing orange is significantly less, wearing orange appears to lessen the chances of being mistaken for game.

NOTE: While the true percentage who actually wear orange is likely less than the percentage who say they routinely wear orange [see exercise #30 for a comparable situation], the differences here are too great to be explained away in that manner.

43. $\alpha = .05$ and $\hat{p} = x/n = 21/100 = .210$
$\hat{p} \pm z_{.025}\sqrt{\hat{p}\hat{q}/n}$
$.210 \pm 1.960\sqrt{(.210)(.790)/100}$
$.210 \pm .080$
$.130 < p < .290$
$13.0\% < p < 29.0\%$
Yes; this result is consistent with the 20% rate reported by the candy maker.

45. For n=829, any x value from 419 [419/829 = .5054] to 426 [426/829 = .5138] rounds to 51%. The two confidence intervals associated with those extreme possibilities are as follows.
$\alpha = .05$ and $\hat{p} = x/n = 419/829 = .505$
$\quad \hat{p} \pm z_{.025}\sqrt{\hat{p}\hat{q}/n}$
$\quad .505 \pm 1.96\sqrt{(.505)(.495)/829}$
$\quad .505 \pm .034$
$\quad .471 < p < .539$
$\alpha = .05$ and $\hat{p} = x/n = 426/829 = .514$
$\quad \hat{p} \pm z_{.025}\sqrt{\hat{p}\hat{q}/n}$
$\quad .514 \pm 1.96\sqrt{(.514)(.486)/829}$
$\quad .514 \pm .034$
$\quad .480 < p < .548$
While the results do not differ substantially from the $.476 < p < .544$ given in the text using $\hat{p} = .51$, they do illustrate the folly of giving an interval with 3 decimal "accuracy" when the original problem was stated to the nearest percent. In truth, even the second decimal is not certain – as the two intervals above round to $.47 < p < .54$ and $.48 < p < .55$.

47. $\alpha = .05$ and $\hat{p} = x/n = 630/750 = .840$
$\quad \hat{p} - z_{.05}\sqrt{\hat{p}\hat{q}/n}$
$\quad .840 - 1.645\sqrt{(.840)(.160)/750}$
$\quad .840 - .022$
$\quad .818 < p$
The interval is expressed as $p > .818$. The desired figure is 81.8%

49. $\alpha = .01$ and $\hat{p} = x/n = 95/100 = .950$
$\quad \hat{p} \pm z_{.005}\sqrt{\hat{p}\hat{q}/n}$
$\quad .950 \pm 2.575\sqrt{(.950)(.050)/100}$
$\quad .950 \pm .056$
$\quad .894 < p < 1.006$
This interval is noteworthy because the upper limit is greater than 1, the maximum possible value for p in any problem. This occurs because of using the normal approximation to the binomial – which is just barely appropriate since $n(1-p) \approx 100(.05) = 5$, the minimum acceptable value to approximate the binomial distribution with the normal. In such cases the interval should be reported as $.894 < p < 1$ [and not as $.894 < p \le 1$, because the presence of 5 tails indicates that p=1 is not true].

51. Use the normal distribution with $\mu = 63.6$ and $\sigma = 2.5$ to estimate p.
$$\hat{p} = P(x > 60)$$
$$= P(z > -1.44)$$
$$= 1 - P(z < -1.44)$$
$$= 1 - .0749$$
$$= .9251$$
$$n = [(z_{.01})^2 \hat{p}\hat{q}]/E^2$$
$$= [(2.33)^2(.9251)(.0749)]/(.025)^2 = 601.87, \text{ rounded up to } 602$$

6-3 Estimating a Population Mean: σ Known

1. For .98 confidence, $\alpha = 1-.98 = .02$ and $\alpha/2 = .02/2 = .01$.
   For the upper .01, A = .9900 [.9901] and z = 2.33.
   $z_{\alpha/2} = z_{.01} = 2.33$

3. For .96 confidence, $\alpha = 1-.96 = .04$ and $\alpha/2 = .04/2 = .02$.
   For the upper .02, A = .9800 [.9798] and z = 2.05.
   $z_{\alpha/2} = z_{.02} = 2.05$

5. Yes, if the sample is a simple random sample.

7. Yes, if the sample is a simple random sample.

NOTE: When the sample mean is given (i.e., not calculated from raw data), the accuracy with which $\bar{x}$ is reported determines the accuracy for the endpoints of the confidence interval.

9. $\alpha = .05$, $\alpha/2 = .025$, $z_{.025} = 1.96$
   a. $E = z_{.025} \cdot \sigma/\sqrt{n}$
      $= 1.96 \cdot (12,345)/\sqrt{100}$
      $= 2419.62$ dollars
   b. $\bar{x} \pm E$
      $95,000 \pm 2,420$
      $92,580 < \mu < 97,420$ (dollars)

11. $\alpha = .10$, $\alpha/2 = .05$, $z_{.05} = 1.645$
    a. $E = z_{.05} \cdot \sigma/\sqrt{n}$
       $= 1.645 \cdot (2.50)/\sqrt{25}$
       $= .8225$ seconds
    b. $\bar{x} \pm E$
       $5.24 \pm .82$
       $4.42 < \mu < 6.06$ (seconds)

13. $n = [z_{.025} \cdot \sigma/E]^2$
    $= [1.96 \cdot (500)/125]^2 = 61.47$, rounded up to 62

15. $n = [z_{.05} \cdot \sigma/E]^2$
    $= [1.645 \cdot (48)/5]^2 = 249.38$, rounded up to 250

17. $\bar{x} = 318.1$

19. First find the value of E as follows.
$$\text{width} = 2 \cdot E$$
$$374.11 - 262.09 = 2 \cdot E$$
$$112.02 = 2 \cdot E$$
$$56.01 = E$$
Then use $\bar{x}$ and E to give the interval as: 318.1 ± 56.01

21. $\bar{x} \pm z_{.025} \cdot \sigma/\sqrt{n}$
$30.4 \pm 1.96 \cdot (1.7)/\sqrt{61}$
30.4 ± 0.4
$30.0 < \mu < 30.8$
It is unrealistic to know the value of σ.

23. $\bar{x} \pm z_{.05} \cdot \sigma/\sqrt{n}$
$172.5 \pm 1.645 \cdot (119.5)/\sqrt{40}$
172.5 ± 31.1
$141.4 < \mu < 203.6$
It is unrealistic to know the value of σ.

25. $n = [z_{.025} \cdot \sigma/E]^2$
$= [1.96 \cdot (15)/2]^2 = 216.09$, rounded up to 217

27. $n = [z_{.025} \cdot \sigma/E]^2$
$= [1.96 \cdot (6250)/500]^2 = 600.25$, rounded up to 601

29. Using the range rule of thumb, $\sigma \approx (\text{range})/4 = (70,000-12,000)/4 = 58,000/4 = 14,500$.
$n = [z_{.025} \cdot \sigma/E]^2 = [1.96 \cdot (14,500)/100]^2 = 80,769.64$, rounded up to 80,770
No, this sample size is not practical. Of the three terms in the formula for n, $z_{\alpha/2}$ and E are set by the expectations of the researcher while σ is determined by the variable under consideration and not under control of the researcher. Either lower the level confidence (which will make $z_{\alpha/2}$ less than 1.96, thus decreasing n) or increase the margin of error (which will make E larger than 100, thus decreasing n).

31. The minimum and maximum values in the data set are 56 and 96 respectively.
Using the range rule of thumb, $\sigma \approx (\text{range})/4 = (96-56)/4 = 40/4 = 10$ and
$n = [z_{.025} \cdot \sigma/E]^2$
$= [1.96 \cdot (10)/2]^2 = 96.04$, rounded up to 97.
Using the sample standard deviation for the male pulse values [11.297] to estimate σ,
$n = [z_{.025} \cdot \sigma/E]^2$
$= [1.96 \cdot (11.297)/2]^2 = 122.56$, rounded up to 123.
The two values are relatively close. Since s (which considers all the data) is a better estimator for σ than R/4 (which is based entirely on the extreme values), the sample size of 123 should be preferred.

33. $\bar{x} \pm [z_{.025} \cdot \sigma/\sqrt{n}] \cdot \sqrt{(N-n)/(N-1)}$
$110 \pm [1.96 \cdot (15)/\sqrt{35}] \cdot \sqrt{(250-35)/(250-1)}$
$110 \pm [4.9695] \cdot [.9292]$
110 ± 4.617
$105 < \mu < 115$

6-4 Estimating a Population Mean: σ Not Known

IMPORTANT NOTE: This manual uses the following conventions.
(1) The designation "df" stands for "degrees of freedom."
(2) Since the t value depends on both the degrees of freedom and the probability lying beyond it, double subscripts are used to identify points on t distributions. The t distribution with 15 degrees of freedom and .025 beyond it, for example, is designated $t_{15,.025} = 2.132$.
(3) Always use the closest entry in Table A-3. When the desired df is exactly halfway between the two nearest tabled values, be conservative and choose the one with the lower df.
(4) As the degrees of freedom increase, the t distribution approaches the standard normal distribution – and the "large" row of the t table actually gives z values. Consequently, the z scores for certain "popular" α and $\alpha/2$ values may be found by reading Table A-3 "frontwards" instead of reading Table A-2 "backwards." This is not only easier but also more accurate, since Table A-3 includes one more decimal place. Note, for example, that $t_{large,.05} = 1.645 = z_{.05}$ (as found from the z table) and $t_{large,.01} = 2.326 = z_{.01}$ (more accurate than the 2.33 as found from the z table). The manual uses this technique from this point on. [That $t_{large,.005} = 2.576 \neq 2.575 = z_{.005}$ is discrepancy caused by using different mathematical approximation techniques to construct the tables, and not a true difference. The manual will continue to use $z_{.005} = 2.575$.]

1. σ unknown and population approximately normal:
    If the sample is a simple random sample, use t.
    $n = 5$; $df = 4$
    $\alpha = .05$; $\alpha/2 = .025$
    $t_{4,.025} = 2.776$

3. σ known and the population is not approximately normal
    Since $n < 30$, neither the z distribution nor the t distribution applies.

5. σ unknown and population approximately normal:
    If the sample is a simple random sample, use t.
    $n = 92$; $df = 91$ [90]
    $\alpha = .10$; $\alpha/2 = .05$
    $t_{91,.05} = 1.662$

7. σ known and population approximately normal:
    If the sample is a simple random sample, use z.
    $\alpha = .02$; $\alpha/2 = .01$
    $z_{.01} = 2.236$

9. $\alpha = .05$, $\alpha/2 = .025$; $n = 15$, $df = 14$
    a. $E = t_{14,.025} \cdot s/\sqrt{n}$
       $= 2.145 \cdot (108)/\sqrt{15}$
       $= 59.8$
    b. $\overline{x} \pm E$
       496 ± 60
       $436 < \mu < 556$

11. $112.84 < \mu < 121.56$
    We are 95% confident that the interval from 112.84 to 121.56 contains μ, the true value of the population mean.

13. σ unknown and distribution approximately normal, use t

$\bar{x} \pm t_{11,.025} \cdot s/\sqrt{n}$

$26,227 \pm 2.201 \cdot (15,873)/\sqrt{12}$

$26,227 \pm 10,085$

$16,142 < \mu < 36,312$ (dollars)

We are 95% certain that the interval from \$16,142 to \$36,312 contains the true mean repair cost for repairing Dodge Vipers under the specified conditions.

15. a. σ unknown and n > 30, use t

$\bar{x} \pm t_{30,.005} \cdot s/\sqrt{n}$

$-.419 \pm 2.750 \cdot (3.704)/\sqrt{31}$

$-.419 \pm 1.829$

$-2.248 < \mu < 1.410$ (°F)

b. Yes, the confidence interval includes 0. No; since the confidence interval includes 0, the true difference could be 0 – i.e., there is not evidence to conclude the three-day forecasts are too high or too low.

17. preliminary values: n = 7, $\Sigma x = .85$, $\Sigma x^2 = .1123$

$\bar{x} = (\Sigma x)/n$

$\quad = (.85)/7 = .121$

$s^2 = [n(\Sigma x^2) - (\Sigma x)^2]/[n(n-1)]$

$\quad = [7(.1123) - (.85)^2]/[7(6)]$

$\quad = .00151$

$s = .039$

σ unknown (and assuming the distribution is approximately normal), use t

$\bar{x} \pm t_{6,.01} \cdot s/\sqrt{n}$

$.121 \pm 3.143 \cdot (.039)/\sqrt{7}$

$.121 \pm .046$

$.075 < \mu < .168$ (grams/mile)

No; since the confidence interval includes values greater than .165, there is a reasonable possibility that the requirement is not being met.

19. a. σ unknown (and assuming the distribution is approximately normal), use t

$\bar{x} \pm t_{9,.025} \cdot s/\sqrt{n}$

$175 \pm 2.262 \cdot (15)/\sqrt{10}$

175 ± 11

$164 < \mu < 186$ (beats per minute)

b. σ unknown (and assuming the distribution is approximately normal), use t

$\bar{x} \pm t_{9,.025} \cdot s/\sqrt{n}$

$124 \pm 2.262 \cdot (18)/\sqrt{10}$

124 ± 13

$111 < \mu < 137$ (beats per minute)

c. The maximum likely value for the true mean heart rate for hand-shovelers, 186.

d. Since the two confidence intervals do not overlap, there is evidence that hand-shovelers experience significantly higher heart rates than those using electric snow blowers.

21. <u>4000 BC</u>

$n = 12$

$\Sigma x = 1544$

$\Sigma x^2 = 198898$

$\bar{x} = (\Sigma x)/n = (1544)/12 = 128.67$

$s^2 = [n(\Sigma x^2) - (\Sigma x)^2]/[n(n-1)]$

$= [12(198898) - (1544)^2]/[12(11)]$

$= 21.515$

$s = 4.638$

σ unknown (and assuming normality), use t

$\bar{x} \pm t_{11,.025} \cdot s/\sqrt{n}$

$128.67 \pm 2.201 \cdot (4.638)/\sqrt{12}$

128.67 ± 2.95

$125.7 < \mu < 131.6$

<u>150 AD</u>

$n = 12$

$\Sigma x = 1600$

$\Sigma x^2 = 213610$

$\bar{x} = (\Sigma x)/n = (1600)/12 = 133.33$

$s^2 = [n(\Sigma x^2) - (\Sigma x)^2]/[n(n-1)]$

$= [12(213610) - (1600)^2]/[12(11)]$

$= 25.152$

$s = 5.015$

σ unknown (and assuming normality), use t

$\bar{x} \pm t_{11,.025} \cdot s/\sqrt{n}$

$133.33 \pm 2.201 \cdot (5.015)/\sqrt{12}$

133.33 ± 3.19

$130.1 < \mu < 136.5$

Since the two 95% confidence intervals overlap, the two samples could from populations with the same mean. This is not evidence to conclude that the head sizes have changed. NOTE: The level of confidence was not specified. While 95% is the standard level of confidence when no specific value is given, it is possible that using less than 95% confidence would shorten the intervals to the point where there is no overlap.

23. a. preliminary values: $n = 36$, $\Sigma x = 29.6677$, $\Sigma x^2 = 24.45037155$

$\bar{x} = (\Sigma x)/n$

$= (29.6677)/36 = .82410$

$s^2 = [n(\Sigma x^2) - (\Sigma x)^2]/[n(n-1)]$

$= [36(24.45037155) - (29.6677)^2]/[36(35)] = .00003249$

$s = .005700$

σ unknown and $n > 30$, use t

$\bar{x} \pm t_{35,.025} \cdot s/\sqrt{n}$

$.82410 \pm 2.032 \cdot (.005700)/\sqrt{36}$

$.82410 \pm .00193$

$.82217 < \mu < .82603$ (lbs)

b. preliminary values: $n = 36$, $\Sigma x = 28.2189$, $\Sigma x^2 = 22.12028574$

$\bar{x} = (\Sigma x)/n$

$= (28.2189)/36 = .78386$

$s^2 = [n(\Sigma x^2) - (\Sigma x)^2]/[n(n-1)]$

$= [36(22.12028574) - (28.2189)^2]/[36(35)] = .00001900$

$s = .004359$

σ unknown and $n > 30$, use t

$\bar{x} \pm t_{35,.025} \cdot s/\sqrt{n}$

$.78386 \pm 2.032 \cdot (.004359)/\sqrt{36}$

$.78386 \pm .00148$

$.78238 < \mu < .78533$ (lbs)

c. The confidence intervals do not overlap. Cans of diet Pepsi weigh significantly less than cans of regular Pepsi.

25. preliminary values: $n = 7$, $\Sigma x = 60.79$, $\Sigma x^2 = 3600.1087$

$\bar{x} = (\Sigma x)/n$

$= (60.79)/7 = 8.684$

$s^2 = [n(\Sigma x^2) - (\Sigma x)^2]/[n(n-1)]$

$= [7(3600.1087) - (60.79)^2]/[7(6)] = 512.0318$

$s = 22.628$

σ unknown (and assuming the distribution is approximately normal), use t

$\bar{x} \pm t_{6,.025} \cdot s/\sqrt{n}$

$8.684 \pm 2.447 \cdot (22.628)/\sqrt{7}$

8.684 ± 20.928

$-12.244 < \mu < 29.613$ (grams/mile)

This is considerably different from the $.075 < \mu < .168$ interval of exercise #17. Confidence intervals appear to be very sensitive to outliers. Outliers found in sample data should be examined for two reasons: (1) to determine whether they are errors, (2) to determine whether their presence in small samples indicates that the data fail to meet the normal distribution requirement.

27. Applying the principles from chapter 2, if $y = a \cdot x + b$
    then $\bar{y} = a \cdot \bar{x} + b$ and $s_y = a \cdot s_x$
    Applying this to C and F, where $C = 5(F-32)/9 = (5/9) \cdot F - 160/9$,
        then $\bar{C} = (5/9) \cdot \bar{F} - 160/9$ and $s_C = (5/9) \cdot s_F$
    a. $E_C = t_{df,\alpha/2} s_C/\sqrt{n}$
        $= t_{df,\alpha/2}(5/9) \cdot s_F/\sqrt{n}$
        $= (5/9) \cdot t_{df,\alpha/2} s_F/\sqrt{n}$
        $= (5/9) \cdot E_F$
    b. $L = \bar{C} - E_C$ $U = \bar{C} + E_C$
        $= [(5/9) \cdot \bar{F} - 160/9] - (5/9) \cdot E_F$ $= [(5/9) \cdot \bar{F} - 160/9] + (5/9) \cdot E_F$
        $= (5/9) \cdot (\bar{F} - E_F) - 160/9$ $= (5/9) \cdot (\bar{F} + E_F) - 160/9$
        $= (5/9) \cdot a - 160/9$ $= (5/9) \cdot b - 160/9$
        $= (5/9) \cdot (a - 32)$ $= (5/9) \cdot (b - 32)$
    c. Yes.

6-5 Estimating a Population Variance

1. $\chi_L^2 = \chi_{15,.975}^2 = 6.262$; $\chi_R^2 = \chi_{15,.025}^2 = 27.488$

3. $\chi_L^2 = \chi_{79,.995}^2 = 51.172$; $\chi_R^2 = \chi_{79,.005}^2 = 116.321$
    NOTE: Use df = 80, the closest entry.

5. $(n-1)s^2/\chi_{19,.025}^2 < \sigma^2 < (n-1)s^2/\chi_{19,.975}^2$
    $(19)(12345)^2/32.852 < \sigma^2 < (19)(12345)^2/8.907$
                $88,140,188 < \sigma^2 < 325,090,544$
                    $9,388 < \sigma < 18,030$ (dollars)

7. $(n-1)s^2/\chi_{29,.05}^2 < \sigma^2 < (n-1)s^2/\chi_{29,.95}^2$
    $(29)(2.50^2/42.557 < \sigma^2 < (29)(2.50^2/17.708$
                $4.2590 < \sigma^2 < 10.2355$
                    $2.06 < \sigma < 3.20$ (seconds)

9. From the upper right section of Table 6-2, n = 191.

11. From the lower left section of Table 6-2, n = 133,448.
    No, for most applications this is not a practical sample size.

NOTE: When raw scores are available, $\bar{x}$ and s should be calculated as the primary descriptive statistics – but use the unrounded value of s^2 in the confidence interval formula. In addition, always make certain that the confidence interval for σ includes the calculated value of s.

13. $(n-1)s^2/\chi^2_{11,.025} < \sigma^2 < (n-1)s^2/\chi^2_{11,.975}$
$(11)(15873)^2/21.920 < \sigma^2 < (11)(15873)^2/3.816$
$126{,}435{,}831 < \sigma^2 < 726{,}277{,}101$
$11{,}244 < \sigma < 26{,}950$ (dollars)
We are 95% confident that the interval from \$11,244 to \$26,950 contains σ, the true standard deviation for the repair amounts of all such cars.

15. summary information
   $n = 6$ $\bar{x} = 1.538$
   $\Sigma x = 9.23$ $s^2 = 3.664$
   $\Sigma x^2 = 32.5197$ $s = 1.914$
       $(n-1)s^2/\chi^2_{5,.025} < \sigma^2 < (n-1)s^2/\chi^2_{5,.975}$
   $(5)(3.664)/12.833 < \sigma^2 < (5)(3.664)/.831$
       $1.4276 < \sigma^2 < 22.0468$
         $1.195 < \sigma < 4.695$ (micrograms/cubic meter)
Yes; the fact that 4 of the 5 sample values are below $\bar{x}$ raises a question about whether the data meets the requirement that the underlying population distribution is normal.

17. a. $(n-1)s^2/\chi^2_{9,.025} < \sigma^2 < (n-1)s^2/\chi^2_{9,.975}$
      $(9)(15)^2/19.023 < \sigma^2 < (9)(15)^2/2.700$
          $106.5 < \sigma^2 < 750.0$
           $10 < \sigma < 27$ (beats per minute)
    b. $(n-1)s^2/\chi^2_{9,.025} < \sigma^2 < (n-1)s^2/\chi^2_{9,.975}$
      $(9)(18)^2/19.023 < \sigma^2 < (9)(18)^2/2.700$
          $153.3 < \sigma^2 < 1080.0$
           $12 < \sigma < 33$ (beats per minute)
    c. No; the two groups do not appear to differ much in variation.

19. a. summary information
      $n = 10$ $\bar{x} = 7.15$
      $\Sigma x = 71.5$ $s^2 = .2272$
      $\Sigma x^2 = 513.27$ $s = .48$
          $(n-1)s^2/\chi^2_{9,.025} < \sigma^2 < (n-1)s^2/\chi^2_{9,.975}$
      $(9)(.2272)/19.023 < \sigma^2 < (9)(.2272)/2.700$
           $.1075 < \sigma^2 < .7573$
             $.33 < \sigma < .87$ (minutes)
    b. summary information
      $n = 10$ $\bar{x} = 7.15$
      $\Sigma x = 71.5$ $s^2 = 3.3183$
      $\Sigma x^2 = 541.09$ $s = 1.82$
          $(n-1)s^2/\chi^2_{9,.025} < \sigma^2 < (n-1)s^2/\chi^2_{9,.975}$
      $(9)(3.3183)/19.023 < \sigma^2 < (9)(3.3183)/2.700$
             $1.5699 < \sigma^2 < 11.0610$
             $1.25 < \sigma < 3.33$ (minutes)
    c. Yes, there is a difference. The multiple line system exhibits more variability among the waiting times. The greater consistency among the single line waiting times seems fairer to the customers and more professional.

21. a. The given interval $2.8 < \sigma < 6.0$

$$7.84 < \sigma^2 < 36.00$$

and the usual calculations $(n-1)s^2/\chi^2_{19,\alpha/2} < \sigma^2 < (n-1)s^2/\chi^2_{19,1-\alpha/2}$

$$(19)(3.8)^2/\chi^2_{19,\alpha/2} < \sigma^2 < (19)(3.8)^2/\chi^2_{19,1-\alpha/2}$$

$$274.36/\chi^2_{19,\alpha/2} < \sigma^2 < 274.36/\chi^2_{19,1-\alpha/2}$$

imply that $7.84 = 274.37/\chi^2_{19,\alpha/2}$ and $36.00 = 274.36/\chi^2_{19,1-\alpha/2}$

$\chi^2_{19,\alpha/2} = 274.36/7.84$ $\chi^2_{19,1-\alpha/2} = 274.36/36.00$

$= 34.99$ $= 7.62$

The closest entries in Table A-4 are $\chi^2_{19,\alpha/2} = 34.805$ and $\chi^2_{19,1-\alpha/2} = 7.633$

which imply $\alpha/2 = .01$ $1 - \alpha/2 = .99$

$\alpha = .02$ $\alpha/2 = .01$

$\alpha = .02$

The level of confidence is therefore is $1-\alpha = 98\%$.

b. $(n-1)s^2/\chi^2_{11,.025} < \sigma^2 < (n-1)s^2/\chi^2_{11,.975}$

using the lower endpoint   OR   using the upper endpoint
$(11)s^2/21.920 = (19.1)^2$ $(11)s^2/3.816 = (45.8)^2$
$s^2 = 726.97$ $s^2 = 727.69$
$s = 27.0$ $s = 27.0$

Review Exercises

1. a. $\hat{p} = x/n = 111/1233 = .0900 = 9.00\%$
   b. $\hat{p} \pm z_{.05}\sqrt{\hat{p}\hat{q}/n}$
   $.0900 \pm 1.96\sqrt{(.0900)(.9100)/1233}$
   $.0900 \pm .0160$
   $.0740 < p < .1060$
   $7.40\% < p < 10.60\%$
   c. $\hat{p}$ unknown, use $\hat{p} = .5$
   $n = [(z_{.005})^2\hat{p}\hat{q}]/E^2 = [(2.575)^2(.5)(.5)]/(.025)^2 = 2652.25$, rounded up to 2653

2. a. σ unknown and population normal, use t
   $\bar{x} \pm t_{24,.025} \cdot s/\sqrt{n}$
   $7.01 \pm 2.064 \cdot 3.741/\sqrt{25}$
   7.01 ± 1.54
   $5.47 < \mu < 8.55$ (years)
   b. $(n-1)s^2/\chi^2_{24,.025} < \sigma^2 < (n-1)s^2/\chi^2_{24,.975}$
   $(24)(3.74)^2/39.364 < \sigma^2 < (24)(3.74)^2/12.401$
   $8.528 < \sigma^2 < 27.071$
   $2.92 < \sigma < 5.20$ (years)
   c. $n = [z_{.005} \cdot \sigma/E]^2 = [2.575 \cdot (3.74)/.25]^2 = 1483.94$, rounded up to 1484
   d. No; those who purchased a General Motors car are not necessarily representative of all car owners.

3. a. $\hat{p} = x/n = 308/611 = .504 = 50.4\%$
   b. $\hat{p} \pm z_{.01}\sqrt{\hat{p}\hat{q}/n}$
   $.5041 \pm 2.326\sqrt{(.5041)(.4959)/611}$
   $.5041 \pm .0470$
   $.4570 < p < .5511$
   $45.7\% < p < 55.1\%$
   c. No; since 43% is not within the confidence interval, the survey results are not consistent with the facts. People might not want to admit they voted for a losing candidate,

especially if the winner is turning out to be a good president. [If the winner is turning out to be a bad president, the percentage in the survey who said they voted for the winner might actually be lower than the true value.]

4. a. summary information

n = 12	$\bar{x}$ = 6.500
Σx = 78.0	s^2 = 5.9945
Σx^2 = 572.94	s = 2.448

σ unknown (and assuming normality), use t
$\bar{x} \pm t_{11,.025} \cdot s/\sqrt{n}$
$6.500 \pm 2.201 \cdot (2.448)/\sqrt{12}$
6.500 ± 1.556
$4.94 < \mu < 8.06$

b. summary information

n = 12	$\bar{x}$ = 5.075
Σx = 60.9	s^2 = 1.3639
Σx^2 = 324.07	s = 1.168

σ unknown (and assuming normality), use t
$\bar{x} \pm t_{11,.025} \cdot s/\sqrt{n}$
$5.075 \pm 2.201 \cdot (1.168)/\sqrt{12}$
5.075 ± 0.742
$4.33 < \mu < 5.82$

c. summary information

n = 12	$\bar{x}$ = 8.433
Σx = 101.2	s^2 = 4.0333
Σx^2 = 897.82	s = 2.008

σ unknown (and assuming normality), use t
$\bar{x} \pm t_{11,.025} \cdot s/\sqrt{n}$
$8.433 \pm 2.201 \cdot (2.008)/\sqrt{12}$
8.433 ± 1.276
$7.16 < \mu < 8.71$

d. The mean grade-level rating appears to be significantly higher for Tolstoy than for Clancy or Rowling.

5. $n = [z_{.05} \cdot \sigma/E]^2$
$= [1.645 \cdot (2.45)/.5]^2 = 64.97$, rounded up to 65

6. $(n-1)s^2/\chi^2_{11,.025} < \sigma^2 < (n-1)s^2/\chi^2_{11,.975}$
$(11)(1.17)^2/21.920 < \sigma^2 < (11)(1.17)^2/3.816$
$.687 < \sigma^2 < 3.946$
$.83 < \sigma < 1.99$

7. $\hat{p}$ unknown, use $\hat{p}$ = .5
$n = [(z_{.015})^2 \hat{p}\hat{q}]/E^2$
$= [(2.17)^2(.5)(.5)]/(.02)^2 = 2943.06$, rounded up to 2944

8. $\hat{p}$ = .93
$n = [(z_{.01})^2 \hat{p}\hat{q}]/E^2$
$= [(2.326)^2(.93)(.07)]/(.04)^2 = 220.32$, rounded up to 221

Cumulative Review Exercises

1. Begin by making a stem-and-leaf plot and calculating summary statistics.

   10 | 5
   11 |
   11 | 599 $n = 9$
   12 | 3 $\Sigma x = 1089$
   12 | 5788 $\Sigma x^2 = 132223$

   a. $\overline{x} = (\Sigma x)/n = (1089)/9 = 121.0$ lbs

   b. $\tilde{x} = 123.0$ lbs

   c. $M = 119, 128$ (bi-modal)

   d. m.r. $= (105 + 128)/2 = 116.5$ lbs

   e. $R = 128 - 105 = 23$ lbs

   f. $s^2 = [n(\Sigma x^2) - (\Sigma x)^2]/[n(n-1)]$
      $= [9(132223) - (1089)^2]/[9(8)]$
      $= 56.75$ lbs^2

   g. $s = 7.5$ lbs

   h. for $Q_1 = P_{25}$, $L = (25/100)(9) = 2.25$, round up to 3
      $Q_1 = x_3 = 119$ lbs

   i. for $Q_2 = P_{50}$, $L = (50/100)(9) = 4.50$, round up to 5
      $Q_2 = x_5 = 123$ lbs

   j. for $Q_3 = P_{75}$, $L = (75/100)(9) = 6.75$, round up to 7
      $Q_3 = x_7 = 127$ lbs

   k. ratio, since differences are consistent and there is a meaningful zero

   l. The boxplot is at the right.

   m. σ unknown (and assuming normality), use t
      $\overline{x} \pm t_{8,.005} \cdot s/\sqrt{n}$
      $121.0 \pm 3.355 \cdot 7.5/\sqrt{9}$
      121.0 ± 8.4
      $112.6 < \mu < 129.4$ (lbs)

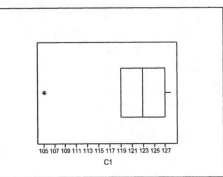

   NOTE: Since both the stem-and-leaf plot and the
   boxplot indicate that the weights do not appear to
   come from a normal population, the methods of the
   chapter are not appropriate. It appears the distribution of the weights of supermodels is
   truncated at some upper limit and skewed to the left. Even though the confidence intervals
   for μ and σ should not be constructed, the intervals in part (m) above and part (n) below are
   given as a review of the techniques.

   n. $(n-1)s^2/\chi^2_{8,.005} < \sigma^2 < (n-1)s^2/\chi^2_{8,.995}$
         $(8)(56.75)/21.955 < \sigma^2 < (8)(56.75)/1.344$
               $20.68 < \sigma^2 < 337.80$
                $4.5 < \sigma < 18.4$ (lbs)

   o. $E = 2$ and $\alpha = .01$
      $n = [z_{.005} \cdot \sigma/E]^2$
        $= [2.575 \cdot 7.5/2]^2 = 94.07$, rounded up to 95

   p. For the general female population, an unusually low weight would be one below 85 lbs
      (i.e., more than 2σ below μ). Individually, none of the supermodels has an unusually
      low weight. As a group, however, they are each well below the general population
      mean weight and appear to weigh substantially less than the general population – as
      evidenced by their mean weight as given by the point estimate in part (a) and the
      interval estimate in part (m).

2. a. binomial: n = 200 and p = .25

a normal approximation is appropriate since

$np = 200(.25) = 50 \geq 5$

$n(1-p) = 200(.75) = 150 \geq 5$

use $\mu = np = 200(.25) = 50$

$\sigma = \sqrt{n(p)(1-p)} = \sqrt{200(.25)(.75)}$

$= 6.214$

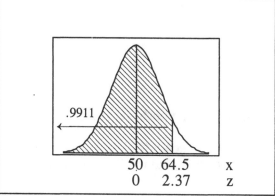

$P(x \geq 65)$

$= P_c(x > 64.5)$

$= P(z > 2.37)$

$= 1 - P(z < 2.37)$

$= 1 - .9911$

$= .0089$

b. $\hat{p} = 65/200 = .325$

$\alpha = .05$

$\hat{p} \pm z_{.025}\sqrt{\hat{p}\hat{q}/n}$

$.325 \pm 1.960\sqrt{(.325)(.675)/200}$

$.325 \pm .065$

$.260 < p < .390$

c. No; the expert's value of .25 does not seem correct for two reasons.

From part (a): if the expert is correct, then the probability of getting the sample obtained is very small – less than 1%.

From part (b): according to the sample obtained, there is 95% confidence that the interval from .26 to .39 includes the true value – that is, there is 95% confidence that the value is not .25.

Chapter 7

Hypothesis Testing

7-2 Basics of Hypothesis testing

1. There is not sufficient evidence to support the claim. Under the assumption that the claim is not true, having 26 girls among 50 babies is the type of result you expect to occur.

3. There is sufficient evidence to support the claim. Under the assumption that the claim is not true, having 475 of 500 persons like pizza would be a rare event. Therefore the assumption that the claim is not true is probably not correct.

5. original statement: $\mu > 50,000$ (does not contain the equality; must be H_1)
   competing idea: $\mu \leq 50,000$
   H_o: $\mu = 50,000$
   H_1: $\mu > 50,000$

7. original statement: $p > .5$ (does not contain the equality; must be H_1)
   competing idea: $p \leq .5$
   H_o: $p = .5$
   H_1: $p > .5$

9. original statement: $\sigma < 2.8$ (does not contain the equality; must be H_1)
   competing idea: $\sigma \geq 2.8$
   H_o: $\sigma = 2.8$
   H_1: $\sigma < 2.8$

11. original statement: $\mu \geq 12$
    competing idea: $\mu < 12$ (does not contain the equality; must be H_1)
    H_o: $\mu = 12$
    H_1: $\mu < 12$

NOTE: Recall that z_α is the z with α <u>above</u> it. By the symmetry of the normal distribution, the z with α <u>below</u> it is $z_{1-\alpha} = -z_\alpha$.

13. Two-tailed test; place $\alpha/2$ in each tail.
    Use $A = 1-\alpha/2 = 1-.0250 = .9750$ and $z = 1.96$.
    critical values are $\pm z_{\alpha/2} = \pm z_{.025} = \pm 1.96$

15. Right-tailed test; place α in the upper tail.
    Use $A = 1-\alpha = 1-.0100 = .9900$ [closest entry $= .9901$] and $z = 2.33$.
    critical value is $+z_\alpha = +z_{.01} = +2.33$ [or 2.326 from the "large" row of the t table]

17. Two-tailed test; place $\alpha/2$ in each tail.
    Use $A = 1-\alpha/2 = 1-.0500 = .9500$ and $z = 1.645$.
    critical values are $\pm z_{\alpha/2} = \pm z_{.05} = \pm 1.645$

19. Left-tailed test; place α in the lower tail.
    Use $A = \alpha = .0200$ [closest entry $= .0202$] and $z = -2.05$.
    critical value is $-z_\alpha = -z_{.02} = -2.05$

21. $\hat{p} = x/n$
     $= x/1025 = .29$
     $z_{\hat{p}} = (\hat{p} - p)/\sqrt{pq/n}$
     $= (.29-.50)/\sqrt{(.50)(.50)/1025} = -.21/.0156 = -13.45$

23. $\hat{p} = x/n$
     $= x/400 = .290$
     $z_{\hat{p}} = (\hat{p} - p)/\sqrt{pq/n}$
     $= (.290-.25)/\sqrt{(.25)(.75)/400} = .04/.0217 = 1.85$

25. P-value $= P(z > .55)$
     $= 1 - P(z < .55)$
     $= 1 - .7088$
     $= .2912$

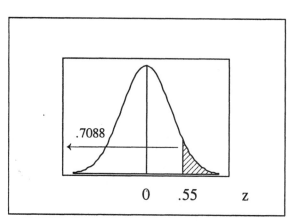

27. P-value $= 2{\cdot}P(z > 1.95)$
     $= 2 \cdot [1 - P(z < 1.95)]$
     $= 2 \cdot [1 - .9744]$
     $= 2 \cdot [.0256]$
     $= .0512$

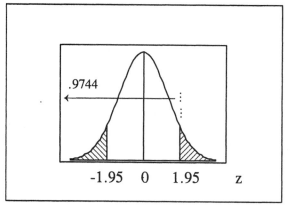

29. P-value $= P(z > 1.97)$
     $= 1 - P(z < 1.97)$
     $= 1 - .9756$
     $= .0244$

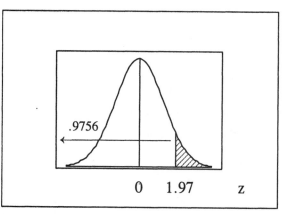

31.   P-value = $2 \cdot P(z > .77)$
       $= 2 \cdot [1 - P(z < .77)]$
       $= 2 \cdot [1 - .7794]$
       $= 2 \cdot [.2206]$
       $= .4412$

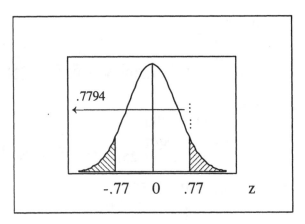

33.   original claim: $p > .5$ (does not contain the equality; must be H_1)
       competing idea: $p \leq .5$
           H_o: $p = .5$
           H_1: $p > .5$
       initial conclusion: Reject H_o.
       final conclusion: There is sufficient evidence to support the claim that the proportion of
           married women is greater than .5.

35.   original claim: $p \neq .038$ (does not contain the equality; must be H_1)
       competing idea: $p = .038$
           H_o: $p = .038$
           H_1: $p \neq .038$
       initial conclusion: Fail to reject H_o.
       final conclusion: There is not sufficient evidence to support the claim that the proportion of
           commercial aviation crashes that are fatal is different from .038.

37.   type I error: supporting the claim $p > .5$ when $p = .5$ is true
       type II error: failing to support the claim $p > .5$ when $p > .5$ is true

39.   type I error: supporting the claim $p \neq .038$ when $p = .038$ is true
       type II error: failing to support the claim $p \neq ,038$ when $p \neq .038$ is true

41. $\hat{p} = x/n = x/491 = .27$
       $z_{\hat{p}} = (\hat{p} - p)/\sqrt{pq/n}$
           $= (.27 - .50)/\sqrt{(.50)(.50)/491}$
           $= -.23/.0226 = -10.19$
       P-value $= P(z > -10.19)$
           $= 1 - (z < -10.19)$
           $= 1 - .0001 = .9999$
       The claim is that $p > .5$. Only $\hat{p}$ values so much larger than .5 that they are unlikely to
       occur by chance if $p = .5$ is true give statistical support to the claim. A $\hat{p}$ smaller than .5
       does not give any support to the claim.

43.   Prefer .01, the lowest of the given values. The P-value is the probability of getting by
       chance alone less than or equal to the number of defects observed if the old rate is still in
       effect. The lower the P-value, the less likely your results could have occurred by chance
       alone – and the more likely your new process truly reduces the rate of defects.

45.   Mathematically, in order for α to equal 0 the magnitude of the critical value would have to
       be infinite. Practically, the only way never to make a type I error is to always fail to reject
       H_o. From either perspective, the only way to achieve $\alpha = 0$ is to never reject H_o no
       matter how extreme the sample data might be.

7-3 Testing a Claim about a Proportion

NOTE: To reinforce the concept that all z scores are standardized rescalings obtained by subtracting the mean and dividing by the standard deviation, the manual uses the "usual" z formula written to apply to $\hat{p}$'s

$$z_{\hat{p}} = (\hat{p} - \mu_{\hat{p}})/\sigma_{\hat{p}}.$$

When the normal approximation to the binomial applies, the $\hat{p}$'s are normally distributed with $\mu_{\hat{p}} = p$ and $\sigma_{\hat{p}} = \sqrt{pq/n}$.

And so the formula for the z statistic may also be written as

$$z_{\hat{p}} = (\hat{p} - p)/\sqrt{pq/n} .$$

1. a. $z_{\hat{p}} = (\hat{p} - \mu_{\hat{p}})/\sigma_{\hat{p}}$
        $= (\hat{p} - p)/\sqrt{pq/n}$
        $= (.2494 - .25)/\sqrt{(.25)(.75)/8023}$
        $= -.0006/.0048$
        $= -.12$
    b. $z = \pm 1.96$
    c. P-value $= 2 \cdot P(z < -.12) = 2 \cdot (.4522) = .9044$
    d. Do not reject H_o; there is not sufficient evidence to conclude $p \neq .25$.
    e. No, the hypothesis test will either "reject" or "fail to reject" the claim that a population parameter is equal to a specified value.

3. original claim: $p < .62$
        $\hat{p} = x/n = x/2500 = .60$
    H_o: $p = .62$
    H_1: $p < .62$
    $\alpha = .01$
    C.R. $z < -z_{.01} = -2.326$
    calculations:
        $z_{\hat{p}} = (\hat{p} - \mu_{\hat{p}})/\sigma_{\hat{p}}$
        $= (.60 - .62)/\sqrt{(.62)(.38)/2500}$
        $= -.02/.00971$
        $= -2.06$
    P-value $= P(z < -2.06) = .0197$
conclusion:
    Do not reject H_o; there is not sufficient evidence to conclude that $p < .62$.
If the data came from self-selected voluntary respondents, the test is not valid.

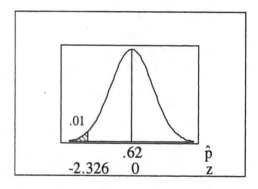

5. original claim: $p > .15$
        $\hat{p} = x/n = 149/880 = .169$
    H_o: $p = .15$
    H_1: $p > .15$
    $\alpha = .05$
    C.R. $z > z_{.05} = 1.645$
    calculations:
        $z_{\hat{p}} = (\hat{p} - \mu_{\hat{p}})/\sigma_{\hat{p}}$
        $= (.169 - .15)/\sqrt{(.15)(.85)/880}$
        $= .019/.0120$
        $= 1.60$
    P-value $= P(z > 1.60) = 1 - P(z < 1.60) = 1 - .9452 = .0548$
conclusion:
    Do not reject H_o; there is not sufficient evidence to conclude that $p > .15$.
No; since technology and the use of technology is changing so rapidly, any figures from 1997 would no longer be valid today. [By the time you are reading this, e-mail could be so common that essentially 100% of the households use it – or it could have been replaced by something newer and be so out-dated that essentially 0% of the households still use it!]

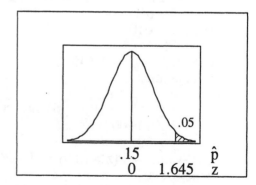

7.   original claim: p > .50
       $\hat{p} = x/n = x/829 = .51$
H_o: p = .50
H_1: p > .50
α = .10
C.R. z > $z_{.10}$ = 1.282
calculations:
       $z_{\hat{p}} = (\hat{p} - \mu_{\hat{p}})/\sigma_{\hat{p}}$
            $= (.51 - .50)/\sqrt{(.50)(.50)/829}$
            $= .01/.0174$
            $= .58$
       P-value = P(z > .58) = 1 - P(z < .58) = 1 - .7190 = .2810
conclusion:
       Do not reject H_o; there is not sufficient evidence to conclude that p > .50.
No; since there are state and regional differences (in both circumstances and opinions) the
conclusions reached for any one state can not be generalized to the country as a whole.

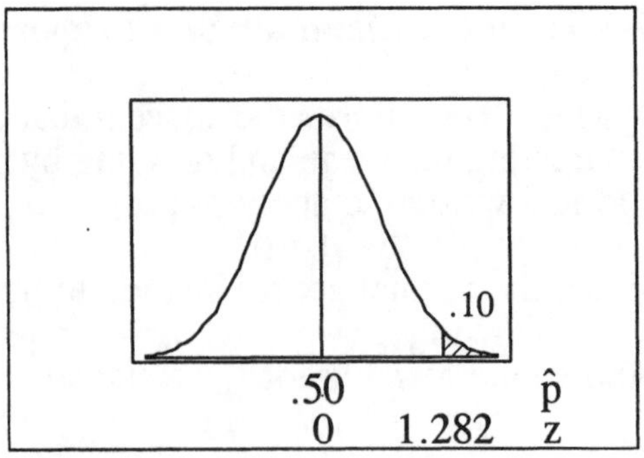

9.   original claim: p = .01
       $\hat{p} = x/n = 20/1234 = .016$
H_o: p = .01
H_1: p ≠ .01
α = .05
C.R. z < -$z_{.025}$ = -1.96
       z > $z_{.025}$ = 1.96
calculations:
       $z_{\hat{p}} = (\hat{p} - \mu_{\hat{p}})/\sigma_{\hat{p}}$
            $= (.016 - .01)/\sqrt{(.01)(.99)/1234}$
            $= .006/.00283$
            $= 2.19$
       P-value = 2·P(z > 2.19) = 2·[1 - P(z < 2.19)] = 2·[1 - .9857] = 2·[.0143] = .0286
conclusion:
       Reject H_o; there is sufficient evidence reject the claim that p = .01 and to conclude that
       p ≠ .01 (in fact, that p > .01).
No; based on these results, consumers appear to be subjected to more overcharges than
under the old pre-scanner system.

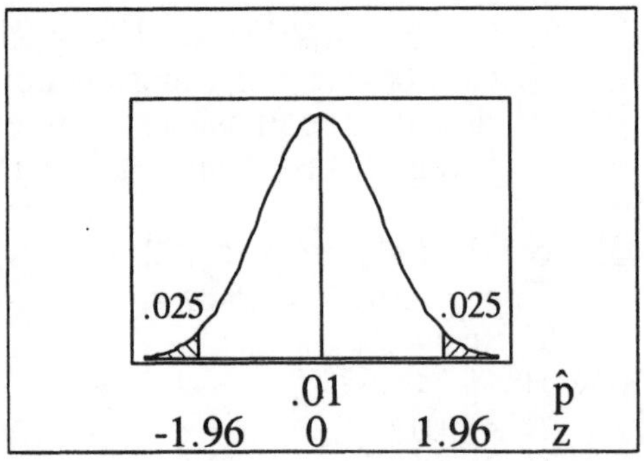

11.   original claim: p > .610
       $\hat{p} = x/n = 2231/3581 = .623$
H_o: p = .610
H_1: p > .610
α = .05
C.R. z > $z_{.05}$ = 1.645
calculations:
       $z_{\hat{p}} = (\hat{p} - \mu_{\hat{p}})/\sigma_{\hat{p}}$
            $= (.623 - .610)/\sqrt{(.610)(.390)/3581}$
            $= .013/.00815$
            $= 1.60$
       P-value = P(z > 1.60) = 1 - P(z < 1.60) = 1 - .9452 = .0548
conclusion:
       Do not reject H_o; there is not sufficient evidence to conclude that p > .610.

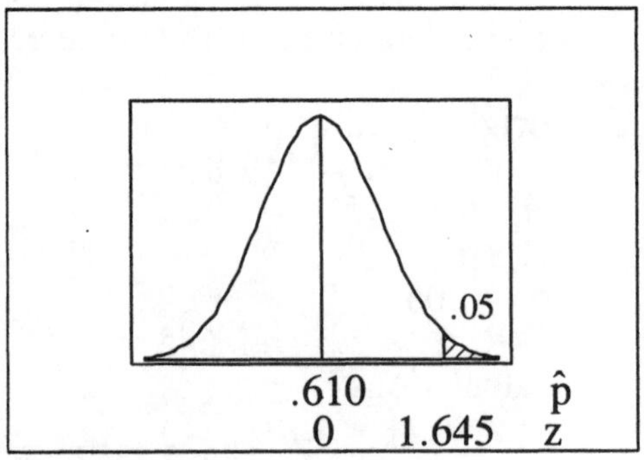

13. original claim: p = .000340
        $\hat{p}$ = x/n = 135/420,095 = .000321
      H_o: p = .000340
      H_1: p ≠ .000340
      α = .005
      C.R. z < $-z_{.0025}$ = -2.81
            z > $z_{.0025}$ = 2.81
      calculations:
          $z_{\hat{p}}$ = ($\hat{p}$ - $\mu_{\hat{p}}$)/$\sigma_{\hat{p}}$
              = (.000321 - .000340)/$\sqrt{\dfrac{(.000340)(.999660)}{420,095}}$
              = -.000019/.0000284
              = -.66

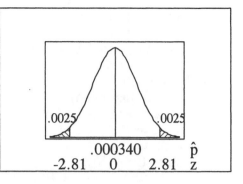

      P-value = 2·P(z < -.66) = 2·(.2546) = .5092
      conclusion:
        Do not reject H_o; there is not sufficient evidence reject the claim that p = .000340.
      No; based on these results, cell phone users have no reason for such concern.

15. original claim: p < .27
        $\hat{p}$ = x/n = 144/785 = .183
      H_o: p = .27
      H_1: p < .27
      α = .01
      C.R. z < $-z_{.01}$ = -2.326
      calculations:
          $z_{\hat{p}}$ = ($\hat{p}$ - $\mu_{\hat{p}}$)/$\sigma_{\hat{p}}$
              = (.183 - .27)/$\sqrt{(.27)(.73)/785}$
              = -.087/.0158
              = -5.46

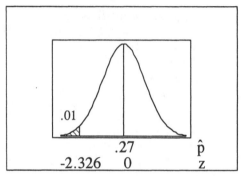

      P-value = P(z < -5.46) = .0001
      conclusion:
        Reject H_o; there is sufficient evidence to conclude that p < .27.
      Since not smoking is the better decision from both a health and a financial perspective, one
      would expect it to be the choice of the better decision-makers – and presumably the
      additional education college graduates have makes them better decision-makers than the
      general population.

17. original claim: p > .75
        $\hat{p}$ = x/n = x/500 = .91
      H_o: p = .75
      H_1: p > .75
      α = .01
      C.R. z > $z_{.01}$ = 2.326
      calculations:
          $z_{\hat{p}}$ = ($\hat{p}$ - $\mu_{\hat{p}}$)/$\sigma_{\hat{p}}$
              = (.91 - .75)/$\sqrt{(.75)(.25)/500}$
              = .16/.0194
              = 8.26

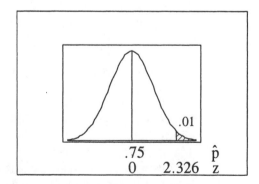

      P-value = P(z > 8.26) = 7.22x10^{-17} = .0000000000000000722 [from TI-83 Plus]
      conclusion:
        Reject H_o; there is sufficient evidence to conclude that p > .75.
      The TI-83 Plus values agree with those calculated above. Yes; based on these sample
      results, the funding will be approved.

19. There are 100 M&M candies listed in Data Set 19, and 5 of them are blue.
     original claim: p = .10
         $\hat{p} = x/n = 5/100 = .05$
     H_0: p = .10
     H_1: p ≠ .10
     α = .05 [assumed]
     C.R. z < $-z_{.025}$ = -1.96
          z > $z_{.025}$ = 1.96
     calculations:
         $z_{\hat{p}} = (\hat{p} - \mu_{\hat{p}})/\sigma_{\hat{p}}$
              $= (.05 - .10)/\sqrt{(.10)(.90)/100}$
              $= -.05/.03$
              $= -1.67$
     P-value = 2·P(z < -1.67) = 2·(.0475) = .0950
     conclusion:
         Do not reject H_0; there is not sufficient evidence reject the claim that p = .10.

21. original claim: p = .10
         $\hat{p} = x/n = 119/1000 = .119$
     H_0: p = .10
     H_1: p ≠ .10
     α = .05
     C.R. z < $-z_{.025}$ = -1.96
          z > $z_{.025}$ = 1.96
     calculations:
         $z_{\hat{p}} = (\hat{p} - \mu_{\hat{p}})/\sigma_{\hat{p}}$
              $= (.119 - .10)/\sqrt{(.10)(.90)/1000}$
              $= .019/.00949$
              $= 2.00$
     P-value = 2·P(z > 2.00) = 2·[1 - P(z < 2.00)] = 2·[1 - .9772] = 2·[.0228] = .0456
     conclusion:
         Reject H_0; there is sufficient evidence reject the claim that p = .10 and conclude that
         p ≠ .10 (in fact, that p > .10).

     a. As seen above, the traditional method leads to rejection of the claim that p = .10 because
        the calculated z = 2.00 is greater than the critical value of 1.96.
     b. As seen above, the P-value method leads to rejection of the claim that p = .10 because
        the calculated P-value = .0456 is less than the level of significance of .05.
     c. α = .05 and $\hat{p} = x/n = 119/1000 = .119$
            $\hat{p} \pm z_{.025}\sqrt{\hat{p}\hat{q}/n}$
            $.119 \pm 1.96\sqrt{(.119)(.881)/1000}$
            .119 ± .020
            .099 < p < .139
        Since .10 is inside the confidence interval, this suggests that p = .10 is a reasonable
        claim that should not be rejected.
     d. The traditional method and the P-value method are mathematically equivalent and will
        always agree. As seen by this example, the confidence interval method does not always
        lead to the same conclusion as the other two methods.

23. original claim: p ≤ c
     competing claim: p > c (does not contain the equality; must be H_1)
     H_0: p = c
     H_1: p > c
     The possible conclusions are "reject p = c, and conclude p > c" or "fail to reject p = c, and
     say there is not enough evidence to conclude p > c" – but the test cannot conclude in favor
     of or support the null hypothesis or the original claim.

25. original claim: p = .10 [normal approximation to the binomial, use z]
    $\hat{p}$ = x/n = 0/50 = 0
    Yes, the methods of this section can be used. In general, the appropriateness of a test depends on the design of the experiment and not the particular results. In particular, for this problem the normal approximation applies because
    np = (50)(.1) = 5 ≥ 5
    n(1-p) = (50)(.9) = 45 ≥ 5
    H_o: p = .10
    H_1: p ≠ .10
    α = .01
    C.R. z < $-z_{.005}$ = -2.575
         z > $z_{.005}$ = 2.575
    calculations:
    $z_{\hat{p}}$ = ($\hat{p}$ - $\mu_{\hat{p}}$)/$\sigma_{\hat{p}}$
       = (0 - .10)/$\sqrt{(.10)(.90)/50}$
       = -2.357

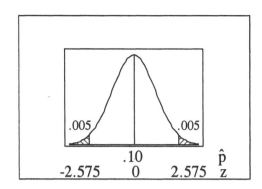

    P-value = 2·P(z < -2.36) = 2·(.0091) = .0182
    conclusion:
    Do no reject H_o; there is not sufficient evidence reject the claim that p = .10.

27. Assuming that each mouse is either a success or not a success, the number of successes must be a whole number between 0 and 20 includive. The possible success rates are all multiples of 5%: 0/20 = 0%, 1/20 = 5%, 2/20 = 10%, 3/20 = 15%, etc. A success rate of 47% is not a possibility.

7-4 Testing a Claim about a Mean: σ Known

1. Yes, since σ is known and the original population is normally distributed.

3. No, since σ is unknown.

5. original claim: μ > 118
   $\bar{x}$ = 120
   H_o: μ = 118
   H_1: μ > 118
   α = .05
   C.R. z > $z_{.05}$ = 1.645
   calculations:
   $z_{\bar{x}}$ = ($\bar{x}$ - μ)/$\sigma_{\bar{x}}$
      = (120 - 118)/(12$\sqrt{50}$)
      = 2/1.697
      = 1.18

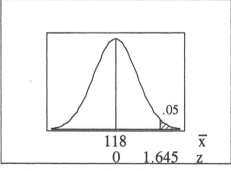

   P-value = P(z > 1.18) = 1 - P(z < 1.18) = 1 - .8810 = .1190
   conclusion:
   Do not reject H_o; there is not sufficient evidence to conclude that μ > 118.

7. original claim: $\mu = 5.00$
    $\overline{x} = 5.25$
    $H_o: \mu = 5.00$
    $H_1: \mu \neq 5.00$
    $\alpha = .01$
    C.R. $z < -z_{.005} = -2.575$
    　　$z > z_{.005} = 2.575$
    calculations:
    　　$z_{\overline{x}} = (\overline{x} - \mu)/\sigma_{\overline{x}}$
    　　　$= (5.25 - 5.00)/(2.50/\sqrt{80})$
    　　　$= .25/.2795$
    　　　$= .89$
    P-value $= 2 \cdot P(z > .89) = 2 \cdot [1 - P(z < .89)] = 2 \cdot [1 - .8133] = 2 \cdot [.1867] = .3734$
    conclusion:
    　　Do not reject H_o; there is not sufficient evidence reject the claim that $\mu = .10$.

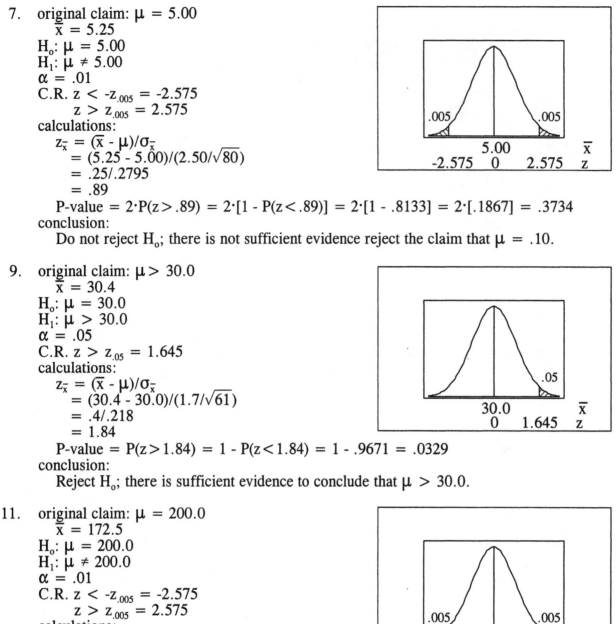

9. original claim: $\mu > 30.0$
    $\overline{x} = 30.4$
    $H_o: \mu = 30.0$
    $H_1: \mu > 30.0$
    $\alpha = .05$
    C.R. $z > z_{.05} = 1.645$
    calculations:
    　　$z_{\overline{x}} = (\overline{x} - \mu)/\sigma_{\overline{x}}$
    　　　$= (30.4 - 30.0)/(1.7/\sqrt{61})$
    　　　$= .4/.218$
    　　　$= 1.84$
    P-value $= P(z > 1.84) = 1 - P(z < 1.84) = 1 - .9671 = .0329$
    conclusion:
    　　Reject H_o; there is sufficient evidence to conclude that $\mu > 30.0$.

11. original claim: $\mu = 200.0$
    $\overline{x} = 172.5$
    $H_o: \mu = 200.0$
    $H_1: \mu \neq 200.0$
    $\alpha = .01$
    C.R. $z < -z_{.005} = -2.575$
    　　$z > z_{.005} = 2.575$
    calculations:
    　　$z_{\overline{x}} = (\overline{x} - \mu)/\sigma_{\overline{x}}$
    　　　$= (172.5 - 200.0)/(119.5/\sqrt{40})$
    　　　$= -27.5/18.895$
    　　　$= -1.46$
    P-value $= 2 \cdot P(z < -1.46) = 2 \cdot (.0721) = .1442$
    conclusion:
    　　Do not reject H_o; there is not sufficient evidence reject the claim that $\mu = 200.0$.

13. original claim: $\mu \neq .9085$
     $\bar{x} = .91470$
     H_o: $\mu = .9085$
     H_1: $\mu \neq .9085$
     $\alpha = .05$ [assumed]
     C.R. $z < -z_{.025} = -1.96$
     $\quad\quad z > z_{.025} = 1.96$
     calculations:
     $\quad z_{\bar{x}} = (\bar{x} - \mu)/\sigma_{\bar{x}}$
     $\quad\quad = (.91470 - .9085)/(.03691/\sqrt{100})$
     $\quad\quad = .0062/.003691$
     $\quad\quad = 1.68$
     P-value $= 2 \cdot P(z > 1.68) = 2 \cdot [1 - P(z < 1.68)] = 2 \cdot [1 - .9535] = 2 \cdot (.0465) = .0930$
     conclusion:

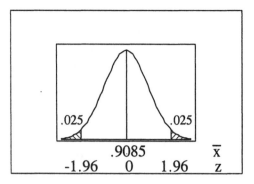

Do not reject H_o; there is not sufficient evidence to conclude that $\mu \neq .9085$
Both the calculated z and the P-value agree with the Minitab output. No; since $\bar{x} >$
.9085, it appears that any deviance is in favor of the consumer. No; since the company
presumably figures its costs based on what is actually produced rather than on what is
advertised, they are not losing money by making the candy slightly heavier than necessary.

15. original claim: $\mu = 0$
     $\bar{x} = -.419$
     H_o: $\mu = 0$
     H_1: $\mu \neq 0$
     $\alpha = .05$ [assumed]
     C.R. $z < -z_{.025} = -1.96$
     $\quad\quad z > z_{.025} = 1.96$
     calculations:
     $\quad z_{\bar{x}} = (\bar{x} - \mu)/\sigma_{\bar{x}}$
     $\quad\quad = (-.419 - 0)/(3.704/\sqrt{31})$
     $\quad\quad = -.419/.6653$
     $\quad\quad = -.63$
     P-value $= 2 \cdot P(z < -.63) = 2 \cdot (.2643) = .5286$
     conclusion:

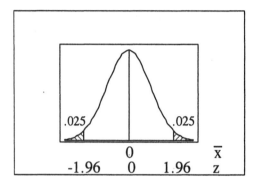

Do not reject H_o; there is not sufficient evidence reject the claim that $\mu = 0$.
Both the calculated z and the P-value agree with the TI-83 Plus output. The mean
difference does not appear to be significantly different form 0. This suggests that the 3-
day forecast high temperatures are correct <u>on the average</u> – but that does not necessarily
instill confidence in any one single prediction, as the $\sigma = 3.7$ suggests that it's not unusual
for the predictions to be off by $2 \cdot (3.7) = 7.4$ degrees in either direction.

17. a. It is unrealistic that the value of σ would be known.
     b. original claim: $\mu \neq 98.6$; $\bar{x} = 98.20$
        H_o: $\mu = 98.6$
        H_1: $\mu \neq 98.6$
        $\alpha = .05$
        C.R. $z < -z_{.025} = -1.96$
        $\quad\quad z > z_{.025} = 1.96$
        calculations:
        $z_{\bar{x}} = (\bar{x} - \mu)/\sigma_{\bar{x}}$
        $-1.96 < (98.20 - 98.6)/(\sigma/\sqrt{106})$
        $\quad\quad \sigma < (98.20 - 98.6)/(-1.96/\sqrt{106})$
        $\quad\quad \sigma < (-.4)/(-.1904)$
        $\quad\quad \sigma < 2.101$; The largest possible standard deviation is 2.10

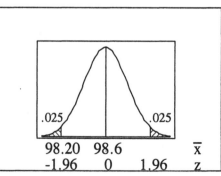

c. No; if $s=.62$ for $n=106$ sample values, it seems unlikely that the true value of σ in this particular instance would be greater than 2.10. The assumption that $\sigma=.62$ seems appropriate in that the conclusion would be the same for any other value reasonably close to .62.

19. original claim: $\mu \neq 98.6$
H_o: $\mu = 98.6$
H_1: $\mu \neq 98.6$
$\alpha = .05$
C.R. $z < -z_{.025} = -1.960$
$\quad z > z_{.025} = 1.960$

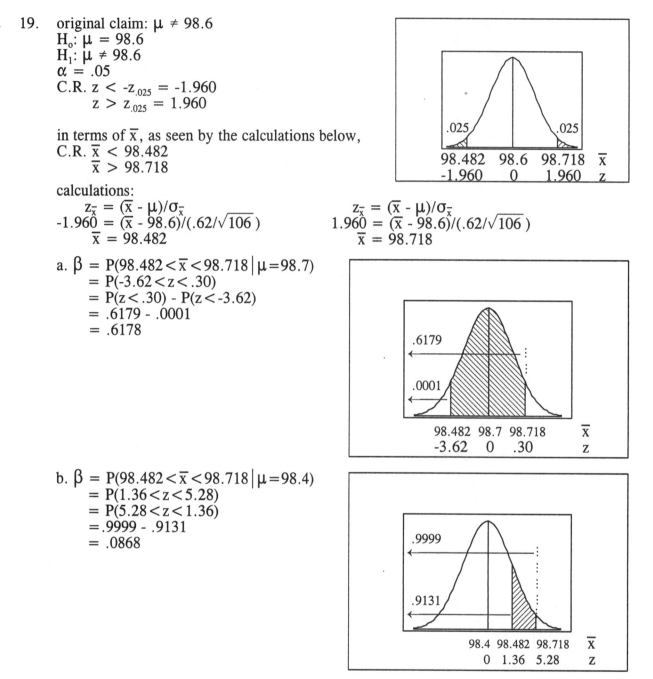

in terms of $\overline{x}$, as seen by the calculations below,
C.R. $\overline{x} < 98.482$
$\quad \overline{x} > 98.718$

calculations:
$z_{\overline{x}} = (\overline{x} - \mu)/\sigma_{\overline{x}}$
$-1.960 = (\overline{x} - 98.6)/(.62/\sqrt{106})$
$\overline{x} = 98.482$

$z_{\overline{x}} = (\overline{x} - \mu)/\sigma_{\overline{x}}$
$1.960 = (\overline{x} - 98.6)/(.62/\sqrt{106})$
$\overline{x} = 98.718$

a. $\beta = P(98.482 < \overline{x} < 98.718 \,|\, \mu = 98.7)$
$\quad = P(-3.62 < z < .30)$
$\quad = P(z < .30) - P(z < -3.62)$
$\quad = .6179 - .0001$
$\quad = .6178$

b. $\beta = P(98.482 < \overline{x} < 98.718 \,|\, \mu = 98.4)$
$\quad = P(1.36 < z < 5.28)$
$\quad = P(5.28 < z < 1.36)$
$\quad = .9999 - .9131$
$\quad = .0868$

7-5 Testing a Claim About a Mean: σ Not Known

1. σ unknown and distribution approximately normal, use t

3. σ known and distribution approximately normal, use z

5. $t_{11,.01} = 2.718 < 2.998 < 3.106 = t_{11,.005}$
    $.01 > $ P-value $< .005$
    $.005 < $ P-value $< .01$

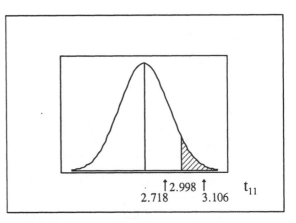

7. $t_{15,.005} = 2.947 < 4.629$
    $.005 > \frac{1}{2}$(P-value)
    P-value $< .01$

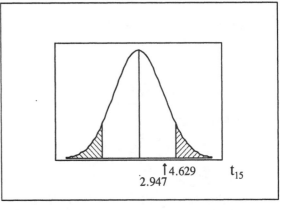

9. original claim: $\mu > 118$
    H_o: $\mu = 118$
    H_1: $\mu > 118$
    $\alpha = .05$
    C.R. $t > t_{19,.05} = 1.729$
    calculations:
    $\quad t_{\bar{x}} = (\bar{x} - \mu)/s_{\bar{x}}$
    $\quad\quad = (120 - 118)/(12\sqrt{20})$
    $\quad\quad = 2/2.683$
    $\quad\quad = .745$
    P-value $= P(t_{19} > .745) > .10$
    conclusion:
    Do not reject H_o; there is not sufficient evidence to conclude that $\mu > 118$.

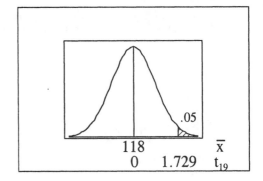

11. original claim: $\mu = 5.00$
    H_o: $\mu = 5.00$
    H_1: $\mu \neq 5.00$
    $\alpha = .01$
    C.R. $t < -t_{80,.005} = -2.639$
    $\quad\quad t > t_{80,.005} = 2.639$
    calculations:
    $\quad t_{\bar{x}} = (\bar{x} - \mu)/t_{\bar{x}}$
    $\quad\quad = (5.25 - 5.00)/(2.50/\sqrt{81})$
    $\quad\quad = .25/.2778$
    $\quad\quad = .900$
    P-value $= 2 \cdot P(t_{19} > .900) > .20$
    conclusion:
    Do not reject H_o; there is not sufficient evidence reject the claim that $\mu = .10$.

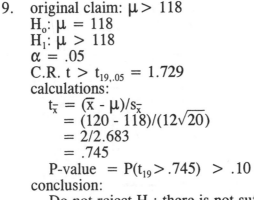

13. original claim: $\mu > 4$
     H_o: $\mu = 4$
     H_1: $\mu > 4$
     $\alpha = .05$
     C.R. $t > t_{11,.05} = 1.796$
     calculations:
     $$t_{\bar{x}} = (\bar{x} - \mu)/s_{\bar{x}}$$
     $$= (5.075 - 4)/(1.168/\sqrt{12})$$
     $$= 1.075/.337$$
     $$= 3.188$$
     P-value $= P(t_{11} > 3.188) < .005$
     conclusion:
     Reject H_o; there is sufficient evidence to conclude that $\mu > 4$.
     Yes; based on this sample, the teachers will use the book.

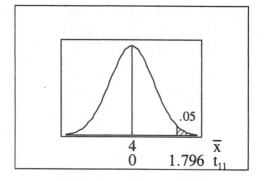

15. original claim: $\mu \neq 0$
     H_o: $\mu = 0$
     H_1: $\mu \neq 0$
     $\alpha = .05$
     C.R. $t < -t_{30,.025} = -2.042$
     $\quad\quad t > t_{30,.025} = 2.042$
     calculations:
     $$t_{\bar{x}} = (\bar{x} - \mu)/s_{\bar{x}}$$
     $$= (-.419 - 0)/(3.704/\sqrt{31})$$
     $$= -.419/.665$$
     $$= -.630$$
     P-value $= 2 \cdot P(t_{30} < -.630) > .20$
     conclusion:
     Do not reject H_o; there is not sufficient evidence to reject the claim that $\mu = 0$.
     Yes; based on this result, the forecasts do seem to be reasonably accurate.

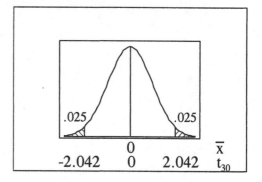

17. original claim: $\mu = 0$
     H_o: $\mu = 0$
     H_1: $\mu \neq 0$
     $\alpha = .01$
     C.R. $t < -t_{39,.005} = -2.712$
     $\quad\quad t > t_{39,.005} = 2.712$
     calculations:
     $$t_{\bar{x}} = (\bar{x} - \mu)/s_{\bar{x}}$$
     $$= (117.3 - 0)/(185.0/\sqrt{40})$$
     $$= 117.3/29.251$$
     $$= 4.010$$
     P-value $= 2 \cdot P(t_{39} > 4.010) < .01$
     conclusion:
     Reject H_o; there is sufficient evidence to reject the claim that $\mu = 0$ and to conclude that $\mu \neq 0$ (in fact, that $\mu > 0$).
     One conclude that on the average, people's wrist watches tend to be almost two minutes fast – which is probably by deliberate choice.

19. original claim: $\mu > 69.5$
     H_o: $\mu = 69.5$
     H_1: $\mu > 69.5$
     $\alpha = .05$
     C.R. $t > t_{34,.05} = 1.691$
     calculations:
       $t_{\bar{x}} = (\bar{x} - \mu)/s_{\bar{x}}$
          $= (73.4 - 69.5)/(8.7/\sqrt{35})$
          $= 3.9/1.471$
          $= 2.652$
       $.005 < \text{P-value} = P(t_{34} > 2.652) < .01$
     conclusion:

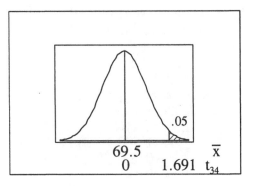

     Reject H_o; there is sufficient evidence to conclude that $\mu > 69.5$.
     Yes, it does appear that male symphony conductors live longer than males form the general
     population – but in one sense that is like saying "male centenarians live longer than males
     from the general population." Male conductors should not be compared to "males from
     the general population" – but to "males from general population who have reached, say,
     age 35." Males who die in their teens, for example, are included in the general male
     population but (since we have no foreknowledge about what vocation they might have
     chosen had they lived) not in the population of male symphony orchestra conductors.

21. original claim: $\mu < 1000$
     H_o: $\mu = 1000$
     H_1: $\mu < 1000$
     $\alpha = .05$ [assumed]
     C.R. $t < -t_{4,.05} = -2.132$
     calculations:
       $t_{\bar{x}} = (\bar{x} - \mu)/s_{\bar{x}}$
          $= (767 - 1000)/(285/\sqrt{5})$
          $= -233/127.46$
          $= -1.828$
       P-value $= .071$
     conclusion:

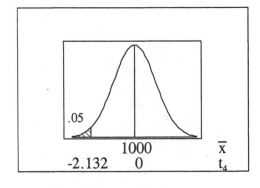

     Do not reject H_o; there is not sufficient evidence to conclude that $\mu < 1000$.
     No; while the sample mean was less than \$1000, the company cannot be 95% certain that
     the population mean is less than that amount.

23. original claim: $\mu = 3.39$
     H_o: $\mu = 3.39$
     H_1: $\mu \neq 3.39$
     $\alpha = .05$ [assumed]
     C.R. $t < -t_{15,.025} = -2.131$
          $t > t_{15,.025} = 2.131$
     calculations:
       $t_{\bar{x}} = (\bar{x} - \mu)/s_{\bar{x}}$
          $= (3.675 - 3.39)/(.6573/\sqrt{16})$
          $= .285/.1643$
          $= 1.734$
       P-value $= .1034$
     conclusion:

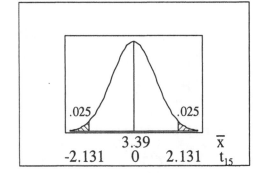

     Do not reject H_o; there is not sufficient evidence to conclude that $\mu \neq 3.39$.
     No; based on these results the supplement does not appear to have a significant effect on
     birth weight.

25. summary statistics: n = 6, Σx = 9.23, Σx^2 = 32.5197, $\bar{x}$ = 1.538, s = 1.914
    original claim: $\mu > 1.5$
    H_o: $\mu = 1.5$
    H_1: $\mu > 1.5$
    $\alpha = .05$
    C.R. t > $t_{5,.05}$ = 2.015
    calculations:
    $t_{\bar{x}} = (\bar{x} - \mu)/s_{\bar{x}}$
    $= (1.538 - 1.5)/(1.914/\sqrt{6})$
    $= .0383/.78147$
    $= .049$

    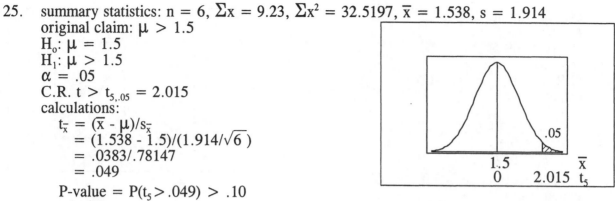

    P-value = $P(t_5 > .049) > .10$
    conclusion:
    Do not reject H_o; there is not sufficient evidence to conclude that $\mu > 1.5$.
    Yes; since 5 of the 6 values are below the sample mean, there is reason to doubt that the population values are normally distributed.

27. summary statistics: n = 23, Σx = 241.27, Σx^2 = 2536.5762, $\bar{x}$ = 10.490 s = .507
    original claim: $\mu < 10.5$
    H_o: $\mu = 10.5$
    H_1: $\mu < 10.5$
    $\alpha = .05$ [assumed]
    C.R. t < $-t_{22,.05}$ = -1.717
    calculations:
    $t_{\bar{x}} = (\bar{x} - \mu)/s_{\bar{x}}$
    $= (10.490 - 10.5)/(.507/\sqrt{23})$
    $= -.01/.1057$
    $= -.095$
    P-value = $P(t_{22} < -.095) > .10$
    conclusion:
    Do not reject H_o; there is not sufficient evidence to conclude that $\mu < 10.5$.
    The precision of the numbers changed from 1 to 2 decimal accuracy, presumably as improved technology produced more accurate timing devices. The test does not consider the sequential nature of the data, and would give the same results no matter what the order of the data. Because the test fails to consider the significant order effect, it only summarizes the past and cannot accurately project into the future. No; the test does not address the issue about future times being close to 10.5, and such a conclusion would not be valid.

29. summary statistics: n = 12, Σx = 78.0, Σx^2 = 572.94, $\bar{x}$ = 6.50, s = 2.448
    original claim: $\mu > 6$
    H_o: $\mu = 6$
    H_1: $\mu > 6$
    $\alpha = .05$ [assumed]
    C.R. t > $t_{11,.05}$ = 1.796
    calculations:
    $t_{\bar{x}} = (\bar{x} - \mu)/s_{\bar{x}}$
    $= (6.50 - 6)/(2.448/\sqrt{12})$
    $= .50/.7068$
    $= .707$
    P-value = $P(t_{11} > .707) > .10$
    conclusion:
    Do not reject H_o; there is not sufficient evidence to conclude that $\mu > 6$.
    While the sample mean is above 6, the teacher cannot be 95% certain that the population mean is above 6.

31. summary statistics: n = 36, Σx = 439.0, Σx^2 = 5353.82, $\bar{x}$ = 12.194, s = .1145
original claim: $\mu > 12$
H_o: $\mu = 12$
H_1: $\mu > 12$
$\alpha = .01$
C.R. t > $t_{35,.01}$ = 2.441
calculations:
$\quad t_{\bar{x}} = (\bar{x} - \mu)/s_{\bar{x}}$
$\qquad = (12.194 - 12)/(.1145/\sqrt{36})$
$\qquad = .1944/.01908$
$\qquad = 10.189$
$\quad$ P-value = $P(t_{35} > 10.189) < .005$

conclusion:
$\quad$ Reject H_o; there is sufficient evidence to conclude that $\mu > 12$.
While the population mean appears to be above 12, the volume should not be reduced unless the standard deviation can be lowered. If $\mu \approx 12.194$ and $\sigma \approx .1145$ as the data suggests, then the limits $\mu \pm 2\sigma$ for "usual" occurrences already include values below 12 – i.e., it would not be unusual to get a can with less than 12 ounces. Reducing the volume would further increase the number of sub-standard cans, possibly to the point of serious legal or public relations problems.

33. The P-value is changed. The summary values n, $\bar{x}$ and s do not change. The standard error of the mean, which is $s_{\bar{x}} = s/\sqrt{n}$, does not change. The calculated t, which is $t_{\bar{x}} = (\bar{x} - \mu)/s_{\bar{x}}$, does not change. Since the test is now two-tailed, the P-value doubles and is now $2 \cdot P(t_{14} > 1.954) = 2 \cdot (.035) = .070$. In this instance, changing the P-value changes the conclusion and (at the assumed .05 level of significance) there is not enough evidence to reject H_o: $\mu = 420$.

35. The new test of hypothesis is given below. In this instance the presence of a drastic outlier outlier drastically changed the summery statistics but did not have much of an impact on the overall test of hypothesis. In general, however, the effects of an outlier can be substantial. One could argue that the effect is not so great in this exercise because the original test already involved an outlier: by the $\bar{x} \pm s$ standard, 5.40 is an "unusual" result – and it is so much larger than the others that it is the only value above $\bar{x}$.

summary statistics: n=6, Σx = 543.83, Σx^2 = 291,603.3597, $\bar{x}$ = 90.638, s = 220.142
original claim: $\mu > 1.5$
H_o: $\mu = 1.5$
H_1: $\mu > 1.5$
$\alpha = .05$
C.R. t > $t_{5,.05}$ = 2.015
calculations:
$\quad t_{\bar{x}} = (\bar{x} - \mu)/s_{\bar{x}}$
$\qquad = (90.638 - 1.5)/(220.142/\sqrt{6})$
$\qquad = 89.138/89.872$
$\qquad = .992$
$\quad$ P-value = $P(t_5 > .992) > .10$

conclusion:
$\quad$ Do not reject H_o; there is not sufficient evidence to conclude that $\mu > 1.5$.

37. original claim: $\mu > 40$
     $H_o: \mu = 40$
     $H_1: \mu > 40$
     $\alpha = .01$
     C.R. $t > t_{9,.01} = 2.821$
     calculations:
     $$t_{\bar{x}} = (\bar{x} - \mu)/s_{\bar{x}}$$
     $$2.821 = (\bar{x} - 40)/(3.801/\sqrt{10})$$
     $$3.390 = \bar{x} - 40$$
     $$\bar{x} = 43.3899$$

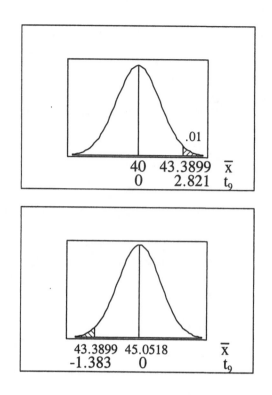

β = P(failing to reject $H_o | H_o$ is false)
  = $P(\bar{x} < 43.3899 | \mu = 45.0518)$
  = $P(t_9 < -1.383)$
  = .10 [since $-t_{9,.10} = -1.383$]

in the above calculation of β,
  $t_{\bar{x}} = (\bar{x} - \mu)/s_{\bar{x}}$
    = $(43.3899 - 45.0518)/(3.801/\sqrt{10})$
    = $-1.6619/1.202$
    = -1.383

7-6 Testing a Claim about a Standard Deviation or Variance

NOTE: Following the pattern used with the z and t distributions, this manual uses the closest entry from Table A-4 for χ^2 as if it were the precise value necessary and does not use interpolation. This procedure sacrifices very little accuracy – and even interpolation does not yield precise values. When extreme accuracy is needed in practice, statisticians refer either to more accurate tables or to computer-produced values.

ADDITIONAL NOTE: The χ^2 distribution depends upon n, and it "bunches up" around df = n-1. In addition, the formula $\chi^2 = (n-1)s^2/\sigma^2$ used in the calculations contains df = n-1. When the exact df needed in the problem does not appear in the table and the closest χ^2 value is used to determine the critical region, some instructors recommend using the same df in the calculations that were used to determine the C.R. This manual typically uses the closest entry to determine the C.R. and the n from the problem in the calculations, even though this introduces a slight discrepancy.

1. $H_o: \sigma = 15$
    $H_1: \sigma \neq 15$
    $\alpha = .05$
    C.R. $\chi^2 < \chi^2_{19,.975} = 8.907$
         $\chi^2 > \chi^2_{19,.025} = 32.852$
    calculations:
    $$\chi^2 = (n-1)s^2/\sigma^2$$
    $$= (19)(10)^2/(15)^2$$
    $$= 8.444$$
    .02 < P-value < .05

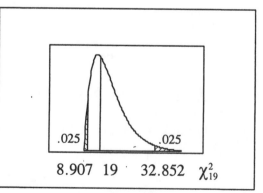

conclusion:
    Reject H_o; there is sufficient evidence to conclude that $\sigma \neq 15$ (in fact, that $\sigma < 15$).

GRAPHICS NOTE: To illustrate χ^2 tests of hypotheses, this manual uses a "generic" figure, resembling a chi-squared distribution with approximately 4 degrees of freedom – chi-squared distributions with 1 and 2 degrees of freedom actually have no upper limit and approach the y axis asymptotically, while chi-squared distributions with more than 30 degrees of freedom are essentially symmetric and normal-looking. The expected value of the chi-squared distribution is the degrees of freedom for the problem, typically n-1. Since the distribution is positively skewed, this manual indicates that value slightly to the right of the figure's "peak." Loosely speaking, the distribution "bunches up" around the degrees of freedom.

3. H_o: $\sigma = 50$
    H_1: $\sigma < 50$
    $\alpha = .01$
    C.R. $\chi^2 < \chi^2_{29,.99} = 14.257$
    calculations:
    $\chi^2 = (n-1)s^2/\sigma^2$
    $\quad = (29)(30)^2/(50)^2$
    $\quad = 10.440$
    P-value $< .005$

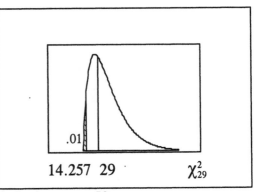

conclusion:
    Reject H_o; there is sufficient evidence to conclude that $\sigma < 50$.

5. original claim: $\sigma > .04$
    H_o: $\sigma = .04$
    H_1: $\sigma > .04$
    $\alpha = .01$
    C.R. $\chi^2 > \chi^2_{39,.01} = 63.691$
    calculations:
    $\chi^2 = (n-1)s^2/\sigma^2$
    $\quad = (39)(.31)^2/(.04)^2$
    $\quad = 2342.4375$
    P-value $< .005$

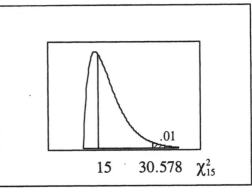

conclusion:
    Reject H_o; there is sufficient evidence to conclude that $\sigma > .04$.
Peanut M&M's vary more than plain M&M's since lack of uniformity in peanut sizes introduces a significant additional source of variation. Plain M&M's are produced to specifications with little variation from piece to piece. Peanut M&M's start with a peanut – not produced to specifications, but occurring with varying sizes. After rejecting very large and very small peanuts, there is still more variation than in a controlled process.

7. original claim: $\sigma \neq 43.7$
    H_o: $\sigma = 43.7$
    H_1: $\sigma \neq 43.7$
    $\alpha = .05$
    C.R. $\chi^2 < \chi^2_{80,.975} = 57.153$
    $\quad\,\, \chi^2 > \chi^2_{80,.025} = 106.629$
    calculations:
    $\chi^2 = (n-1)s^2/\sigma^2$
    $\quad = (80)(52.3)^2/(43.7)^2$
    $\quad = 114.586$
    $.01 < $ P-value $< .02$

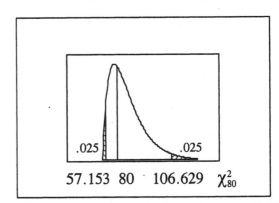

conclusion:
    Reject H_o; there is sufficient evidence to conclude that $\sigma \neq 43.7$ (in fact, that $\sigma > 43.7$). Since it appears that the standard deviation has increased, the new production method seems to be worse (at least so far as product consistency is concerned) than the old one.

9. original claim: $\sigma < 6.2$
H_o: $\sigma = 6.2$
H_1: $\sigma < 6.2$
$\alpha = .05$
C.R. $\chi^2 < \chi^2_{24,.95} = 13.848$
calculations:
$\quad \chi^2 = (n-1)s^2/\sigma^2$
$\quad\quad = (24)(3.8)^2/(6.2)^2$
$\quad\quad = 9.016$
$\quad$ P-value $< .005$

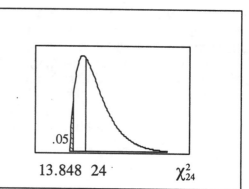

conclusion:
$\quad$ Reject H_o; there is sufficient evidence to conclude that $\sigma < 6.2$.
Customers prefer waiting times with less variation because eliminating very long or very short waits depending only on which line is chosen creates a more predictable and more professional business climate. No; the smaller variability does not mean a smaller mean time.

NOTE: When dealing with raw data, use the calculated value of s^2 directly in the formula $\chi^2 = (n-1)s^2/\sigma^2$. It is good to take the square root to find the value of s to see whether it is larger or smaller than the hypothesized σ, but do not introduce unnecessary rounding errors by entering the value for s and then squaring it.

11. summary statistics: $n = 9$, $\Sigma x = 1089$, $\Sigma x^2 = 132223$, $\bar{x} = 121.0$, $s^2 = 56.75$
original claim: $\sigma < 29$
H_o: $\sigma = 29$
H_1: $\sigma < 29$
$\alpha = .01$
C.R. $\chi^2 < \chi^2_{8,.99} = 1.646$
calculations:
$\quad \chi^2 = (n-1)s^2/\sigma^2$
$\quad\quad = (8)(56.75)/(29)^2$
$\quad\quad = .540$
P-value $< .005$

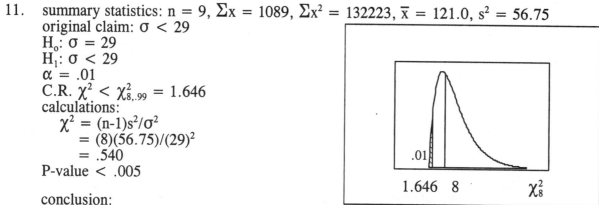

conclusion:
$\quad$ Reject H_o; there is sufficient evidence to conclude that $\sigma < 29$.

13. Usually we use the closest table entry – or, when the desired df is exactly halfway between the closest two tabled values, we act conservatively and choose the smaller df. For this exercise only (and related exercise #17), use interpolation to estimate
$\quad \chi^2_{35,.95} = (18.493 + 26.509)/2 = 22.501$.
summary statistics: $n = 36$, $\Sigma x = 442.5000$, $\Sigma x^2 = 5439.35004672$
$\quad \bar{x} = 12.2917$, $s^2 = .008216$ [$s = .0907$]
original claim: $\sigma < .10$
H_o: $\sigma = .10$
H_1: $\sigma < .10$
$\alpha = .05$
C.R. $\chi^2 < \chi^2_{35,.95} = 22.501$
calculations:
$\quad \chi^2 = (n-1)s^2/\sigma^2$
$\quad\quad = (35)(.008216)/(.10)^2$
$\quad\quad = 28.755$
$\quad$ $.10 <$ P-value $< .90$

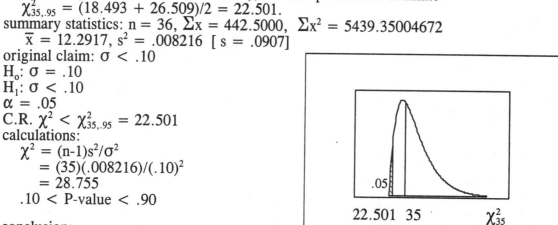

conclusion:
$\quad$ Do not reject H_o; there is not sufficient evidence to conclude that $\sigma < .10$.

If the mean is less than 12 ounces, then the company is guilty of cheating the consumers. If the mean is greater than 12 ounces but the standard deviation is high, there still may be a significant number of individual cans containing less than 12 ounces.

15. summary statistics: $n = 40$, $\Sigma x = 6902.0$, $\Sigma x^2 = 1,217,971.76$
     $\bar{x} = 172.55$, $s^2 = 693.119$ [$s = 26.327$]
     original claim: $\sigma = 28.7$
     $H_0: \sigma = 28.7$
     $H_1: \sigma \neq 28.7$
     $\alpha = .05$
     C.R. $\chi^2 < \chi^2_{39,.975} = 24.433$
     $\chi^2 > \chi^2_{39,.025} = 59.342$
     calculations:
     $\chi^2 = (n-1)s^2/\sigma^2$
     $= (39)(693.110)/(28.7)^2$
     $= 32.818$
     P-value $> .20$

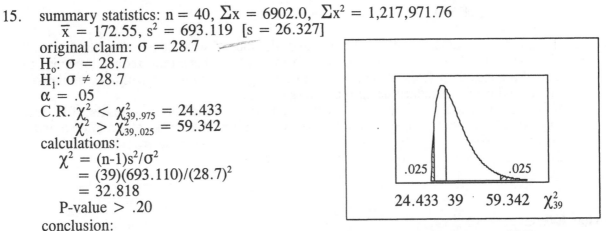

conclusion:
     Do not reject H_0; there is not sufficient evidence to conclude that $\sigma \neq 28.7$. Underestimating the variation in the weight could lead to designing an elevator that handled the average expected weight per load, but that had a relatively high risk of being overloaded.

17. Usually we use the closest table entry – or, when the desired df is exactly halfway between the closest two tabled values, we act conservatively and choose the smaller df. For this exercise only (and related exercise #13), use interpolation to estimate
     $\chi^2_{35,.95} = (18.493 + 26.509)/2 = 22.501$.
     original claim: $\sigma < .10$
     $H_0: \sigma = .10$
     $H_1: \sigma < .10$
     $\alpha = .05$
     C.R. $\chi^2 < \chi^2_{35,.95} = 22.501$
     calculations:
     $(n-1)s^2/\sigma^2 < 22.501$
     $(35)s^2/(.10)^2 < 22.501$
     $s^2 < .006428$
     $s < .0802$
     The largest s that would lead to rejection of H_0 (and support of $\sigma < .10$) is $s = .08$

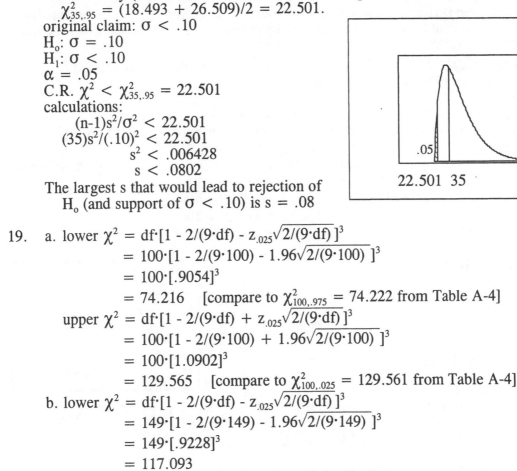

19. a. lower $\chi^2 = df \cdot [1 - 2/(9 \cdot df) - z_{.025}\sqrt{2/(9 \cdot df)}\,]^3$
     $= 100 \cdot [1 - 2/(9 \cdot 100) - 1.96\sqrt{2/(9 \cdot 100)}\,]^3$
     $= 100 \cdot [.9054]^3$
     $= 74.216$ [compare to $\chi^2_{100,.975} = 74.222$ from Table A-4]
     upper $\chi^2 = df \cdot [1 - 2/(9 \cdot df) + z_{.025}\sqrt{2/(9 \cdot df)}\,]^3$
     $= 100 \cdot [1 - 2/(9 \cdot 100) + 1.96\sqrt{2/(9 \cdot 100)}\,]^3$
     $= 100 \cdot [1.0902]^3$
     $= 129.565$ [compare to $\chi^2_{100,.025} = 129.561$ from Table A-4]
     b. lower $\chi^2 = df \cdot [1 - 2/(9 \cdot df) - z_{.025}\sqrt{2/(9 \cdot df)}\,]^3$
     $= 149 \cdot [1 - 2/(9 \cdot 149) - 1.96\sqrt{2/(9 \cdot 149)}\,]^3$
     $= 149 \cdot [.9228]^3$
     $= 117.093$

$$\begin{aligned}
\text{upper } \chi^2 &= df \cdot [1 - 2/(9 \cdot df) - z_{.025}\sqrt{2/(9 \cdot df)}\,]^3 \\
&= 149 \cdot [1 - 2/(9 \cdot 149) - 1.96\sqrt{2/(9 \cdot 149)}\,]^3 \\
&= 149 \cdot [1.0742]^3 \\
&= 184.690
\end{aligned}$$

21. a. If there are disproportionately more 0's and 5's, there would be an overabundance of digits less than 6 – i.e., at the lower end of the scale. This would reduce the mean to something less than 4.5 and (since most of the digits are bunched within these narrower limits) reduce the standard deviation to less than 3.

   b. If H_o were true and $\sigma = 3$, we would expect a uniform distribution of digits; if H_1 were true and $\sigma < 3$, we would expect a distribution of digits that peaks at 0 and/or 5 and is positively skewed. In either case, the data would not be coming from a normal distribution – which is one of the assumptions of the test.

Review Exercises

1. a. No; because the sample consisted only of those who chose to reply, the respondents are not necessarily representative of the general population.

   b. No; While the 0.2 lb loss is statistically significant, it is not of practical significance. In truth, there is essentially no chance that any H_o is <u>exactly</u> true to an infinite number of decimal places – and so given a large enough sample, virtually any H_o will be rejected. When the sample is so large that it is able to detect minute differences of no useful value, some statisticians call this "an insignificant significant difference."

   c. Choose .001, the smallest of the suggested P-values. This corresponds to the smallest chance that such results could have occurred by chance alone, and hence gives the most support to the conclusion that the cure was effective.

   d. Do not reject $H_o: \mu = 12$, there is not sufficient evidence to conclude that $\mu > 12$.

   e. "…rejecting the null hypothesis when it is true."

2. a. original claim: $\mu < 10,000 \Rightarrow H_o: \mu = 10,000$ and $H_1: \mu < 10,000$
   asks about μ, with σ unknown and $n > 30$: use the t distribution [with df=749]

   b. original claim: $\sigma > 1.8 \Rightarrow H_o: \sigma = 1.8$ and $H_1: \sigma < 1.8$
   asks about σ: assuming the population is approximately normal, use the χ^2 distribution

   c. original claim: $p > .50 \Rightarrow H_o: p = .50$ and $H_1: p > .50$
   asks about p, with $np \geq 5$ and $n(1-p) \geq 5$: use the z distribution

   d. original claim: $\mu = 100 \Rightarrow H_o: \mu = 100$ and $H_1: \mu \neq 100$
   asks about μ, with σ known and $n > 30$: use the z distribution

3. a. original claim: $\mu = 100$
   $\bar{x} = 98.4$
   $H_o: \mu = 100$
   $H_1: \mu \neq 100$
   $\alpha = .10$
   C.R. $z < -z_{.05} = -1.645$
   $\quad\quad z > z_{.05} = 1.645$
   calculations:

   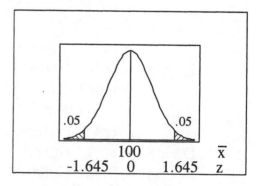

   $$\begin{aligned}
   z_{\bar{x}} &= (\bar{x} - \mu)/\sigma_{\bar{x}} \\
   &= (98.4 - 100)/(15/\sqrt{50}) \\
   &= -1.6/2.121 = -.75
   \end{aligned}$$
   P-value $= 2 \cdot P(z < -.75) = 2 \cdot (.2266) = .4532$
   conclusion:
   Do not reject H_o; there is not sufficient evidence reject the claim that $\mu = 100$.

b. original claim: $\mu = 100$
  H_o: $\mu = 100$
  H_1: $\mu \neq 100$
  $\alpha = .10$
  C.R. $t < -t_{49,.05} = -1.676$
      $t > t_{49,.05} = 1.676$
  calculations:
    $t_{\bar{x}} = (\bar{x} - \mu)/s_{\bar{x}}$
      $= (98.4 - 100)/(16.3/\sqrt{50})$
      $= -1.6/2.305 = -.694$
    P-value $> .20$
  conclusion:
    Do not reject H_o; there is not sufficient evidence to reject the claim that $\mu = 100$.

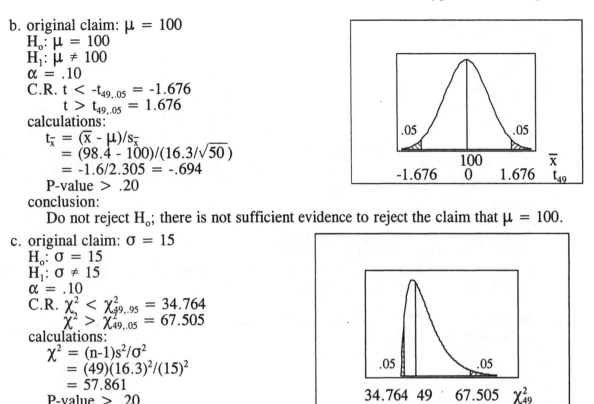

c. original claim: $\sigma = 15$
  H_o: $\sigma = 15$
  H_1: $\sigma \neq 15$
  $\alpha = .10$
  C.R. $\chi^2 < \chi^2_{49,.95} = 34.764$
      $\chi^2 > \chi^2_{49,.05} = 67.505$
  calculations:
    $\chi^2 = (n-1)s^2/\sigma^2$
      $= (49)(16.3)^2/(15)^2$
      $= 57.861$
    P-value $> .20$
  conclusion:
    Do not reject H_o; there is not sufficient evidence to reject the claim that $\sigma = 15$.

d. Yes; based on the preceding results, the random number generator appears to be working correctly.

4. original claim: $p < .50$
  $\hat{p} = x/n = x/15 = = .44$
  H_o: $p = .50$
  H_1: $p < .50$
  $\alpha = .05$
  C.R. $z < -z_{.05} = -1.645$
  calculations:
    $z_{\hat{p}} = (\hat{p} - \mu_{\hat{p}})/\sigma_{\hat{p}}$
      $= (.44 - .50)/\sqrt{(.50)(.50)/150}$
      $= -.06/.0408$
      $= -1.47$
    P-value $= P(z < -1.47) = .0708$
  conclusion:
    Do not reject H_o; there is not sufficient evidence to conclude that $p < .50$.

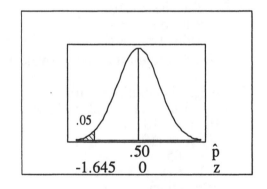

5. original claim: $\mu = 5.670$
    H_o: $\mu = 5.670$
    H_1: $\mu \neq 5.670$
    $\alpha = .01$
    C.R. $t < -t_{49,.005} = -2.678$
    $\quad\quad t > t_{49,.005} = 2.678$
    calculations:
    $\quad t_{\bar{x}} = (\bar{x} - \mu)/s_{\bar{x}}$
    $\quad\quad = (5.622 - 5.670)/(.068/\sqrt{50})$
    $\quad\quad = -.048/.009617$
    $\quad\quad = -4.991$

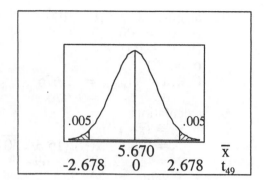

    P-value $< .01$
    conclusion:
    Reject H_o; there is sufficient evidence to reject the claim that $\mu = 5.670$ and conclude
    that $\mu \neq 5.670$ (in fact, that $\mu < 5.670$).

The quarters may have had a mean weight of 5.670 g when they were minted, but wear
from being in circulation will reduce the weight of coins.

6. There are n=5 blue M%M's listed in Data Set 19.
    original claim: $\mu \geq .9085$
    H_o: $\mu = .9085$
    H_1: $\mu < .9085$
    $\alpha = .05$
    C.R. $t < -t_{4,.05} = -2.132$
    calculations:
    $\quad t_{\bar{x}} = (\bar{x} - \mu)/s_{\bar{x}}$
    $\quad\quad = (.9014 - .9085)/(.0573/\sqrt{5})$
    $\quad\quad = -.0071/.0256$
    $\quad\quad = -.277$
    P-value $> .10$
    conclusion:

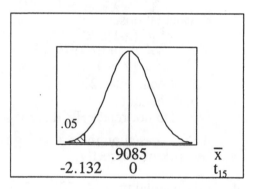

    Do not reject H_o; there is not sufficient evidence to reject the claim that $\mu \geq .9085$.
No, this result does not provide evidence of such a discrepancy.

7. original claim: $p < .10$
    $\hat{p} = x/n = 111/1233 = .090$
    H_o: $p = .10$
    H_1: $p < .10$
    $\alpha = .05$
    C.R. $z < -z_{.05} = -1.645$
    calculations:
    $\quad z_{\hat{p}} = (\hat{p} - \mu_{\hat{p}})/\sigma_{\hat{p}}$
    $\quad\quad = (.090 - .10)/\sqrt{(.10)(.90)/1233}$
    $\quad\quad = -.010/.00854$
    $\quad\quad = -1.17$
    P-value $= P(z<-1.17) = .1210$
    conclusion:

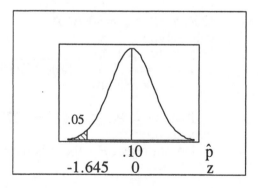

    Do not reject H_o; there is not sufficient evidence to conclude that $p < .10$.
She can use that claim if she is willing to run a 12.10% risk of being wrong.

8. original claim: p = .43
$\hat{p} = x/n = 308/611 = .504$
H_o: p = .43
H_1: p ≠ .43
α = .04
C.R. z < $-z_{.02}$ = -2.05
    z > $z_{.02}$ = 2.05
calculations:
$z_{\hat{p}} = (\hat{p} - \mu_{\hat{p}})/\sigma_{\hat{p}}$
$= (.0504 - .43)/\sqrt{(.43)(.57)/611}$
$= .074/.0200$
$= 3.70$

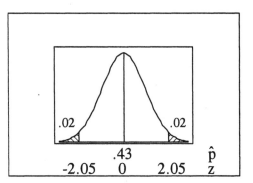

P-value = 2·P(z > 3.70) = 2·[1 - P(z < 3.70)] = 2·[1 - .9999] = 2·[.0001] = .0002
conclusion:
Reject H_o; there is sufficient evidence to reject the claim that p = .43 and conclude that p ≠ .43 (in fact, that p > .43).

The result indicates that the proportion who say they voted for the winner is greater than proportion that actually did so vote. It appears that people are not responding honestly to the question – and this should raise a concern about the honesty of respondents in general, in situations where there is no independent reliable knowledge about the true value.

9. original claim: μ < 12
H_o: μ = 12
H_1: μ < 12
α = .05 [assumed]
C.R. t < $-t_{23,.05}$ = -1.714
calculations:
$t_{\bar{x}} = (\bar{x} - \mu)/s_{\bar{x}}$
$= (11.4 - 12)/(.62/\sqrt{24})$
$= -.6/.1266$
$= -4.741$
P-value < .005

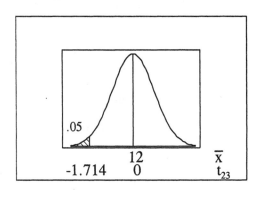

conclusion:
Reject H_o; there is sufficient evidence to conclude that μ < 12.

No; assuming that proper random sampling techniques were used, his claim that the sample is too small to be meaningful is not valid. The t distribution adjusts for the various sample sizes to produce a valid test for any sample size n.

10. original claim: p < .10
$\hat{p} = x/n = x/1248 = .08$
H_o: p = .10
H_1: p < .10
α = .01
C.R. z < $-z_{.01}$ = -2.326
calculations:
$z_{\hat{p}} = (\hat{p} - \mu_{\hat{p}})/\sigma_{\hat{p}}$
$= (.08 - .10)/\sqrt{(.10)(.90)/1248}$
$= -.02/.00849$
$= -2.36$

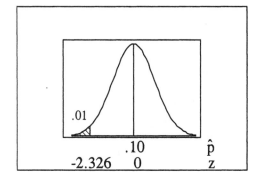

P-value = P(z < -2.36) = .0091
conclusion:
Reject H_o; there is sufficient evidence to conclude that p < .10.

Yes; based on this result, the phrase "almost 1 out of 10" is justified. While "almost" is not well-defined, 10% is the closest round number to 8%. The test indicates we are confident that p < .10, but does not specify how much less. NOTE: While the point $x=(.08)(1248)=99.84$; any $94 \leq x \leq 106$ rounds to 8%. Using the upper possibility $\hat{p} = x/n = 106/1248 = .0849$ $z_{\hat{p}} = -1.774$, which is not enough evidence to reject H_o: $p=.10$. This not only supports the opinion that "almost 1 out of 10" is justified, but it also shows how sensitive tests may be to rounding procedures – and that in this case the results are actually not reported with sufficient accuracy to conduct the test.

11. original claim: $\sigma < .15$
   H_o: $\sigma = .15$
   H_1: $\sigma < .15$
   $\alpha = .05$
   C.R. $\chi^2 < \chi^2_{70,.95} = 51.739$
   calculations:
      $\chi^2 = (n-1)s^2/\sigma^2$
         $= (70)(.12)^2/(.15)^2$
         $= 44.800$
      $.005 < \text{P-value} < .01$

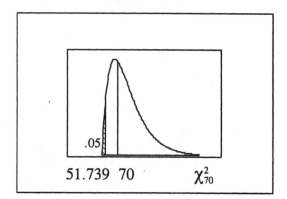

conclusion:
   Reject H_o; there is sufficient evidence to conclude that $\sigma < .15$.
   Yes; assuming the costs and any other factors do not suggest otherwise, the company should consider purchasing the machine.

12. summary statistics: $n = 70$, $\Sigma x = 251.023$, $\Sigma x^2 = 900.557440$, $\bar{x} = 3.586$, $s = .0740$
   original claim: $\mu = 3.5$
   H_o: $\mu = 3.5$
   H_1: $\mu \neq 3.5$
   $\alpha = .05$ [assumed]
   C.R. $t < -t_{69,.025} = -1.994$
      $t > t_{69,.025} = 1.994$
   calculations:
      $t_{\bar{x}} = (\bar{x} - \mu)/s_{\bar{x}}$
         $= (3.586 - 3.5)/(.0740/\sqrt{70})$
         $= .086/.008849$
         $= 9.724$
   P-value $< .01$
conclusion:
   Reject H_o; there is sufficient evidence to reject the claim that $\mu = 3.5$ and conclude that $\mu \neq 3.5$ (in fact, that $\mu > 3.5$).
It appears the true mean weight is greater that the 3.5 g claimed on the label.

Cumulative Review Exercises

1. scores in order of magnitude: .018 .0268 .0281 .0320 .0440 .0524 .161 .175 .176
   summary statistics: $n = 9$, $\Sigma x = .7133$, $\Sigma x^2 = .09505961$
   a. $\bar{x} = (\Sigma x)/n$
      $= .7133/9 = .0793$
   b. $\tilde{x} = .044$
   c. $s^2 = [n(\Sigma x^2) - (\Sigma x)^2]/[n(n-1)]$
      $= [9(.09505961) - (.7133)^2]/[9(8)] = .004816$
      $s = .0694$
   d. $s^2 = .004816$
   e. $R = .176 - .018 = .158$

f. $\overline{x} = \pm t_{8,.025}s_{\overline{x}}$
   $.0793 \pm 2.306 \cdot (.0694/\sqrt{9})$
   $.0793 \pm .0533$
   $.0259 < \mu < .1326$

g. original claim: $\mu < .16$

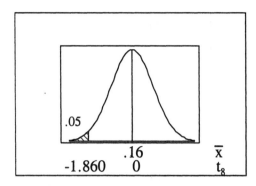

   H_o: $\mu = .16$
   H_1: $\mu < .16$
   $\alpha = .05$
   C.R. $t < -t_{8,.05} = -1.860$
   calculations:
       $t_{\overline{x}} = (\overline{x} - \mu)/s_{\overline{x}}$
          $= (.0793 - .16)/(.0694/\sqrt{9})$
          $= -.0807/.0231 = -3.491$
       P-value $< .005$
   conclusion:
       Reject H_o; there is sufficient evidence to conclude that $\mu < .16$.

h. Yes; the values were listed in chronological order, and there seems to be a dereasing trend over time.

2. a. normal distribution
       $\mu = 496$
       $\sigma = 108$
       $P(x > 500)$
          $= P(z > .04)$
          $= 1 - P(z < .04)$
          $= 1 - .5160$
          $= .4840$

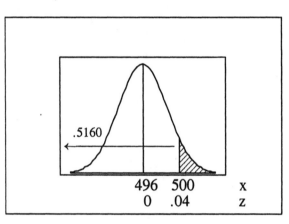

    b. let A = a selected score is above 500
       $P(A) = .4840$, for each selection
       $P(A_1$ and A_2 and A_3 and A_4 and $A_5)$
          $= P(A_1) \cdot P(A_2) \cdot P(A_3) \cdot P(A_4) \cdot P(A_5)$
          $= (.4840)^5$
          $= .0266$

    c. normal distribution,
          since the original distribution is so
       $\mu_{\overline{x}} = \mu = 496$
       $\sigma_{\overline{x}} = \sigma/\sqrt{n} = 108/\sqrt{5} = 48.30$
       $P(\overline{x} > 500)$
          $= P(z > .08)$
          $= 1 - P(z < .08)$
          $= 1 - .5319$
          $= .4681$

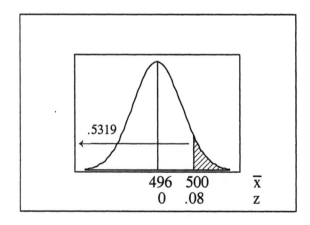

d. For P_{90}, A = .9000 [.8997] and z = 1.28.
 [or, $z_{.10} = t_{large,.10} = 1.282$]
 $x = \mu + z\cdot\sigma$
  $= 496 + (1.282)(108)$
  $= 496 + 138$
  $= 634$

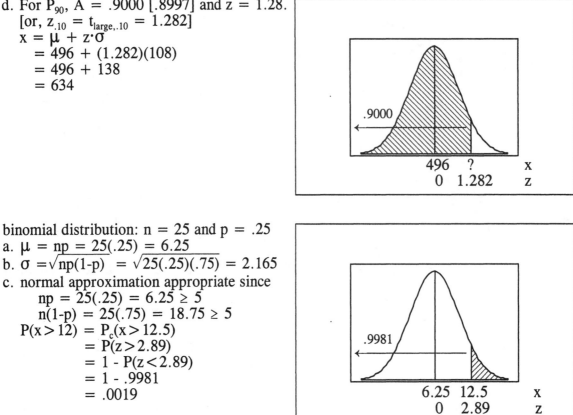

3. binomial distribution: n = 25 and p = .25
 a. $\mu = np = 25(.25) = 6.25$
 b. $\sigma = \sqrt{np(1-p)} = \sqrt{25(.25)(.75)} = 2.165$
 c. normal approximation appropriate since
   $np = 25(.25) = 6.25 \geq 5$
   $n(1-p) = 25(.75) = 18.75 \geq 5$
  $P(x > 12) = P_c(x > 12.5)$
    $= P(z > 2.89)$
    $= 1 - P(z < 2.89)$
    $= 1 - .9981$
    $= .0019$

d. original claim: p = .25
 $\hat{p} = x/n = $ (more than 12)/25 = 13/25 = .52
 H_o: p = .25
 H_1: p $\neq$ .25
 $\alpha = .05$
 C.R. z < $-z_{.025}$ = -1.96
   z > $z_{.025}$ = 1.96
 calculations:
   $z_{\hat{p}} = (\hat{p} - \mu_{\hat{p}})/\sigma_{\hat{p}}$
    $= (.52 - .25)/\sqrt{(.25)(.75)/25}$
    $= .27/.0866$
    $= 3.12$

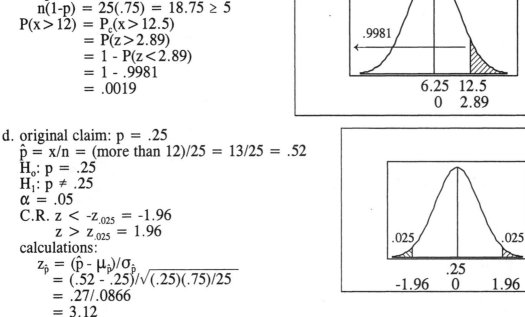

   P-value = $2\cdot P(z > 3.12) = 2\cdot[1 - P(z < 3.12)] = 2\cdot[1 - .9991] = 2\cdot[.0009] = .0018$
 conclusion:
   Reject H_o; there is sufficient evidence to reject the claim that p = .25 and conclude
   that p $\neq$ .25 (in fact, that p > .25).
 NOTE: The above test of hypothesis follows the form and notation used in this chapter.
 One could also consider the .0019 of part (c) above, calculate P-value = $2\cdot(.0019)$ =
 .0038, and reject H_o because .0038 < .05. The difference in P-values arises because
 (c) uses the correction for continuity while the section on testing hypotheses about p and
 part (d) ignore the correction.
 e. $\hat{p}$ unknown, use $\hat{p}$ = .5
  $n = [(z_{.05})^2\hat{p}\hat{q}]/E^2 = [(1.645)^2(.5)(.5)]/(.04)^2 = 422.8$, rounded up to 423

Chapter 8

Inferences from Two Samples

8-2 Inferences about Two Proportions

NOTE: To be consistent with the notation of the previous chapters, reinforcing the patterns and concepts presented there, the manual uses the "usual" z formula written to apply to $\hat{p}_1 - \hat{p}_2$'s:

$$z_{\hat{p}_1 - \hat{p}_2} = (\hat{p}_1 - \hat{p}_2 - \mu_{\hat{p}_1 - \hat{p}_2}) / \sigma_{\hat{p}_1 - \hat{p}_2}$$

with $\mu_{\hat{p}_1 - \hat{p}_2} = p_1 - p_2$ and $\sigma_{\hat{p}_1 - \hat{p}_2} = \sqrt{\bar{p} \cdot \bar{q}/n_1 + \bar{p} \cdot \bar{q}/n_2}$ [when H_o includes $p_1 = p_2$]

where $\bar{p} = (x_1 + x_2)/(n_1 + n_2)$

And so the formula for the z statistic may also be written as

$$z_{\hat{p}_1 - \hat{p}_2} = ((\hat{p}_1 - \hat{p}_2) - (p_1 - p_2)) / \sqrt{\bar{p} \cdot \bar{q}/n_1 + \bar{p} \cdot \bar{q}/n_2}$$

1. $x = \hat{p} \cdot n = .81 \cdot 37 = 29.97$, rounded to 30

3. $x = \hat{p} \cdot n = .289 \cdot 294 = 84.966$, rounded to 85

5. $\hat{p}_1 = x_1/n_1 = 192/436 = .440$ $\qquad\qquad$ $\hat{p}_2 = x_2/n_2 = 40/121 = .331$
    $\hat{p}_1 - \hat{p}_2 = .440 - .331 = .110$
    a. $\bar{p} = (x_1 + x_2)/(n_1 + n_2) = (192 + 40)/(436 + 121) = 232/557 = .417$
    b. $z_{\hat{p}_1 - \hat{p}_2} = (\hat{p}_1 - \hat{p}_2 - \mu_{\hat{p}_1 - \hat{p}_2}) / \sigma_{\hat{p}_1 - \hat{p}_2}$
       $= (.110 - 0)/\sqrt{(.417)(.583)/436 + (.417)(.583)/121}$
       $= .100/.0507 = 2.17$
    c. $\pm z_{.025} = \pm 1.960$
    d. P-value $= 2 \cdot P(z > 2.17) = 2 \cdot [1 - P(z < 2.17)] = 2 \cdot [1 - .9850] = 2 \cdot [.0150] = .0300$

7. Let the employees be group 1.
    original claim: $p_1 - p_2 > 0$
       $\hat{p}_1 = x_1/n_1 = 192/436 = .440$
       $\hat{p}_2 = x_2/n_2 = 40/121 = .331$
       $\hat{p}_1 - \hat{p}_2 = .440 - .331 = .110$
       $\bar{p} = (x_1 + x_2)/(n_1 + n_2)$
          $= (192 + 40)/(436 + 121)$
          $= 232/557 = .417$
    H_o: $p_1 - p_2 = 0$
    H_1: $p_1 - p_2 > 0$
    $\alpha = .05$
    C.R. $z > z_{.05} = 1.645$
    calculations:

       $z_{\hat{p}_1 - \hat{p}_2} = (\hat{p}_1 - \hat{p}_2 - \mu_{\hat{p}_1 - \hat{p}_2}) / \sigma_{\hat{p}_1 - \hat{p}_2}$
          $= (.110 - 0)/\sqrt{(.417)(.583)/436 + (.417)(.583)/121}$
          $= .110/.0507 = 2.17$
       P-value $= P(z > 2.17) = 1 - P(z < 2.17) = 1 - .9850 = .0150$
    conclusion:
       Reject H_o; there is sufficient evidence to conclude that $p_1 - p_2 > 0$.

NOTE: Since $\bar{p}$ is the weighted average of $\hat{p}_1$ and $\hat{p}_2$, it must always fall between those two values. If it does not, then an error has been made that must be corrected before proceeding. Calculation of $\sigma_{\hat{p}_1 - \hat{p}_2} = \sqrt{\bar{p} \cdot \bar{q}/n_1 + \bar{p} \cdot \bar{q}/n_2}$ can be accomplished with no round-off loss on most calculators by calculating $\bar{p}$ and proceeding as follows: STORE 1-RECALL = * RECALL = STORE RECALL $\div$ n_1 + RECALL $\div$ n_2 = $\sqrt{\ }$. The quantity $\sigma_{\hat{p}_1-\hat{p}_2}$ may then be STORED for future use. Each calculator is different -- learn how your calculator works, and do the homework on the same calculator you will use for the exam. If you have any questions about performing/storing calculations on your calculator, check with your instructor or class assistant.

9. Let the women with low activity be group 1.
    $\hat{p}_1 = x_1/n_1 = 101/10239 = .00986$ $\hat{p}_2 = x_2/n_2 = 56/9877 = .00567$
    $\hat{p}_1-\hat{p}_2 = .00986 - .00567 = .00419$
    $(\hat{p}_1-\hat{p}_2) \pm z_{.05}\sqrt{\hat{p}_1\hat{q}_1/n_1 + \hat{p}_2\hat{q}_2/n_2}$
    $.00419 \pm 1.645\cdot\sqrt{(.00986)(.99014)/10239 + (.00567)(.99433)/9877}$
    $.00419 \pm .00203$
    $.00216 < p_1-p_2 < .00623$
    Yes; since the confidence interval does not include 0, there does seem to be a significant difference – and that physical activity corresponds to a lower rate of the disease. Whether the difference is "substantial" is subjective, as the difference is only about 0.4%.

11. Let the 2000 season be group 1.
    original claim: $p_1-p_2 = 0$
         $\hat{p}_1 = x_1/n_1 = 83/247 = .336$
         $\hat{p}_2 = x_2/n_2 = 89/258 = .345$
         $\hat{p}_1-\hat{p}_2 = .336 - .345 = -.00893$
         $\bar{p} = (x_1 + x_2)/(n_1 + n_2)$
             $= (83 + 89)/(247 + 258)$
             $= 172/505 = .341$
    H_o: $p_1-p_2 = 0$
    H_1: $p_1-p_2 \neq 0$
    $\alpha = .05$ [assumed]
    C.R. $z < -z_{.025} = -1.96$
          $z > z_{.025} = 1.96$
    calculations:
        $z_{\hat{p}_1-\hat{p}_2} = (\hat{p}_1-\hat{p}_2 - \mu_{\hat{p}_1-\hat{p}_2})/\sigma_{\hat{p}_1-\hat{p}_2}$
             $= (-.00893 - 0)/\sqrt{(.341)(.659)/247 + (.341)(.659)/258}$
             $= -.00893/.0422 = -.21$
       P-value $= 2\cdot P(z<-.21) = 2\cdot(.4168) = .8336$
    conclusion:
         Do not reject H_o; there is not sufficient evidence to reject the claim that $p_1-p_2 = 0$.
    Yes, the reversal rate appears to be the same for both years.

13. Let those who received the vaccine be group 1.
    original claim: $p_1-p_2 < 0$
         $\hat{p}_1 = x_1/n_1 = 14/1070 = .013$
         $\hat{p}_2 = x_2/n_2 = 95/532 = .179$
         $\hat{p}_1-\hat{p}_2 = .013 - .179 = -.165$
         $\bar{p} = (x_1 + x_2)/(n_1 + n_2)$
             $= (14 + 95)/(1070 + 532)$
             $= 109/1602 = .068$
    H_o: $p_1-p_2 = 0$
    H_1: $p_1-p_2 < 0$
    $\alpha = .05$ [assumed]
    C.R. $z < -z_{.05} = -1.645$
    calculations:
        $z_{\hat{p}_1-\hat{p}_2} = (\hat{p}_1-\hat{p}_2 - \mu_{\hat{p}_1-\hat{p}_2})/\sigma_{\hat{p}_1-\hat{p}_2}$
           $= (-.165 - 0)/\sqrt{(.068)(.932)/1070 + (.068)(.932)/532}$
           $= -.165/.0134 = -12.39$

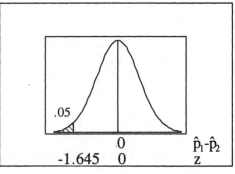

P-value $= P(z < -12.39) = .0001$
conclusion:
   Reject H_o; there is sufficient evidence to conclude that $p_1-p_2 < 0$.

15.   Let those not wearing seat belts be group 1.
      original claim: $p_1-p_2 > 0$
         $\hat{p}_1 = x_1/n_1 = 50/290 = .1724$
         $\hat{p}_2 = x_2/n_2 = 16/123 = .1301$
         $\hat{p}_1-\hat{p}_2 = .1724 - .1301 = .0423$
         $\bar{p} = (x_1 + x_2)/(n_1 + n_2) = (50 + 16)/(290 + 123) = 66/413 = .160$
      H_o: $p_1-p_2 = 0$
      H_1: $p_1-p_2 > 0$
      $\alpha = .05$
      C.R. $z > z_{.05} = 1.645$

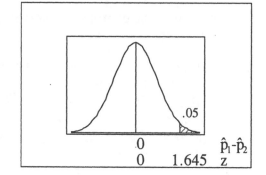

      calculations:
         $z_{\hat{p}_1-\hat{p}_2} = (\hat{p}_1-\hat{p}_2 - \mu_{\hat{p}_1-\hat{p}_2})/\sigma_{\hat{p}_1-\hat{p}_2}$
            $= (.0423 - 0)/\sqrt{(.160)(.840)/290 + (.160)(.840)/123}$
            $= .0423/.0394 = 1.07$
      P-value $= P(z > 1.07) = 1 - P(z < 1.07) = 1 - .8577 = .1423$
      conclusion:
         Do not reject H_o; there is not sufficient evidence to conclude that $p_1-p_2 > 0$.
      Based on these results, no specific action should be taken.
      NOTE: This test involves only children who were hospitalized – and perhaps a greater
      percentage of children wearing seat belts avoided being hospitalized in the first place.

17.   Let the husband defendants be group 1.
      original claim: $p_1-p_2 > 0$
         $\hat{p}_1 = x_1/n_1 = 277/318 = .871$
         $\hat{p}_2 = x_2/n_2 = 155/222 = .698$
         $\hat{p}_1-\hat{p}_2 = .871 - .692 = .173$
         $\bar{p} = (x_1 + x_2)/(n_1 + n_2)$
            $= (277 + 155)/(318 + 222)$
            $= 432/540 = .80$
      H_o: $p_1-p_2 = 0$
      H_1: $p_1-p_2 > 0$
      $\alpha = .05$ [assumed]
      C.R. $z > z_{.05} = 1.645$

      calculations:
         $z_{\hat{p}_1-\hat{p}_2} = (\hat{p}_1-\hat{p}_2 - \mu_{\hat{p}_1-\hat{p}_2})/\sigma_{\hat{p}_1-\hat{p}_2}$
            $= (.173 - 0)/\sqrt{(.80)(.20)/318 + (.80)(.20)/222}$
            $= .173/.0350 = 4.941$
      P-value $= P(z > 4.94) = 1 - P(z < 4.94) = 1 - .9999 = .0001$
      conclusion:
         Reject H_o; there is sufficient evidence to conclude that $p_1-p_2 > 0$.
      In spouse murder cases the question is usually not about who committed the murder, but
      whether the murder was committed in self-defense. Since women are usually physically
      smaller and weaker than men, a self-defense plea would be more believable from a wife
      defendant.

19.   Let the Autozone stores be group 1.
      original claim: $p_1 - p_2 = 0$
      $\hat{p}_1 = x_1/n_1 = 63/100 = .630$
      $\hat{p}_2 = x_2/n_2 = 30/37 = .811$
      $\hat{p}_1 - \hat{p}_2 = .630 - .811 = -.181$
      $\bar{p} = (x_1 + x_2)/(n_1 + n_2)$
      $= (63 + 30)/(100 + 37)$
      $= 93/137 = .679$
      H_o: $p_1 - p_2 = 0$
      H_1: $p_1 - p_2 \neq 0$
      $\alpha = .05$
      C.R. $z < -z_{.025} = -1.96$
      $\quad z > z_{.025} = 1.96$
      calculations:

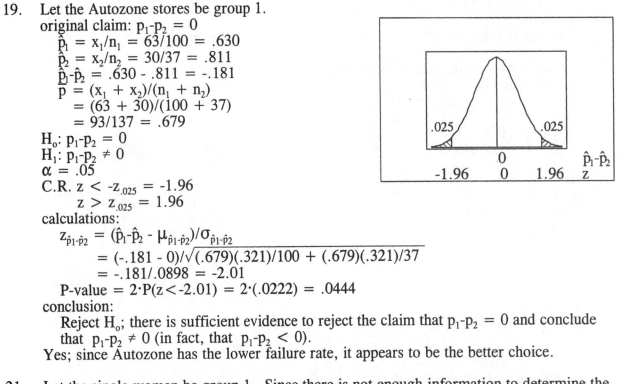

   $z_{\hat{p}_1 - \hat{p}_2} = (\hat{p}_1 - \hat{p}_2 - \mu_{\hat{p}_1 - \hat{p}_2})/\sigma_{\hat{p}_1 - \hat{p}_2}$
   $= (-.181 - 0)/\sqrt{(.679)(.321)/100 + (.679)(.321)/37}$
   $= -.181/.0898 = -2.01$
   P-value $= 2 \cdot P(z < -2.01) = 2 \cdot (.0222) = .0444$
      conclusion:
      Reject H_o; there is sufficient evidence to reject the claim that $p_1 - p_2 = 0$ and conclude
      that $p_1 - p_2 \neq 0$ (in fact, that $p_1 - p_2 < 0$).
      Yes; since Autozone has the lower failure rate, it appears to be the better choice.

21.   Let the single women be group 1. Since there is not enough information to determine the
      exact values of x_1 and x_2, the problem is limited to precision given below.
      $\hat{p}_1 = x_1/n_1 = [49 \text{ or } 50]/205 = .24 \qquad \hat{p}_2 = x_2/n_2 = [70 \text{ or } 71]/260 = .27$
      $\hat{p}_1 - \hat{p}_2 = .24 - .27 = -.03$
      $(\hat{p}_1 - \hat{p}_2) \pm z_{.005}\sqrt{\hat{p}_1\hat{q}_1/n_1 + \hat{p}_2\hat{q}_2/n_2}$
      $-.03 \pm 2.575 \cdot \sqrt{(.24)(.76)/205 + (.27)(.73)/260}$
      $-.03 \pm .10$
      $-.13 < p_1 - p_2 < .07$
      No; since the confidence interval includes 0, there does not appear to be a gender gap on
      this issue.

23.   Let those younger than 21 be group 1.
      $\hat{p}_1 = x_1/n_1 = .0425 \qquad\qquad \hat{p}_2 = x_2/n_2 = .0455$
      $\hat{p}_1 - \hat{p}_2 = .0425 - .0455 = -.0029$
      $(\hat{p}_1 - \hat{p}_2) \pm z_{.025}\sqrt{\hat{p}_1\hat{q}_1/n_1 + \hat{p}_2\hat{q}_2/n_2}$
      $-.0029 \pm 1.96 \cdot \sqrt{(.0425)(.9575)/2750 + (.0455)(.9545)/2200}$
      $-.0029 \pm .0115$
      $-.0144 < p_1 - p_2 < .0086$
      Yes; the interval contains zero, indicating no significant difference between the two rates
      of violent crimes.

25.   Let those given the written survey be group 1.
      $\hat{p}_1 = x_1/n_1 = 67/850 = .079$ [note: $x_1 = (.079)(850) = 67$]
      $\hat{p}_2 = x_2/n_2 = 105/850 = .124$ [note: $x_2 = (.124)(850) = 105$]
      $\hat{p}_1 - \hat{p}_2 = .079 - .124 = -.0447$
      $\bar{p} = (x_1 + x_2)/(n_1 + n_2)$
      $= (67 + 105)/(850 + 850)$
      $= 172/1700 = .1012$

a. original claim: $p_1 - p_2 = 0$
   H_o: $p_1 - p_2 = 0$
   H_1: $p_1 - p_2 \neq 0$
   $\alpha = .05$ [assumed]
   C.R. $z < -z_{.025} = -1.96$
   $\quad\quad z > z_{.025} = 1.96$

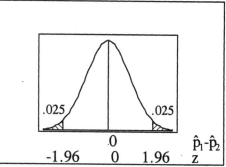

calculations:
$z_{\hat{p}_1 - \hat{p}_2} = (\hat{p}_1 - \hat{p}_2 - \mu_{\hat{p}_1 - \hat{p}_2})/\sigma_{\hat{p}_1 - \hat{p}_2}$
$\quad = (-.0447 - 0)/\sqrt{(.1012)(.8988)/850 + (.1012)(.8988)/850}$
$\quad = -.0447/.0146 = -3.06$
P-value $= 2 \cdot P(z < -3.06) = 2 \cdot (.0011) = .0022$
conclusion:
Reject H_o; there is sufficient evidence to reject the claim that $p_1 - p_2 = 0$ and to conclude that $p_1 - p_2 \neq 0$ (in fact, that $p_1 - p_2 < 0$ – i.e., that fewer students receiving the written test admit carrying a gun.)
Yes; based on this result the difference between 7.9% and 12.4% is significant.

b. $(\hat{p}_1 - \hat{p}_2) \pm z_{.005}\sqrt{\hat{p}_1\hat{q}_1/n_1 + \hat{p}_2\hat{q}_2/n_2}$
   $-.0447 \pm 2.575 \cdot \sqrt{(.079)(.921)/850 + (.124)(.876)/850}$
   $-.0447 \pm .0376$
   $-.0823 < p_1 - p_2 < -.0071$
The interval does not contain zero, indicating a significant difference between the two response rates. We are 99% confident the interval from -8.3% to -0.7% contains the true difference between the population percentages.

27. Let the central city be group 1.
    original claim: $p_1 - p_2 = 0$
    $\hat{p}_1 = x_1/n_1 = 85/294 = .289$ [note: $x_1 = (.289)(294) = 85$]
    $\hat{p}_2 = x_2/n_2 = 174/1015 = .171$ [note: $x_2 = (.171)(1015) = 174$]
    $\hat{p}_1 - \hat{p}_2 = .289 - .171 = .118$
    $\bar{p} = (x_1 + x_2)/(n_1 + n_2)$
    $\quad = (84 + 174)/(294 + 1015)$
    $\quad = 259/1309 = .198$

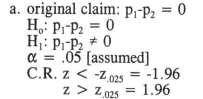

H_o: $p_1 - p_2 = 0$
H_1: $p_1 - p_2 \neq 0$
$\alpha = .01$
C.R. $z < -z_{.005} = -2.575$
$\quad\quad z > z_{.005} = 2.575$
calculations:
$z_{\hat{p}_1 - \hat{p}_2} = (\hat{p}_1 - \hat{p}_2 - \mu_{\hat{p}_1 - \hat{p}_2})/\sigma_{\hat{p}_1 - \hat{p}_2}$
$\quad = (.118 - 0)/\sqrt{(.198)(.802)/294 + (.198)(.802)/1015}$
$\quad = .118/.0264 = 4.47$
P-value $= 2 \cdot P(z > 4.47) = 2 \cdot [1 - P(z < 4.47)] = 2 \cdot [1 - .9999] = 2 \cdot [.0001] = .0002$
conclusion:
Reject H_o; there is sufficient evidence to reject the claim that $p_1 - p_2 = 0$ and to conclude that $p_1 - p_2 \neq 0$ (in fact, $p_1 - p_2 > 0$).

29. Let the movies tested for alcohol use be group 1.
     original claim: $p_1 - p_2 < 0$
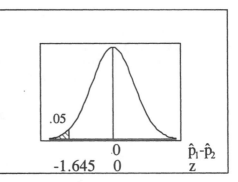
$$\hat{p}_1 = x_1/n_1 = 25/50 = .50$$
$$\hat{p}_2 = x_2/n_2 = 28/50 = .56$$
$$\hat{p}_1 - \hat{p}_2 = .50 - .56 = -.06$$
$$\bar{p} = (x_1 + x_2)/(n_1 + n_2)$$
$$= (25 + 28)/(50 + 50)$$
$$= 53/100 = .53$$
H_o: $p_1 - p_2 = 0$
H_1: $p_1 - p_2 < 0$
$\alpha = .05$ [assumed]
C.R. $z < -z_{.01} = -1.645$
calculations:
$$z_{\hat{p}_1 - \hat{p}_2} = (\hat{p}_1 - \hat{p}_2 - \mu_{\hat{p}_1 - \hat{p}_2})/\sigma_{\hat{p}_1 - \hat{p}_2}$$
$$= (-.06 - 0)/\sqrt{(.53)(.47)/50 + (.53)(.47)/50}$$
$$= -.06/.0998 = -.60$$
P-value $= P(z < -.60) = .2743$
conclusion:
     Do not reject H_o; there is not sufficient evidence to conclude that $p_1 - p_2 < 0$.
No; even though the data match those in Data Set 7, the results do not apply. One
requirement for this section is that the two samples be independent. The data in Data Set 7
are not from independent samples, but from the same sample – analogous to studying
whether the proportion of people who declare bankruptcy is different from the proportion
of people with more than $10,000 of credit card debt, and using the same 50 people.

31. For all parts of this exercise.
     $\hat{p}_1 = x_1/n_1 = 112/200 = .56$ $\hat{p}_2 = x_2/n_2 = 88/200 = .44$
     $\hat{p}_1 - \hat{p}_2 = .56 - .44 = .12$ $\bar{p} = (x_1 + x_2)/(n_1 + n_2) = 200/400 = .50$
     a. $(\hat{p}_1 - \hat{p}_2) \pm z_{.025}\sqrt{\hat{p}_1\hat{q}_1/n_1 + \hat{p}_2\hat{q}_2/n_2}$
         $.120 \pm 1.96 \cdot \sqrt{(.56)(.44)/200 + (.44)(.56)/200}$
         $.120 \pm .097$
         $.023 < p_1 - p_2 < .217$
         Since the interval does not include zero, the implication is that p_1 and p_2 are different.
         Since the interval lies entirely above, conclude that $p_1 - p_2 > 0$.
     b. for group 1   for group 2
         $\hat{p} \pm z_{.025}\sqrt{\hat{p}\hat{q}/n}$ $\hat{p} \pm z_{.025}\sqrt{\hat{p}\hat{q}/n}$
         $.560 \pm 1.96\sqrt{(.56)(.44)/200}$ $.440 \pm 1.96\sqrt{(.44)(.56)/200}$
         $.560 \pm .069$ $.440 \pm .069$
         $.491 < p < .629$ $.371 < p < .509$
         Since the intervals overlap, the implication is that p_1 and p_2 could be the same.
     c. original claim: $p_1 - p_2 = 0$
         H_o: $p_1 - p_2 = 0$
         H_1: $p_1 - p_2 \neq 0$
         $\alpha = .05$
         C.R. $z < -z_{.025} = -1.96$
               $z > z_{.025} = 1.96$

         calculations:
             $z_{\hat{p}_1 - \hat{p}_2} = (\hat{p}_1 - \hat{p}_2 - \mu_{\hat{p}_1 - \hat{p}_2})/\sigma_{\hat{p}_1 - \hat{p}_2}$
             $$= (.12 - 0)/\sqrt{(.5)(.5)/200 + (.5)(.5)/200}$$
             $$= .12/.05 = 2.40$$
         P-value $= 2 \cdot P(z > 2.40) = 2 \cdot [1 - P(z < 2.40)] = 2 \cdot [1 - .9918] = 2 \cdot [.0082] = .0164$
         conclusion:
             Reject H_o; there is sufficient evidence to reject the claim that $p_1 - p_2 = 0$ and conclude

that p_1-$p_2 \neq 0$ (in fact, that p_1-$p_2 > 0$).

d. Based on parts (a)-(c), conclude that p_1 and p_2 are unequal, and that $p_1 > p_2$. The overlapping interval method of part (b) appears to be the least effective method for comparing two populations.

33.   Let the black drivers be group 1.
      The new test of hypothesis is as follows.
          original claim: p_1-$p_2 > 0$

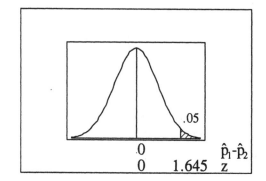

$\hat{p}_1 = x_1/n_1 = 240/2000 = .120$
$\hat{p}_2 = x_2/n_2 = 1470/14,000 = .105$
$\hat{p}_1$-$\hat{p}_2 = .120 - .105 = .015$
$\bar{p} = (x_1 + x_2)/(n_1 + n_2)$
$\quad = (240 + 1470)/(2000 + 14000)$
$\quad = 1710/16,000 = .107$
H_o: p_1-$p_2 = 0$
H_1: p_1-$p_2 > 0$
$\alpha = .05$
C.R. $z > z_{.05} = 1.645$

calculations:
$z_{\hat{p}_1\text{-}\hat{p}_2} = (\hat{p}_1\text{-}\hat{p}_2 - \mu_{\hat{p}_1\text{-}\hat{p}_2})/\sigma_{\hat{p}_1\text{-}\hat{p}_2}$
$\quad = (.015 - 0)/\sqrt{(.107)(.893)/2000 + (.107)(.893)/14000}$
$\quad = .015/.007385 = 2.03$
P-value = $P(z>2.03) = 1 - P(z<2.03) = 1 - .9788 = .0212$
conclusion:
   Reject H_o; there is sufficient evidence to conclude that p_1-$p_2 > 0$.
The new confidence interval is as follows.
$(\hat{p}_1\text{-}\hat{p}_2) \pm z_{.05}\sqrt{\hat{p}_1\hat{q}_1/n_1 + \hat{p}_2\hat{q}_2/n_2}$
$.015 \pm 1.645 \cdot \sqrt{(.120)(.880)/2000 + (.105)(.895)/14000}$
$.015 \pm .013$
$.002 < p_1$-$p_2 < .028$
The test statistic increased from .64 to 2.03, now allowing one to conclude p_1-$p_2 > 0$. The confidence interval grew narrower to not include zero, now allowing one to conclude p_1-$p_2 > 0$. With the same point estimates, the increased sample size provides enough evidence to support the original claim.

35. $\hat{p}_1 = x_1/n_1 = 40/100 = .40$ for groups 1 and 2, $\bar{p} = 70/200 = .35$
      $\hat{p}_2 = x_2/n_2 = 30/100 = .30$ for groups 2 and 3, $\bar{p} = 50/200 = .25$
      $\hat{p}_3 = x_3/n_3 = 20/100 = .20$ for groups 1 and 3, $\bar{p} = 50/200 = .30$

a. H_o: p_1-$p_2 = 0$
   H_1: p_1-$p_2 \neq 0$
   $\alpha = .05$
   C.R. $z < -z_{.025} = -1.96$
       $z > z_{.025} = 1.96$

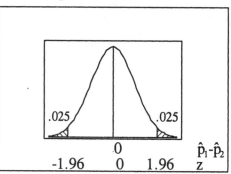

calculations:
   $z_{\hat{p}_1\text{-}\hat{p}_2} = (\hat{p}_1\text{-}\hat{p}_2 - \mu_{\hat{p}_1\text{-}\hat{p}_2})/\sigma_{\hat{p}_1\text{-}\hat{p}_2}$
   $\quad = (.10 - 0)/\sqrt{(.35)(.65)/100 + (.35)(.65)/100}$
   $\quad = .10/.0675 = 1.482$
conclusion:
   Do not reject H_o; there is not sufficient evidence to reject the claim that p_1-$p_2 = 0$.

b. H_o: $p_2-p_3 = 0$
   H_1: $p_2-p_3 \neq 0$
   $\alpha = .05$
   C.R. $z < -z_{.025} = -1.96$
   $\quad\quad z > z_{.025} = 1.96$

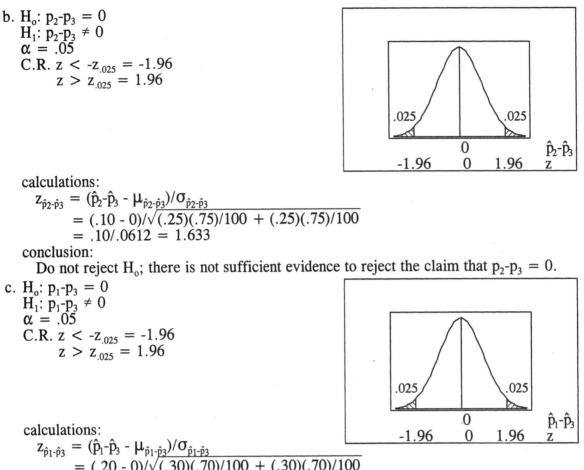

calculations:

$z_{\hat{p}_2-\hat{p}_3} = (\hat{p}_2-\hat{p}_3 - \mu_{\hat{p}_2-\hat{p}_3})/\sigma_{\hat{p}_2-\hat{p}_3}$

$\quad = (.10 - 0)/\sqrt{(.25)(.75)/100 + (.25)(.75)/100}$

$\quad = .10/.0612 = 1.633$

conclusion:

Do not reject H_o; there is not sufficient evidence to reject the claim that $p_2-p_3 = 0$.

c. H_o: $p_1-p_3 = 0$
   H_1: $p_1-p_3 \neq 0$
   $\alpha = .05$
   C.R. $z < -z_{.025} = -1.96$
   $\quad\quad z > z_{.025} = 1.96$

calculations:

$z_{\hat{p}_1-\hat{p}_3} = (\hat{p}_1-\hat{p}_3 - \mu_{\hat{p}_1-\hat{p}_3})/\sigma_{\hat{p}_1-\hat{p}_3}$

$\quad = (.20 - 0)/\sqrt{(.30)(.70)/100 + (.30)(.70)/100}$

$\quad = .20/.0648 = 3.086$

conclusion:

Reject H_o; there is sufficient evidence to reject the claim that $p_1-p_2 = 0$ and to conclude that $p_1-p_3 \neq 0$ (in fact, $p_1-p_3 > 0$).

d. No; failing to find a difference between population 1 and population 2, and between population 2 and population 3, does not necessarily mean one will fail to find a difference between population 1 and population 3. The fact that adjacent values may be equal does not necessarily mean that the possibility of equality extends to non-adjacent values.

37. a. No; for the placebo group, $np \approx (144)(.018) = 2.592 < 5$. The normal approximation to the binomial does not apply.
    b. For the 144 people in the placebo group, 1.8% is not a possible sample result since $2/144 = 1.4\%$ and $3/144 = 2.1\%$

8-2 Inferences about Two Means: Independent Samples

NOTE: To be consistent with the previous notation, reinforcing the patterns and concepts presented in those sections, the manual uses the "usual" t formula written to apply to $\bar{x}_1-\bar{x}_2$'s

$$t_{\bar{x}_1-\bar{x}_2} = (\bar{x}_1-\bar{x}_2 - \mu_{\bar{x}_1-\bar{x}_2})/s_{\bar{x}_1-\bar{x}_2}$$

with $\mu_{\bar{x}_1-\bar{x}_2} = \mu_1 - \mu_2$ and $s_{\bar{x}_1-\bar{x}_2} = \sqrt{s_1^2/n_1 + s_2^2/n_2}$

And so the formula for the t statistic <u>may also be w</u>ritten as

$$t_{\bar{x}_1-\bar{x}_2} = ((\bar{x}_1-\bar{x}_2) - (\mu_1-\mu_2))/\sqrt{s_1^2/n_1 + s_2^2/n_2}$$

1. Independent samples, since two groups are selected and evaluated separately.

3. Matched pairs, since each reported weight is paired with its measured weight.

5. Let the light users be group 1.
    original claim: $\mu_1-\mu_2 > 0$
    $\quad \bar{x}_1-\bar{x}_2 = 53.3 - 51.3 = 2.0$
    H_o: $\mu_1-\mu_2 = 0$
    H_1: $\mu_1-\mu_2 > 0$
    $\alpha = .01$
    C.R. $t > t_{63,.01} = 2.385$
    calculations:

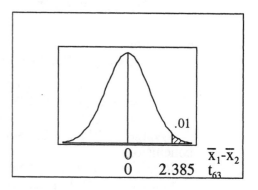

$$t_{\bar{x}_1-\bar{x}_2} = (\bar{x}_1-\bar{x}_2 - \mu_{\bar{x}_1-\bar{x}_2})/s_{\bar{x}_1-\bar{x}_2}$$
$$= (2.0 - 0)/\sqrt{(3.6)^2/64 + (4.5)^2/65}$$
$$= 2.0/.7170$$
$$= 2.790$$

    P-value = $P(t_{63} > 2.790) < .005$
    conclusion:
    Reject H_o; there is sufficient evidence to conclude that $\mu_1-\mu_2 > 0$.
    Yes; based on these results, heavy marijuana use appears to impede performance.

7. Let the placebo users be group 1.
    $\quad \bar{x}_1-\bar{x}_2 = 21.57 - 20.38 = 1.19$
    $(\bar{x}_1-\bar{x}_2) \pm t_{32,.025} \sqrt{s_1^2/n_1 + s_2^2/n_2}$
    $1.19 \pm 2.037 \cdot \sqrt{(3.87)^2/43 + (3.91)^2/33}$
    1.19 ± 1.84
    $-.65 < \mu_1-\mu_2 < 3.03$
    No; based on these results, we cannot be 95% certain that the two populations have different means. There is not enough evidence to make this the generally recommended treatment for bipolar depression.

9. Let those receiving the magnet treatment be group 1.
    original claim: $\mu_1-\mu_2 > 0$
    $\quad \bar{x}_1-\bar{x}_2 = .49 - .44 = .05$
    H_o: $\mu_1-\mu_2 = 0$
    H_1: $\mu_1-\mu_2 > 0$
    $\alpha = .05$
    C.R. $t > t_{19,.05} = 1.729$
    calculations:

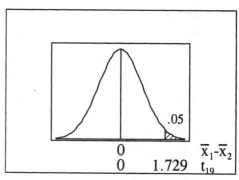

$$t_{\bar{x}_1-\bar{x}_2} = (\bar{x}_1-\bar{x}_2 - \mu_{\bar{x}_1-\bar{x}_2})/s_{\bar{x}_1-\bar{x}_2}$$
$$= (.05 - 0)/\sqrt{(.96)^2/20 + (1.4)^2/20}$$
$$= .05/.3796$$
$$= .132$$

    P-value = $P(t_{19} > .132) > .10$

conclusion:
   Do not reject H_o; there is not sufficient evidence to conclude that $\mu_1 - \mu_2 > 0$.
No, it does not appear that magnets are effective in treating back pain. If much larger sample sizes achieved these same results, the calculated t could fall in the critical region and appear to provide evidence that the treatment is effective – but the observed difference would still be .05, and one would have to decide whether that statistically significant difference is of practical significance.

11. a. original claim: $\mu_1 - \mu_2 = 0$
   $\overline{x}_1 - \overline{x}_2 = .81682 - .78479 = .03203$
   $H_o: \mu_1 - \mu_2 = 0$
   $H_1: \mu_1 - \mu_2 \neq 0$
   $\alpha = .01$
   C.R. $t < -t_{35,.005} = -2.728$
      $t > t_{35,.005} = 2.728$

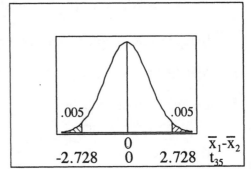

   calculations:
      $t_{\overline{x}_1 - \overline{x}_2} = (\overline{x}_1 - \overline{x}_2 - \mu_{\overline{x}_1 - \overline{x}_2})/s_{\overline{x}_1 - \overline{x}_2}$

      $= (.03203 - 0)/\sqrt{(.007507)^2/36 + (.004391)^2/36}$
      $= .03203/.0014495$
      $= 22.098$
   P-value $= 2 \cdot P(t_{35} > 22.098) < .01$
   conclusion:
      Reject H_o; there is sufficient evidence to reject the claim that $\mu_1 - \mu_2 = 0$ and conclude that $\mu_1 - \mu_2 \neq 0$ (in fact, that $\mu_1 - \mu_2 > 0$).
   Diet Coke probably weighs less because it uses an artificial sweetener that weighs less than real sugar – or perhaps because it simply uses less sweetener, and sweeteners in general are more dense than the rest of the product.

   b. $(\overline{x}_1 - \overline{x}_2) \pm t_{35,.005} \sqrt{s_1^2/n_1 + s_2^2/n_2}$
   $.03203 \pm 2.728 \cdot \sqrt{(.007507)^2/36 + (.004391)^2/36}$
   $.03203 \pm .00395$
   $.02808 < \mu_1 - \mu_2 < .03598$

13. Let the control subjects be group 1.
   $\overline{x}_1 - \overline{x}_2 = .45 - .34 = .11$
   $(\overline{x}_1 - \overline{x}_2) \pm t_{9,.005} \sqrt{s_1^2/n_1 + s_2^2/n_2}$
   $.11 \pm 3.250 \cdot \sqrt{(.08)^2/10 + (.08)^2/10}$
   $.11 \pm .12$
   $-.01 < \mu_1 - \mu_2 < .23$
   The confidence intervals suggests that the two samples could come from populations with the same mean. Based on this result, one cannot be 99% certain that there is such a biological basis for the disorders.

15. Let the treatment subjects be group 1.
   $\overline{x}_1 - \overline{x}_2 = 4.20 - 1.71 = 2.49$
   $(\overline{x}_1 - \overline{x}_2) \pm t_{21,.025} \sqrt{s_1^2/n_1 + s_2^2/n_2}$
   $2.49 \pm 2.080 \cdot \sqrt{(2.20)^2/22 + (.72)^2/22}$
   2.49 ± 1.03
   $1.46 < \mu_1 - \mu_2 < 3.52$
   Yes, since high scores indicate the presence of more errors and the confidence interval falls entirely above zero, the results support the common belief that drinking alcohol is hazardous for operators of passenger vehicles – and for their passengers.

17. original claim: $\mu_1-\mu_2 > 0$
     $\overline{x}_1-\overline{x}_2 = 53.3 - 45.3 = 8.0$
     H_o: $\mu_1-\mu_2 = 0$
     H_1: $\mu_1-\mu_2 > 0$
     $\alpha = .01$
     C.R. $t > t_{39,.01} = 2.429$
     calculations:

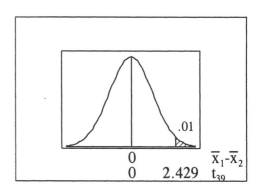

$$t_{\overline{x}_1-\overline{x}_2} = (\overline{x}_1-\overline{x}_2 - \mu_{\overline{x}_1-\overline{x}_2})/s_{\overline{x}_1-\overline{x}_2}$$
$$= (8.0 - 0)/\sqrt{(11.6)^2/40 + (13.2)^2/40}$$
$$= 8.0/2.778$$
$$= 2.879$$

     P-value $= P(t_{39} > 2.879) < .005$
     conclusion:
        Reject H_o; there is sufficient evidence to conclude that $\mu_1-\mu_2 > 0$.

19. Let the westbound stowaways be group 1.
     original claim: $\mu_1-\mu_2 = 0$
     $\overline{x}_1-\overline{x}_2 = 26.71 - 24.84 = 1.87$
     H_o: $\mu_1-\mu_2 = 0$
     H_1: $\mu_1-\mu_2 \neq 0$
     $\alpha = .05$
     C.R. $t < -t_{104,.025} = -1.983$ [from Excel]
        $t > t_{104,.025} = 1.983$
     calculations:

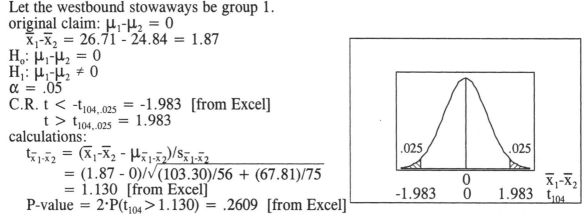

$$t_{\overline{x}_1-\overline{x}_2} = (\overline{x}_1-\overline{x}_2 - \mu_{\overline{x}_1-\overline{x}_2})/s_{\overline{x}_1-\overline{x}_2}$$
$$= (1.87 - 0)/\sqrt{(103.30)/56 + (67.81)/75}$$
$$= 1.130 \text{ [from Excel]}$$

     P-value $= 2 \cdot P(t_{104} > 1.130) = .2609$ [from Excel]
     conclusion:
        Do not reject H_o; there is not sufficient evidence to reject the claim that $\mu_1-\mu_2 = 0$.
     No, there appears to be no significant difference between the ages.

21. Let the filtered cigarettes be group 1

group 1: filtered (n = 21)	group 2: unfiltered (n = 8)
$\Sigma x = 279$	$\Sigma x = 192$
$\Sigma x^2 = 3987$	$\Sigma x^2 = 4628$
$\overline{x} = 13.286$	$\overline{x} = 24.000$
$s = 3.744$	$s = 1.690$

     original claim: $\mu_1-\mu_2 < 0$
     $\overline{x}_1-\overline{x}_2 = 13.286 - 24.000 = -10.714$
     H_o: $\mu_1-\mu_2 = 0$
     H_1: $\mu_1-\mu_2 < 0$
     $\alpha = .05$
     C.R. $t < -t_{7,.05} = -1.895$
     calculations:

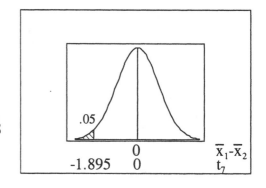

$$t_{\overline{x}_1-\overline{x}_2} = (\overline{x}_1-\overline{x}_2 - \mu_{\overline{x}_1-\overline{x}_2})/s_{\overline{x}_1-\overline{x}_2}$$
$$= (-10.714 - 0)/\sqrt{(3.744)^2/21 + (1.690)^2/8}$$
$$= -.10.714/1.0122$$
$$= -10.585$$

     P-value $= P(t_7 < -10.585) < .005$
     conclusion:
        Reject H_o; there is sufficient evidence to conclude that $\mu_1-\mu_2 < 0$.

23.   Let the men be group 1

group 1: men (n = 40)   group 2: women (n = 40)

$\Sigma x = 1039.9$ $\Sigma x = 1029.6$

$\Sigma x^2 = 27,493.83$ $\Sigma x^2 = 27,984.46$

$\overline{x} = 25.9975$ $\overline{x} = 25.7400$

$s = 3.43$ $s = 6.17$

original claim: $\mu_1 - \mu_2 = 0$

$\overline{x}_1 - \overline{x}_2 = 25.9975 - 25.7400 = .2575$

H_o: $\mu_1 - \mu_2 = 0$

H_1: $\mu_1 - \mu_2 \neq 0$

$\alpha = .05$ [assumed]

C.R. $t < -t_{39,.025} = -2.024$

   $t > t_{39,.025} = 2.024$

calculations:

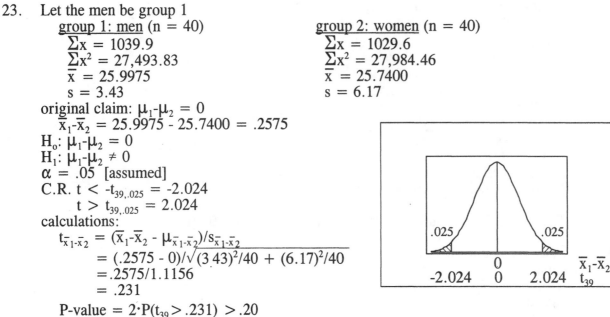

$t_{\overline{x}_1 - \overline{x}_2} = (\overline{x}_1 - \overline{x}_2 - \mu_{\overline{x}_1 - \overline{x}_2})/s_{\overline{x}_1 - \overline{x}_2}$

$= (.2575 - 0)/\sqrt{(3.43)^2/40 + (6.17)^2/40}$

$= .2575/1.1156$

$= .231$

P-value $= 2 \cdot P(t_{39} > .231) > .20$

conclusion:

Do not reject H_o; there is not sufficient evidence to reject the claim that $\mu_1 - \mu_2 = 0$.

25.   Let the placebo users be group 1.

$\overline{x}_1 - \overline{x}_2 = 21.57 - 20.38 = 1.19$

$s_p^2 = [(n_1-1) \cdot s_1^2 + (n_2-1) \cdot s_2^2]/(n_1 + n_2 - 2)$ $df = df_1 + df_2$

$= [42(3.87)^2 + 32(3.91)^2]/74$ $= 42 + 32$

$= 15.111$ $= 74$

$(\overline{x}_1 - \overline{x}_2) \pm t_{74,.025} \sqrt{s_p^2/n_1 + s_p^2/n_2}$

$1.19 \pm 1.992 \cdot \sqrt{15.111/43 + 15.111/33}$

1.19 ± 1.79

$-.60 < \mu_1 - \mu_2 < 2.98$

No; based on these results, we cannot be 95% certain that the two populations have different means. There is not enough evidence to make this the generally recommended treatment for bipolar depression.

▶The conclusion is the same, but the willingness to accept another assumption allows the confidence interval to be slightly narrower.

27.   Let those receiving the magnet treatment be group 1.

original claim: $\mu_1 - \mu_2 > 0$

$\overline{x}_1 - \overline{x}_2 = .49 - .44 = .05$

$s_p^2 = [(n_1-1) \cdot s_1^2 + (n_2-1) \cdot s_2^2]/(n_1 + n_2 - 2)$ $df = df_1 + df_2$

$= [19(.96)^2 + 19(1.41)^2]/38 = 1.4408$ $= 19 + 19 = 38$

H_o: $\mu_1 - \mu_2 = 0$

H_1: $\mu_1 - \mu_2 > 0$

$\alpha = .05$

C.R. $t > t_{38,.05} = 1.686$

calculations:

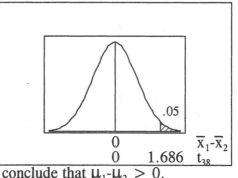

$t_{\overline{x}_1 - \overline{x}_2} = (\overline{x}_1 - \overline{x}_2 - \mu_{\overline{x}_1 - \overline{x}_2})/s_{\overline{x}_1 - \overline{x}_2}$

$= (.05 - 0)/\sqrt{1.4408/20 + 1.4408/20}$

$= .05/.3796$

$= .132$

P-value $= P(t_{38} > .132) > .10$

conclusion:

Do not reject H_o; there is not sufficient evidence to conclude that $\mu_1 - \mu_2 > 0$.

No, it does not appear that magnets are effective in treating back pain. If much larger sample sizes achieved these same results, the calculated t could fall in the critical region and appear to provide evidence that the treatment is effective – but the observed difference would still be .05, and one would have to decide whether that statistically significant difference is of practical significance.

▸The conclusion is the same, but the willingness to accept another assumption makes rejection slightly more likely – the absolute value of the critical value is slightly smaller and the absolute value of the calculated test statistic is (since $n_1 = n_2$) the same.

29.   NOTE: Exercise #19 used df=104 as determined by Excel, to produce CV = ± 1.983. Even though changing the sample size and the variance of one of the groups will change these values, this manual re-uses them unchanged – both to avoid a discussion of Excel's algorithm for determining df, and to better compare the new result to the original one.

a. Let the westbound stowaways be group 1.

new group 1 values based on n = 57

$\Sigma x = 1496 + 90 = 1586$

$\Sigma x^2 = 45646 + 8100 = 53746$

$\bar{x} = 27.825 \quad s^2 = 171.72$

original claim: $\mu_1 - \mu_2 = 0$

$\bar{x}_1 - \bar{x}_2 = 27.825 - 24.840 \doteq 2.985$

$H_o: \mu_1 - \mu_2 = 0$

$H_1: \mu_1 - \mu_2 \neq 0$

$\alpha = .05$

C.R. $t < -t_{104,.025} = -1.983$ [from Excel]

$\quad\quad t > t_{104,.025} = 1.983$

calculations:

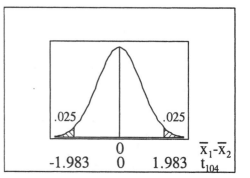

$t_{\bar{x}_1 - \bar{x}_2} = (\bar{x}_1 - \bar{x}_2 - \mu_{\bar{x}_1 - \bar{x}_2})/s_{\bar{x}_1 - \bar{x}_2}$

$\quad = (2.985 - 0)/\sqrt{(171.72)/57 + (67.81)/75}$

$\quad = 2.985/1.979$

$\quad = 1.508$

$.10 < \text{P-value} = 2 \cdot P(t_{104} > 1.508) < .20$

conclusion:

Do not reject H_o; there is not sufficient evidence to reject the claim that $\mu_1 - \mu_2 = 0$.

No, there appears to be no significant difference between the ages.

▸The conclusion is the same, but the P-value is smaller because the calculated t increased from 1.130 to 1.508.

b. Let the westbound stowaways be group 1.

new group 1 values based on n = 57

$\Sigma x = 1496 + 5000 = 6496$

$\Sigma x^2 = 45,646 + 25,000,000 = 25,045,646$

$\bar{x} = 27.825 \quad s^2 = 434023$

original claim: $\mu_1 - \mu_2 = 0$

$\bar{x}_1 - \bar{x}_2 = 113.965 - 24.840 = 89.125$

$H_o: \mu_1 - \mu_2 = 0$

$H_1: \mu_1 - \mu_2 \neq 0$

$\alpha = .05$

C.R. $t < -t_{104,.025} = -1.983$ [from Excel]

$\quad\quad t > t_{104,.025} = 1.983$

calculations:

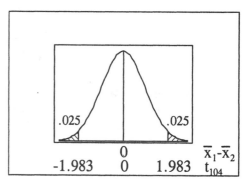

$t_{\bar{x}_1 - \bar{x}_2} = (\bar{x}_1 - \bar{x}_2 - \mu_{\bar{x}_1 - \bar{x}_2})/s_{\bar{x}_1 - \bar{x}_2}$

$\quad = (89.125 - 0)/\sqrt{434023/57 + 67.81/75}$

$\quad = 89.125/87.266 = 1.021$

P-value $= 2 \cdot P(t_{104} > 1.021) > .20$

conclusion:

Do not reject H_o; there is not sufficient evidence to reject the claim that $\mu_1 - \mu_2 = 0$.

No, there appears to be no significant difference between the ages.
▸The conclusion is the same, but the calculated t statistic actually decreased from 1.130 to 1.021. Even though the difference between the sample means increased, the variability in the problem (which in a sense is a measure of uncertainty) increased even more.

31. a. $x = 5,10,15$

$\mu = \Sigma x/n$
$= 30/3 = 10$
$\sigma^2 = \Sigma(x-\mu)^2/n$
$= [(-5)^2 + (0)^2 + (5)^2]/3$
$= 50/3$

b. $y = 1,2,3$

$\mu = \Sigma y/n$
$= 6/3 = 2$
$\sigma^2 = \Sigma(y-\mu)^2/n$
$= [(-1)^2 + (0)^2 + (1)^2]/3$
$= 2/3$

c. $z = x-y = 4,3,2,9,8,7,14,13,12$

$\mu = \Sigma z/n = 72/9 = 8$
$\sigma^2 = \Sigma(z-\mu)^2/n$
$= [(-4)^2 + (-5)^2 + (-6)^2 + (1)^2 + (0)^2 + (-1)^2 + (6)^2 + (5)^2 + (4)^2]/9$
$= 156/9$
$= 52/3$

d. $\sigma^2_{x-y} = \sigma^2_x + \sigma^2_y$

$52/3 = 50/3 + 2/3$
$52/3 = 52/3$

e. Let R stand for range.

$R_{x-y} = highest_{x-y} - lowest_{x-y}$
$= (highest\ x - lowest\ y) - (lowest\ x - highest\ y)$
$= highest\ x - lowest\ y - lowest\ x + highest\ y$
$= (highest\ x - lowest\ x) + (highest\ y - lowest\ y)$
$= R_x + R_y$

The range of all possible x-y values is the sum of the individual ranges of x and y.
NOTE: The problem refers to all possible x-y differences (where n_x and n_y might even be different) and not to x-y differences for paired data.

33. $A = s_1^2/n_1 = .0064/10 = .00064$ $df = (A + B)^2/(A^2/df_1 + B^2/df_2)$
$B = s_2^2/n_2 = .0064/10 = .00064$ $= (.00064 + .00064)^2/(.00064^2/9 + .00064^2/9)$
$= .000001638/.000000091$
$= 18$

When $s_1^2 = s_2^2$ and $n_1 = n_2$ formula 8-1 yields $df = df_1 + df_2$.
In this exercise the tabled t value changes from $t_{9,.005} = 3.250$ to $t_{18,.005} = 2.878$.
In general, the larger df signifies a "tighter" t distribution that is closer to the z distribution. In the test of hypothesis, the sampling distribution will be "tighter" and the critical t value will be smaller – since the calculated t is not affected, the P-value will be smaller. In the confidence interval, the tabled t value will be smaller – since the other values are not affected, the interval will be narrower. Using df = (smaller of df_1 or df_2) is more conservative in that will not reject H_o as often and will lead to wider confidence intervals – i.e., it will allows for a wider range of possible values for the parameter.

8-4 Inferences from Matched Pairs

NOTE: To be consistent with the notation of the previous sections, and thereby reinforcing the patterns and concepts presented in those sections, the manual uses the "usual" t formula written to apply to d's

$$t_{\bar{d}} = (\bar{d} - \mu_{\bar{d}})/s_{\bar{d}}, \quad \text{with } \mu_{\bar{d}} = \mu_d \text{ and } s_{\bar{d}} = s_d/\sqrt{n}$$

And so the formula for the t statistic may also be written as

$$t_{\bar{d}} = (\bar{d} - \mu_d)/(s_d/\sqrt{n})$$

1. $d = x - y$: 1 -1 -2 -3 4
    summary: $n = 5$, $\Sigma d = -1$, $\Sigma d^2 = 31$
    a. $\bar{d} = (\Sigma d)/n = -1/5 = -.2$
    b. $s_d^2 = [n \cdot \Sigma d^2 - (\Sigma d)^2]/[n(n-1)]$
        $= [5 \cdot 31 - (-1)^2]/[5(4)] = 154/20 = 7.7$
        $s_d = 2.8$
    c. $t_{\bar{d}} = (\bar{d} - \mu_{\bar{d}})/s_{\bar{d}}$
        $= (-.2 - 0)/(2.774/\sqrt{5})$
        $= -.2/1.241 = -.161$
    d. $\pm t_{4,.025} = \pm 2.776$

3. $\bar{d} \pm t_{4,.025} \cdot s_d/\sqrt{n}$
    $-.2 \pm 2.776 \cdot 2.775/\sqrt{5}$
    $-.2 \pm 3.4$
    $-3.6 < \mu_d < 3.2$

5. $d = x - y$: -5.1 1.3 -0.1 1.2 0.8 -1.4 0.4 -0.5 -2.8 0.1 1.9 -1.5
    $n = 12$
    $\Sigma d = -5.7$ $\bar{d} = -.475$
    $\Sigma d^2 = 45.87$ $s_d = 1.981$
    a. original claim: $\mu_d = 0$
        H_o: $\mu_d = 0$
        H_1: $\mu_d \neq 0$
        $\alpha = .05$
        C.R. $t < -t_{11,.025} = -2.201$
             $t > t_{11,.025} = 2.201$
        calculations:
            $t_{\bar{d}} = (\bar{d} - \mu_{\bar{d}})/s_{\bar{d}}$
            $= (-.475 - 0)/(1.981/\sqrt{12})$
            $= -.475/.5718 = -.831$
        P-value $= 2 \cdot P(t_{11} < -.831) > .20$
    conclusion:
        Do not reject H_o; there is not sufficient evidence to reject the claim that $\mu_d = 0$.
    b. $\bar{d} \pm t_{11,.025} \cdot s_d/\sqrt{n}$
        $-.475 \pm 2.201 \cdot 1.981/\sqrt{12}$
        $-.475 \pm 1.259$
        $-1.73 < \mu_d < .78$

    Since the confidence interval contains 0, there is no significant difference between the reported and measured heights.

7. d = x - y: -20 0 10 -40 -30 -10 30 -20 -20 -10

n = 10

$\Sigma d = -110$ $\bar{d} = -11.0$

$\Sigma d^2 = 4900$ $s_d = 20.248$

a. original claim: $\mu_d < 0$

H_o: $\mu_d = 0$

H_1: $\mu_d < 0$

$\alpha = .05$

C.R. t < $-t_{9,.05}$ = -1.833

calculations:

$t_{\bar{d}} = (\bar{d} - \mu_{\bar{d}})/s_{\bar{d}}$

$= (-11.0 - 0)/(20.248/\sqrt{10})$

$= -11.0/6.40$

$= -1.718$

$.05 < $ P-value $= P(t_9 < -1.718) < .10$

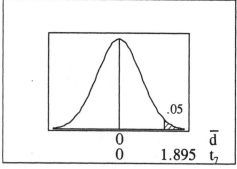

conclusion:

Do not reject H_o; there is not sufficient evidence to conclude that $\mu_d < 0$.

b. $\bar{d} \pm t_{9,.025} \cdot s_d/\sqrt{n}$

$-11.0 \pm 2.262 \cdot 20.248/\sqrt{10}$

-11.0 ± 14.5

$-25.5 < \mu_d < 3.5$

We have 95% confidence that the interval from -25.5 to 3.5 contains the true mean population difference. Since this interval includes 0, the mean before and after scores are not significantly different, and there is not enough evidence to say that the course has any effect.

9. d = $x_B - x_A$: -.2 4.1 1.6 1.8 3.2 2.0 2.9 9.6

n = 8

$\Sigma d = 25.0$ $\bar{d} = 3.125$

$\Sigma d^2 = 137.46$ $s_d = 2.9114$

a. $\bar{d} \pm t_{7,.025} \cdot s_d/\sqrt{n}$

$3.125 \pm 2.365 \cdot 2.9114/\sqrt{8}$

3.125 ± 2.434

$.69 < \mu_d < 5.56$

b. original claim: $\mu_d > 0$

H_o: $\mu_d = 0$

H_1: $\mu_d > 0$

$\alpha = .05$

C.R. t > $t_{7,.05}$ = 1.895

calculations:

$t_{\bar{d}} = (\bar{d} - \mu_{\bar{d}})/s_{\bar{d}}$

$= (3.125 - 0)/(2.9114/\sqrt{8})$

$= 3.125/1.029 = 3.036$

$.005 < $ P-value $= P(t_7 > 3.036) < .01$

conclusion:

Reject H_o; there is sufficient evidence to conclude that $\mu_d > 0$.

c. Yes; hypnotism appears to be effective in reducing pain.

11. $d = x - y$: -106 20 -101 33 -72 36 -62 -38 70 -127 -24
      $n = 11$
      $\Sigma d = -371$ $\bar{d} = -33.73$
      $\Sigma d^2 = 56299$ $s_d = 66.171$

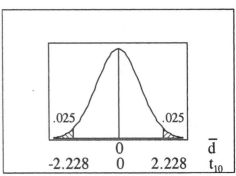

   a. original claim: $\mu_d = 0$
      H_o: $\mu_d = 0$
      H_1: $\mu_d \neq 0$
      $\alpha = .05$
      C.R. $t < -t_{10,.025} = -2.228$
             $t > t_{10,.025} = 2.228$
      calculations:
          $t_{\bar{d}} = (\bar{d} - \mu_{\bar{d}})/s_{\bar{d}}$
             $= (-33.73 - 0)/(66.171/\sqrt{11})$
             $= -33.73/19.951 = -1.690$
          $.10 < $ P-value $ = 2 \cdot P(t_{11} < -1.690) < .20$
      conclusion:
      Do not reject H_o; there is not sufficient evidence to reject the claim that $\mu_d = 0$.

   b. $\bar{d} \pm t_{10,.025} \cdot s_d/\sqrt{n}$
     -33.73 $\pm 2.228 \cdot 66.171/\sqrt{11}$
     -33.73 $\pm$ 44.45
     $-78.2 < \mu_d < 10.7$

   c. We cannot be 95% certain that either type of seed is better. If there are no differences in cost, or any other considerations, choose the kiln dried – even though we can't be sure that it's generally better, it did have the higher yield in this particular trial.

13.   a. original claim: $\mu_d \neq 0$
      H_o: $\mu_d = 0$
      H_1: $\mu_d \neq 0$
      $\alpha = .05$
      C.R. $t < -t_{9,.025} = -2.262$
            $t > t_{9,.025} = 2.262$
      calculations:

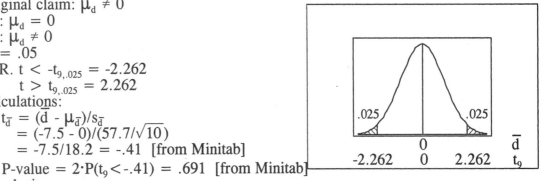

         $t_{\bar{d}} = (\bar{d} - \mu_{\bar{d}})/s_{\bar{d}}$
           $= (-7.5 - 0)/(57.7/\sqrt{10})$
           $= -7.5/18.2 = -.41$ [from Minitab]
        P-value $= 2 \cdot P(t_9 < -.41) = .691$ [from Minitab]
      conclusion:
      Do not reject H_o; there is not sufficient evidence to conclude that $\mu_d \neq 0$.
     No; based on this result, do not spend the money for the drug.

   b. original claim: $\mu_d < 0$
      H_o: $\mu_d = 0$
      H_1: $\mu_d < 0$
      $\alpha = .05$
      C.R. $t < -t_{9,.05} = -1.833$
      calculations:

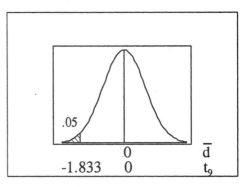

         $t_{\bar{d}} = (\bar{d} - \mu_{\bar{d}})/s_{\bar{d}}$
           $= (-7.5 - 0)/(57.7/\sqrt{10})$
           $= -7.5/18.2 = -.41$ [from Minitab]
        P-value $= P(t_9 < -.41) = \frac{1}{2}(.691) = .3455$
      conclusion:
      Do not reject H_o; there is not sufficient evidence to conclude that $\mu_d < 0$.
     All the calculations remain the same. The alternative hypothesis and critical region change to reflect the one-tailed test. The P-value is ½ the ,691 given by Minitab for the two-tailed test.

15.   original claim: $\mu_d \neq 0$
H_o: $\mu_d = 0$
H_1: $\mu_d \neq 0$
$\alpha = .05$ [assumed]
C.R. $t < -t_{11,.025} = -2.201$
      $t > t_{11,.025} = 2.201$
calculations:
      $t_{\bar{d}} = (\bar{d} - \mu_{\bar{d}})/s_{\bar{d}}$
          $= -.501$ [from Excel]
      P-value $= 2 \cdot P(t_{11} < -.51) = .626$ [from Excel]
conclusion:
      Do not reject H_o; there is not sufficient evidence to conclude $\mu_d \neq 0$.

17. $d = x_8 - x_{12}$: given at the right
      $n = 11$
      $\Sigma d = -8.6$ $\bar{d} = -.782$
      $\Sigma d^2 = 15.06$ $s_d = .913$

sbj#	8am	12am	diff
19	97.0	97.7	-0.7
20	98.0	98.8	-0.8
22	96.4	98.0	-1.6
26	98.2	98.7	-0.5
71	98.8	98.0	0.8
78	98.6	98.5	0.1
80	97.8	98.3	-0.5
81	98.7	98.7	0.0
83	97.8	99.1	-1.3
98	96.4	98.2	-1.8
99	96.9	99.2	-2.3

a. $\bar{d} \pm t_{10,.025} \cdot s_d/\sqrt{n}$
      $-.782 \pm 2.228 \cdot .913/\sqrt{11}$
      $-.782 \pm .613$
      $-1.40 < \mu_d < -.17$
b. original claim: $\mu_d = 0$
      H_o: $\mu_d = 0$
      H_1: $\mu_d \neq 0$
      $\alpha = .05$
      C.R. $t < -t_{10,.025} = -2.228$
          $t > t_{10,.025} = 2.228$
      calculations:
          $t_{\bar{d}} = (\bar{d} - \mu_{\bar{d}})/s_{\bar{d}}$
              $= (-.782 - 0)/(.913/\sqrt{11})$
              $= -.782/.275 = -2.840$
          $.01 < $ P-value $= 2 \cdot P(t_{10} < -2.840) < .02$
      conclusion:
          Reject H_o; there is sufficient evidence to reject the claim that $\mu_d = 0$ and conclude
          that $\mu_d \neq 0$ (in fact, that $\mu_d < 0$.
      No; based on this result. morning and night body temperatures do not appear to be
      about the same. The 8 am (morning) temperatures are significantly lower.

19.   The Minitab output is given below.
      MTB > let c11=c3-c9
      MTB > ttest c11   TEST OF MU = 0.00 VS MU N.E. 0.00

	N	MEAN	STDEV	SE MEAN	T	P VALUE
C11	31	-6.00	11.26	2.02	-2.97	0.0059

a. original claim: $\mu_d = 0$
      H_o: $\mu_d = 0$
      H_1: $\mu_d \neq 0$
      $\alpha = .05$
      C.R. $t < -t_{30,.025} = -2.042$
          $t > t_{30,.025} = 2.042$
      calculations:
          $t_{\bar{d}} = (\bar{d} - \mu_{\bar{d}})/s_{\bar{d}}$
              $= (-6.0 - 0)/(11.26/\sqrt{31})$
              $= -6.0/2.022 = -2.97$ [Minitab]
          P-value $= 2 \cdot P(t_{30} < -2.97) = .0059$ [Minitab]

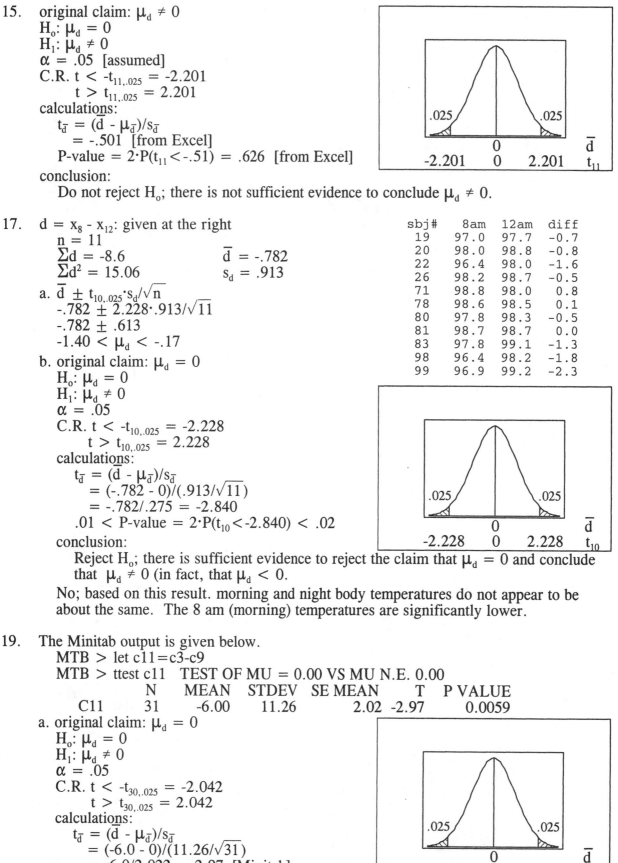

conclusion:
    Reject H_o; there is sufficient evidence to reject the claim that $\mu_d = 0$ and conclude that $\mu_d \neq 0$ (in fact, that $\mu_d < 0$).

b. $\overline{d} \pm t_{30,.025} \cdot s_d/\sqrt{n}$
    $-6.0 \pm 2.042 \cdot 11.26/\sqrt{31}$
    -6.0 ± 4.1
    $-10.1 < \mu_d < -1.9$

c. Even though the conclusions are different for the example in the chapter (using the first 5 values) and this exercise (using all the data), the results are not contradictory. With n=5, there was almost (but not quite) enough evidence to declare a difference; with n=31 (even though the mean difference is smaller), there is enough evidence to declare a difference. It appears that the five-day low forecasts tend to be higher than the low temperatures actually reached.

21. a. Yes, an outlier can have a drastic effect on the test of hypothesis and/or the confidence interval. Depending on the data an outlier could increase the mean difference to create a significant difference where none exists, or an outlier could increase the variability to hide a significant different that does exist.

    b. Since the test statistic is unit free, it will be same no matter what units are used. The confidence interval will be stated in the new units, but it will include precisely the same climatic conditions no matter how those conditions are described in the various units.

23. The following table of summary statistics applies to all parts of this exercise.

	values										n	Σv	Σv^2	$\overline{v}$	s_v
x:	1	3	2	2	1	2	3	3	2	1	10	20	46	2.0	.816
y:	1	2	1	2	1	2	1	2	1	2	10	15	25	1.5	.527
d=x-y:	0	1	1	0	0	0	2	1	1	-1	10	5	9	.5	.850

a. original claim: $\mu_d > 0$
    H_o: $\mu_d = 0$
    H_1: $\mu_d > 0$
    $\alpha = .05$
    C.R. $t > t_{9,.05} = 1.833$
    calculations:
        $t_{\overline{d}} = (\overline{d} - \mu_{\overline{d}})/s_{\overline{d}}$
        $= (.5 - 0)/(.850/\sqrt{10})$
        $= .5/.2687 = 1.861$
        $.025 < $ P-value $ = P(t_9 > 1.861) < .05$

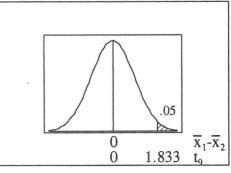

conclusion:
    Reject H_o; there is sufficient evidence to conclude that $\mu_d > 0$.

b. original claim: $\mu_1 - \mu_2 > 0$
    $\overline{x}_1 - \overline{x}_2 = 2.0 - 1.5 = .5$
    H_o: $\mu_1 - \mu_2 = 0$
    H_1: $\mu_1 - \mu_2 > 0$
    $\alpha = .01$
    C.R. $t > t_{9,.05} = 1.833$
    calculations:
        $t_{\overline{x}_1 - \overline{x}_2} = (\overline{x}_1 - \overline{x}_2 - \mu_{\overline{x}_1 - \overline{x}_2})/s_{\overline{x}_1 - \overline{x}_2}$
        $= (.5 - 0)/\sqrt{(.816)^2/10 + (.527)^2/10}$
        $= .5/.3073 = 1.627$
        $.05 < $ P-value $ = P(t_9 > 1.627) < .10$
    conclusion:
        Do not reject H_o; there is not sufficient evidence to conclude that $\mu_1 - \mu_2 > 0$.

c. Yes; since different methods can give different results, it is important that the correct method is used.

8-5 Comparing Variation in Two Samples

NOTE: The following convention are used in this manual regarding the F test.
   * The set if scores with the larger sample variance is designated group 1.
   * Even though always designating the scores with the larger sample variance as group 1 makes lower critical unnecessary in two-tailed tests, the lower critical value are calculated (using the method given in exercise #19) and included for completeness and consistency with the other tests. The F distribution always "bunches up" with expected value 1.0, regardless of the df.
   * The df for group 1 (numerator) and group 2 (denominator) are given with the F as a superscript and subscript respectively.
   * If the desired df does not appear in Table A-5, the closest entry is used. If the desired entry is exactly halfway between two tabled values, the conservative approach of using the smaller df is employed. Since any finite number is closer to 120 than ∞, 120 is used for all df larger than 120.
   * Since all hypotheses in the text question the equality of σ_1^2 and σ_2^2, the calculation of F [which is statistically defined to be $F = (s_1^2/\sigma_1^2)/(s_2^2/\sigma_2^2)$] is shortened to $F = s_1^2/s_2^2$.
   * Some problems are stated in terms of variance, and some are stated in terms of standard deviation. Since $\sigma_1^2 = \sigma_2^2$ is equivalent to $\sigma_1^2 = \sigma_2^2$, the manual simply states all claims and hypotheses and conclusions using the variance.

1. Let the treatment population be group 1.
    original claim: $\sigma_1^2 \neq \sigma_2^2$
    $H_o: \sigma_1^2 = \sigma_2^2$
    $H_1: \sigma_1^2 \neq \sigma_2^2$
    $\alpha = .05$
    C.R. $F < F_{29,.975}^{24} = .4527$
       $F > F_{29,.025}^{24} = 2.1540$
    calculations:
       $F = s_1^2/s_2^2$
         $= (.78)^2/(.52)^2$
         $= 2.25$
    conclusion:

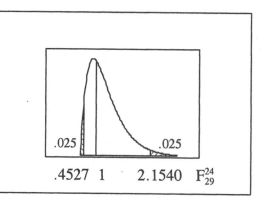

Reject H_o; there is sufficient evidence to conclude that $\sigma_1^2 \neq \sigma_2^2$ (in fact, that $\sigma_1^2 > \sigma_2^2$).

3. Let the sham population be group 1.
    original claim: $\sigma_1^2 > \sigma_2^2$
    $H_o: \sigma_1^2 = \sigma_2^2$
    $H_1: \sigma_1^2 > \sigma_2^2$
    $\alpha = .05$
    C.R. $F > F_{19,.05}^{19} = 2.1555$
    calculations:
       $F = s_1^2/s_2^2$
         $= (1.4)^2/(.96)^2$
         $= 2.1267$

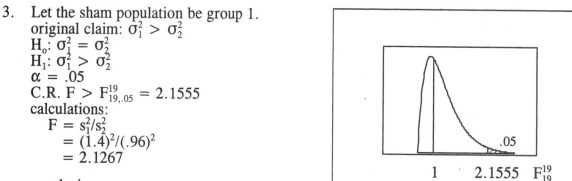

    conclusion:
       Do not reject H_o; there is not sufficient evidence to conclude that $\sigma_1^2 > \sigma_2^2$.

5. Let the regular Coke be group 1.
    original claim: $\sigma_1^2 \neq \sigma_2^2$
    H_o: $\sigma_1^2 = \sigma_2^2$
    H_1: $\sigma_1^2 \neq \sigma_2^2$
    $\alpha = .05$
    C.R. F $< F_{35,.975}^{35} = .4822$
    $\quad\;\;$ F $> F_{35,.025}^{35} = 2.0739$
    calculations:
    $\quad$ F $= s_1^2/s_2^2$
    $\quad\quad = (.007507)^2/(.004391)^2$
    $\quad\quad = 2.9228$
    conclusion:

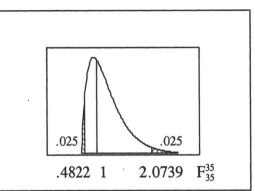

 Reject H_o; there is sufficient evidence to conclude that $\sigma_1^2 \neq \sigma_2^2$ (in fact, that $\sigma_1^2 > \sigma_2^2$). The significant difference in variability between the weights of regular and diet Coke is not necessarily due to differences in quality control efforts. Regular Coke may contain a relatively heavy ingredient (e.g., sugar) that is difficult ro dispense with the same level of consistency as the other ingredients.

7. Let the filtered cigarettes be group 1.
    original claim: $\sigma_1^2 > \sigma_2^2$
    H_o: $\sigma_1^2 = \sigma_2^2$
    H_1: $\sigma_1^2 > \sigma_2^2$
    $\alpha = .05$
    C.R. F $> F_{7,.05}^{20} = 3.4445$
    calculations:
    $\quad$ F $= s_1^2/s_2^2$
    $\quad\quad = (.31)^2/(.16)^2$
    $\quad\quad = 3.7539$

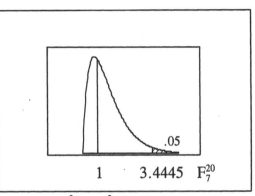

    conclusion:
     Reject H_o; there is sufficient evidence to conclude that $\sigma_1^2 > \sigma_2^2$.

9. Let the students be group 1.
    original claim: $\sigma_1^2 > \sigma_2^2$
    H_o: $\sigma_1^2 = \sigma_2^2$
    H_1: $\sigma_1^2 > \sigma_2^2$
    $\alpha = .05$ [assumed]
    C.R. F $> F_{151,.05}^{216} = 1.3519$
    calculations:
    $\quad$ F $= s_1^2/s_2^2$
    $\quad\quad = (3.67)^2/(3.65)^2$
    $\quad\quad = 1.0110$

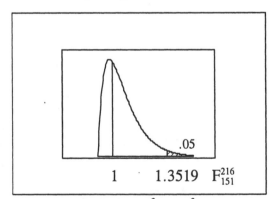

    conclusion:
     Do not reject H_o; there is not sufficient evidence to conclude that $\sigma_1^2 > \sigma_2^2$.

11. NOTE: While Data Set 11 covers a full year of 365 days with 53 Wednesdays and 52 observations for each of the other days, part (a) chooses to use equal sample sizes by dropping December 31 and using the first 52 Wednesdays.

a. Let the Sundays be group 1.
original claim: $\sigma_1^2 = \sigma_2^2$
H_o: $\sigma_1^2 = \sigma_2^2$
H_1: $\sigma_1^2 \neq \sigma_2^2$
$\alpha = .05$
C.R. $F < F_{51,.975}^{51} = .5333$
$F > F_{51,.025}^{51} = 1.8752$
calculations:
$F = s_1^2/s_2^2$
$= (.2000)^2/(.1357)^2$
$= 2.1722$

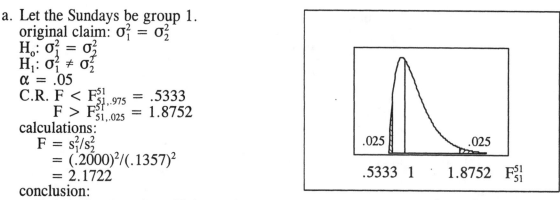

conclusion:
Reject H_o; there is sufficient evidence to reject the claim that $\sigma_1^2 = \sigma_2^2$ and conclude that $\sigma_1^2 \neq \sigma_2^2$ (in fact, that $\sigma_1^2 > \sigma_2^2$).

b. For Sundays, 37 of the 52 observations are 0; for Wednesdays, 37 of the 53 [or 36 of the first 52] observations are 0. The data for both days are very positively skewed. The rainfall amounts do not come from populations with normal distributions.

c. Because the original populations are not normally distributed, the test in part (a) is not valid – which is good news, since there is no reason why rainfall amounts should be more variable on Sundays than on Wednesdays.

13.   The summary statistics are as follows.
placebo: $n=13$ $\Sigma x=1490.5$ $\Sigma x^2=171965.47$ $\bar{x}=114.65$ $s^2=89.493$
calcium: $n=15$ $\Sigma x=1740.4$ $\Sigma x^2=202936.92$ $\bar{x}=116.03$ $s^2=71.722$

Let the placebo population be group 1.
original claim: $\sigma_1^2 = \sigma_2^2$
H_o: $\sigma_1^2 = \sigma_2^2$
H_1: $\sigma_1^2 \neq \sigma_2^2$
$\alpha = .05$
C.R. $F < F_{14,.975}^{12} = .3147$
$F > F_{14,.025}^{12} = 3.0502$
calculations:
$F = s_1^2/s_2^2$
$= 89.493/71.722$
$= 1.2478$ [agrees with TI-83 Plus]

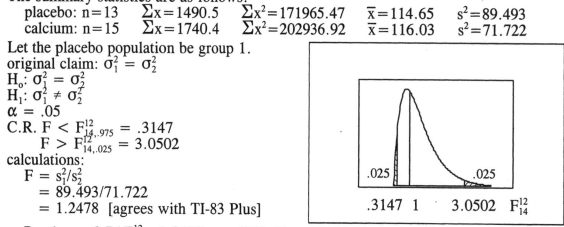

P-value $= 2 \cdot P(F_{14}^{12} > 1.2478) = .6852$ [from TI-83 Plus]
conclusion:
Do not reject H_o; there is not sufficient evidence to reject the claim that $\sigma_1^2 = \sigma_2^2$.
Yes, these two groups appear acceptable for an experiment that requires $\sigma_1^2 = \sigma_2^2$.

15.   The Minirab output is given below.
MTB > describe c1-c2

	N	MEAN	MEDIAN	TRMEAN	STDEV	SEMEAN	MIN	MAX	Q1	Q3
RowlFRE	12	80.75	81.35	81.19	4.68	1.35	70.90	86.20	78.75	84.53
TolsFRE	12	66.15	68.95	66.75	7.86	2.27	51.90	74.40	59.85	72.05

Let the Tolstoy pages be group 1.
original claim: $\sigma_1^2 = \sigma_2^2$
H_o: $\sigma_1^2 = \sigma_2^2$
H_1: $\sigma_1^2 \neq \sigma_2^2$
$\alpha = .05$
C.R. $F < F_{11,.975}^{11} = .2836$
$F > F_{11,.025}^{11} = 3.5257$
calculations:
$F = s_1^2/s_2^2$
$= (7.86)^2/(4.68)^2$
$= 2.8208$

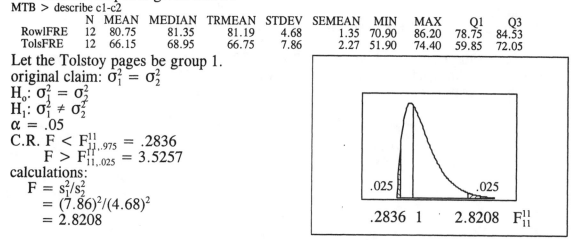

conclusion:
Do not reject H_o; there is not sufficient evidence to reject the claim that $\sigma_1^2 = \sigma_2^2$.

17. Let the .111" cans be group 1.
original claim: $\sigma_1^2 \neq \sigma_2^2$
H_o: $\sigma_1^2 = \sigma_2^2$
H_1: $\sigma_1^2 \neq \sigma_2^2$
$\alpha = .05$
C.R. $F < F_{174,.975}^{173} = .6980$
$F > F_{173,.025}^{174} = 1.4327$
calculations:
$F = s_1^2/s_2^2$
$= (22.1)^2/(22.1)^2$
$= 1.0000$

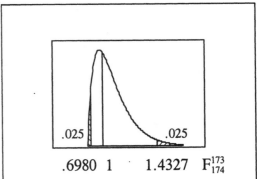

conclusion:
Do not reject H_o; there is not sufficient evidence to conclude that $\sigma_1^2 \neq \sigma_2^2$.
Yes, eliminating the outlier changed the decision.

19. a. $F_L = F_{9,.975}^9 = 1/F_{9,.025}^9 = 1/4.0260 = .2484$
$F_R = F_{9,.025}^9 = 4.0260$
b. $F_L = F_{6,.975}^9 = 1/F_{9,.025}^6 = 1/4.3197 = .2315$
$F_R = F_{6,.025}^9 = 5.5234$
c. $F_L = F_{9,.975}^6 = 1/F_{6,.025}^9 = 1/5.5234 = .1810$
$F_R = F_{9,.025}^6 = 4.3197$

Review Exercises

1. Let those that were warmed be group 1.
a. original claim: $p_1 - p_2 < 0$
$\hat{p}_1 = x_1/n_1 = 6/104 = .0577$
$\hat{p}_2 = x_2/n_2 = 18/96 = .1875$
$\hat{p}_1 - \hat{p}_2 = .0577 - .1875 = -.1298$
$\bar{p} = (x_1 + x_2)/(n_1 + n_2)$
$\phantom{\bar{p} }= (6 + 18)/(104 + 96)$
$\phantom{\bar{p} }= 24/200 = .12$
H_o: $p_1 - p_2 = 0$
H_1: $p_1 - p_2 < 0$
$\alpha = .05$
C.R. $z < -z_{.05} = -1.645$
calculations:
$z_{\hat{p}_1 - \hat{p}_2} = (\hat{p}_1 - \hat{p}_2 - \mu_{\hat{p}_1 - \hat{p}_2})/\sigma_{\hat{p}_1 - \hat{p}_2}$
$\phantom{z_{\hat{p}_1-\hat{p}_2} }= (-.1298 - 0)/\sqrt{(.12)(.88)/104 + (.12)(.88)/96}$
$\phantom{z_{\hat{p}_1-\hat{p}_2} }= -.1298/.0460$
$\phantom{z_{\hat{p}_1-\hat{p}_2} }= -2.82$
P-value $= P(z < -2.82) = .0024$
conclusion:
Reject H_o; there is sufficient evidence to conclude that $p_1 - p_2 < 0$.
Yes; if these results are verified, surgical patients should be routinely warmed.
b. The test in part (a) places 5% in the lower tail. To correspond with this test, use a 90% two-sided confidence interval [or a 95% one-sided confidence interval].

c. $(\hat{p}_1 - \hat{p}_2) \pm z_{.05} \sqrt{\hat{p}_1 \hat{q}_1/n_1 + \hat{p}_2 \hat{q}_2/n_2}$

$-.1298 \pm 1.645 \cdot \sqrt{(.0577)(.9423)/104 + (.1875)(.8125)/96}$

$-.1298 \pm .0756$

$-.205 < p_1 - p_2 < -.054$

d. Since the test of hypothesis and the confidence interval use different estimates for $\sigma_{\hat{p}_1 - \hat{p}_2}$, they are not mathematically equivalent and it is possible that they may might lead to different conclusions.

2. $d = x - y$: -5.75 -1.25 -1.00 -5.00 0.00 0.25 2.25 -0.50 0.75 -1.50 -0.25

$n = 11$

$\Sigma d = -12.00$ $\bar{d} = -1.091$

$\Sigma d^2 = 68.8750$ $s_d = 2.3619$

a. original claim: $\mu_d = 0$

$H_o: \mu_d = 0$

$H_1: \mu_d \neq 0$

$\alpha = .05$

C.R. $t < -t_{10,.025} = -2.228$

$\quad\ t > t_{10,.025} = 2.228$

calculations:

$t_{\bar{d}} = (\bar{d} - \mu_{\bar{d}})/s_{\bar{d}}$

$\quad = (-1.091 - 0)/(2.3619/\sqrt{11})$

$\quad = -1.091/.7121$

$\quad = -1.532$

$.10 < $ P-value $ = 2 \cdot P(t_{11} < -1.532) < .20$

conclusion:

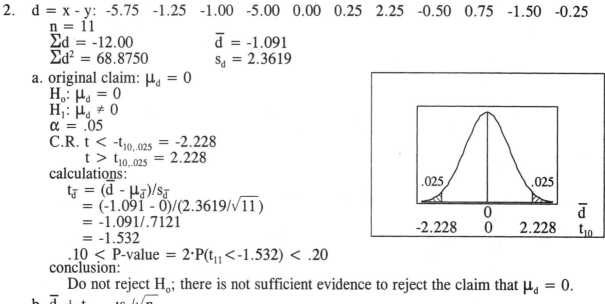

Do not reject H_o; there is not sufficient evidence to reject the claim that $\mu_d = 0$.

b. $\bar{d} \pm t_{10,.025} \cdot s_d/\sqrt{n}$

$-1.091 \pm 2.228 \cdot 2.3619/\sqrt{11}$

-1.091 ± 1.587

$-2.67 < \mu_d < .50$

c. We cannot be 95% certain that either type of seed is better. If there are no differences in cost, or any other considerations, choose the kiln dried – even though we can't be sure that it's generally better, it did have the higher yield in this particular trial.

3. Let the obsessive-compulsive patients be group 1.

$\bar{x}_1 - \bar{x}_2 = 1390.03 - 1268.41 = 121.62$

a. $(\bar{x}_1 - \bar{x}_2) \pm t_{9,.025} \sqrt{s_1^2/n_1 + s_2^2/n_2}$

$121.62 \pm 2.262 \cdot \sqrt{(156.84)^2/10 + (137.97)^2/10}$

121.62 ± 149.42

$-27.80 < \mu_1 - \mu_2 < 271.04$

b. original claim: $\mu_1 - \mu_2 = 0$

$H_o: \mu_1 - \mu_2 = 0$

$H_1: \mu_1 - \mu_2 \neq 0$

$\alpha = .05$

C.R. $t < -t_{9,.025} = -2.262$

$\quad\ t > t_{9,025} = 2.262$

calculations:

$t_{\bar{x}_1 - \bar{x}_2} = (\bar{x}_1 - \bar{x}_2 - \mu_{\bar{x}_1 - \bar{x}_2})/s_{\bar{x}_1 - \bar{x}_2}$

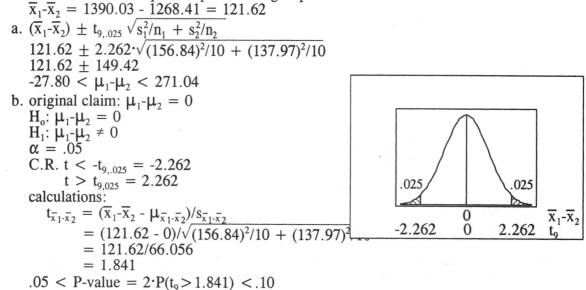

$\quad = (121.62 - 0)/\sqrt{(156.84)^2/10 + (137.97)^2/10}$

$\quad = 121.62/66.056$

$\quad = 1.841$

$.05 < $ P-value $ = 2 \cdot P(t_9 > 1.841) < .10$

conclusion:

Do not reject H_o; there is not sufficient evidence to reject the claim that $\mu_1 - \mu_2 = 0$.

c. No, it does not appear that the total brain volume can be used as a reliable indicator.

4. Let the obsessive-compulsive patients be group 1.
original claim: $\sigma_1^2 \neq \sigma_2^2$
H_o: $\sigma_1^2 = \sigma_2^2$
H_1: $\sigma_1^2 \neq \sigma_2^2$
$\alpha = .05$
C.R. $F < F_{9,.975}^9 = .2484$
　　$F > F_{9,.025}^9 = 4.0260$
calculations:
　$F = s_1^2/s_2^2$
　　$= (156.84)^2/(137.97)^2$
　　$= 1.2922$
conclusion:
　Do not reject H_o; there is not sufficient evidence to conclude that $\sigma_1^2 \neq \sigma_2^2$.

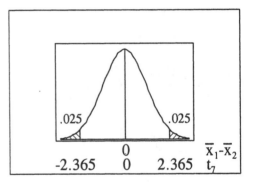

.025　　　　　.025

.6980　1　·　1.4327　F_{174}^{173}

5. Let the filtered cigarettes be group 1

group 1: filtered (n = 21)	group 2: unfiltered (n = 8)
$\Sigma x = 270$	$\Sigma x = 125$
$\Sigma x^2 = 3660$	$\Sigma x^2 = 1963$
$\bar{x} = 12.857$	$\bar{x} = 15.625$
$s = 3.071$	$s = 1.188$

original claim: $\mu_1 - \mu_2 = 0$
　$\bar{x}_1 - \bar{x}_2 = 12.857 - 15.625 = -2.768$
H_o: $\mu_1 - \mu_2 = 0$
H_1: $\mu_1 - \mu_2 \neq 0$
$\alpha = .05$
C.R. $t < -t_{7,.025} = -2.365$
　　$t > t_{7,.025} = 2.365$
calculations:
　$t_{\bar{x}_1-\bar{x}_2} = (\bar{x}_1 - \bar{x}_2 - \mu_{\bar{x}_1-\bar{x}_2})/s_{\bar{x}_1-\bar{x}_2}$
　　$= (-2.768 - 0)/\sqrt{(3.071)^2/21 + (1.188)^2/8}$
　　$= -2.768/.7908$
　　$= -3.500$
　P-value $= 2 \cdot P(t_7 < -3.500) \approx .01$
conclusion:
　Reject H_o; there is sufficient evidence to reject the claim that $\mu_1 - \mu_2 = 0$ and conclude that $\mu_1 - \mu_2 \neq 0$ (in fact, that $\mu_1 - \mu_2 < 0$).
Yes; based on this result, cigarette filters are effective in reducing the amount of carbon monoxide.

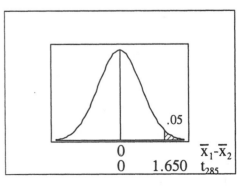

.025　　　　　.025

　　　　0
-2.365　0　2.365　t_7
　　　　　　　　$\bar{x}_1-\bar{x}_2$

6. Let those receiving the zinc supplement be group 1.
original claim: $\mu_1 - \mu_2 > 0$
　$\bar{x}_1 - \bar{x}_2 = 3214 - 3088 = 126$
H_o: $\mu_1 - \mu_2 = 0$
H_1: $\mu_1 - \mu_2 > 0$
$\alpha = .05$
C.R. $t > t_{285,.05} = 1.650$
calculations:
　$t_{\bar{x}_1-\bar{x}_2} = (\bar{x}_1 - \bar{x}_2 - \mu_{\bar{x}_1-\bar{x}_2})/s_{\bar{x}_1-\bar{x}_2}$
　　$= (126 - 0)/\sqrt{(669)^2/294 + (728)^2/286}$
　　$= 126/58.098$
　　$= 2.169$
　$.01 < $ P-value $= P(t_{285} > 2.169) < .025$

.05

0
0　　1.650　t_{285}
　　　　　$\bar{x}_1-\bar{x}_2$

conclusion:
   Reject H_o; there is sufficient evidence to conclude that $\mu_1-\mu_2 > 0$.
Yes, there is sufficient evidence to support the claim that the zinc supplement results in increased birth weights.

7.   Let those who saw the first woman be group 1.
   original claim: $p_1-p_2 > 0$
$$\hat{p}_1 = x_1/n_1 = 58/2000 = .0290$$
$$\hat{p}_2 = x_2/n_2 = 35/2000 = .0175$$
$$\hat{p}_1-\hat{p}_2 = .0290 - .0175 = .0115$$
$$\bar{p} = (x_1 + x_2)/(n_1 + n_2)$$
$$= (58 + 35)/(2000 + 2000)$$
$$= 93/4000 = .02325$$
H_o: $p_1-p_2 = 0$
H_1: $p_1-p_2 > 0$
$\alpha = .05$
C.R. $z > z_{.05} = 1.645$
calculations:

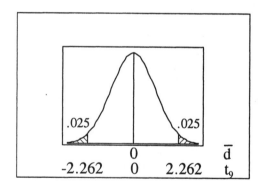

$$z_{\hat{p}1-\hat{p}2} = (\hat{p}_1-\hat{p}_2 - \mu_{\hat{p}_1-\hat{p}_2})/\sigma_{\hat{p}_1-\hat{p}_2}$$
$$= (.0115 - 0)/\sqrt{(.02325)(.97675)/2000 + (.02325)(.97576)/2000}$$
$$= .0115/.004765$$
$$= 2.41$$
   P-value $= P(z>2.41) = 1 - P(z<2.41) = 1 - .9920 = .0080$
conclusion:
   Reject H_o; there is sufficient evidence to conclude that $p_1-p_2 > 0$.

8.   The summary statistics are as follows.
   $d = x_1 - x_2$: 5 0 0 0 8 1 1 4 0 1
      $n = 10$
      $\Sigma d = 20$ $\bar{d} = 2.00$
      $\Sigma d^2 = 108$ $s_d = 2.749$
a. original claim: $\mu_d = 0$
   H_o: $\mu_d = 0$
   H_1: $\mu_d \neq 0$
   $\alpha = .05$ [assumed]
   C.R. $t < -t_{9,.025} = -2.262$
         $t > t_{9,.025} = 2.262$
   calculations:
      $$t_{\bar{d}} = (\bar{d} - \mu_{\bar{d}})/s_{\bar{d}}$$
      $$= (2.00 - 0)/(2.749/\sqrt{10})$$
      $$= 2.00/.869$$
      $$= 2.301$$
   $.02 <$ P-value $< .05$
   conclusion:
      Reject H_o; there is sufficient evidence to reject the claim that $\mu_d = 0$ and to conclude that $\mu_d \neq 0$ (in fact, that $\mu_d > 0$ – i.e., that training reduces the weights).
b. $\bar{d} \pm t_{9,.025} \cdot s_d/\sqrt{n}$
   $2.00 \pm 2.262 \cdot 2.749/\sqrt{10}$
   2.00 ± 1.97
   $0.0 < \mu_d < 4.0$

Cumulative Review Exercises

1. Refer to the summary table at the right.

a. $P(Y) = 53/750 = .071$

b. $P(M \text{ or } Y) = P(M) + P(Y) - P(M \text{ and } Y)$
$= 250/750 + 53/750 - 26/750$
$= 277/750$
$= .369$

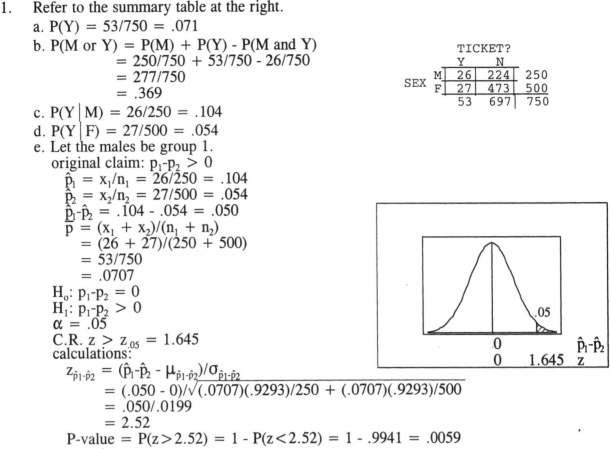

TICKET?

SEX		Y	N	
	M	26	224	250
	F	27	473	500
		53	697	750

c. $P(Y \mid M) = 26/250 = .104$

d. $P(Y \mid F) = 27/500 = .054$

e. Let the males be group 1.
original claim: $p_1-p_2 > 0$
$\hat{p}_1 = x_1/n_1 = 26/250 = .104$
$\hat{p}_2 = x_2/n_2 = 27/500 = .054$
$\hat{p}_1-\hat{p}_2 = .104 - .054 = .050$
$\bar{p} = (x_1 + x_2)/(n_1 + n_2)$
$= (26 + 27)/(250 + 500)$
$= 53/750$
$= .0707$
$H_o: p_1-p_2 = 0$
$H_1: p_1-p_2 > 0$
$\alpha = .05$
C.R. $z > z_{.05} = 1.645$
calculations:
$z_{\hat{p}1-\hat{p}2} = (\hat{p}_1-\hat{p}_2 - \mu_{\hat{p}_1-\hat{p}_2})/\sigma_{\hat{p}_1-\hat{p}_2}$
$= (.050 - 0)/\sqrt{(.0707)(.9293)/250 + (.0707)(.9293)/500}$
$= .050/.0199$
$= 2.52$
P-value $= P(z>2.52) = 1 - P(z<2.52) = 1 - .9941 = .0059$
conclusion:
Reject H_o; there is sufficient evidence to conclude that $p_1-p_2 > 0$.
No; while we can conclude that men receive more speeding tickets than women, we cannot conclude that men speed more often. That conclusion would be valid only if there were a perfect correspondence between speeding and receiving a ticket. It could be that women speed just as often as men but get fewer tickets because they (a) tend to exceed the posted limit by smaller amounts than men do or (b) are more likely to be let off with only a warning than men are. It could also be that men drive more, and so they would be expected to get more tickets even if they did not speed more – consider a population that is ½ male and ½ female in which all people speed equally: if the men do 80% of the driving, for example, then they would receive 80% of the tickets.

2. There is a problem with the reported results, and no statistical analysis would be appropriate. Since there were 100 drivers in each group, and the number in each group owning a cell phone must be a whole number between 0 and 100 inclusive, the sample proportion for each group must be a whole percent. The reported values of 13.7% and 10.6% are not mathematical possibilities for the sample success rates of groups of 100.

3. Let the Viagra users be group 1. The following summary statistics apply to all parts [and, as usual, the unrounded values are used in all subsequent calculations.]
$\hat{p}_1 = x_1/n_1 = 29/734 = .040$
$\hat{p}_2 = x_2/n_2 = 15/725 = .021$
$\hat{p}_1-\hat{p}_2 = .040 - .021 = .019$

a. $\hat{p}_1 \pm z_{.025}\sqrt{\hat{p}_1\hat{q}_1/n_1}$
   $.040 \pm 1.96 \sqrt{(.040)(.960)/734}$
   $.040 \pm .014$
   $.025 < p_1 < .054$

b. $\hat{p}_2 \pm z_{.025}\sqrt{\hat{p}_2\hat{q}_2/n_2}$
   $.021 \pm 1.96 \sqrt{(.021)(.979)/725}$
   $.021 \pm .010$
   $.010 < p_1 < .031$

c. $(\hat{p}_1-\hat{p}_2) \pm z_{.025}\sqrt{\hat{p}_1\hat{q}_1/n_1 + \hat{p}_2\hat{q}_2/n_2}$
   $.019 \pm 1.96 \sqrt{(.040)(.960)/734 + (.021)(.979)/725}$
   $.019 \pm .018$
   $.001 < p_1-p_2 < .036$

d. The method in part (iii) is best. It uses all the information at once and asks whether the data is consistent with the claim that $p_1 = p_2$. Method (ii) constructs an interval with no specific claim in mind. Method (i) does not use all the information at once – in general, methods that incorporate all the information into a single statistical procedure are superior to ones that use smaller sample sizes to give partial answers that need to be combined and/or compared afterward.

4. a. Let p = the proportion of runners who finished the race that were females.
       original claim: p < .50
          $\hat{p} = x/n = 39/150 = .26$
       H_0: p = .50
       H_1: p < .50
       $\alpha = .05$ [assumed]
       C.R. z < $-z_{.05}$ = -1.645
       calculations:
          $z_{\hat{p}} = (\hat{p} - \mu_{\hat{p}})/\sigma_{\hat{p}}$
             $= (.26 - .50)/\sqrt{(.50)(.50)/150}$
             $= -.24/.0408$
             $= -5.88$
       P-value = P(z<-5.88) = .0001
    conclusion:
       Reject H_0; there is sufficient evidence to conclude that p < .50.

    b. The ordered 39 times are as follows.
       | 12047 | 12289 | 13593 | 13704 | 13854 | 14216 | 14235 | 14721 | 15036 | 15077 |
       | 15326 | 15357 | 15402 | 16013 | 16297 | 16352 | 16401 | 16758 | 16771 | 16792 |
       | 16871 | 16991 | 17211 | 17260 | 17286 | 17636 | 17726 | 17799 | 18469 | 18580 |
       | 18647 | 10177 | 20084 | 20675 | 20891 | 21911 | 21983 | 25399 | 25858 |

       The summary statistics are as follows.
          n = 39 $\Sigma x = 670735$ $\Sigma x^2 = 11,902,396,416$
          $\bar{x} = (\Sigma x)/n = 670735/39 = 17198.3$
          $\tilde{x} = x_{20} = 17792$
          $s^2 = [n(\Sigma x^2) - (\Sigma x)^2]/[n(n-1)] = [19(11902396416) - (670735)^2]/[39(38)]$
          s = 3107.2

    The frequency distribution at the right indicates that the times appear to be approximately normally distributed.

time (seconds)	frequency
11000 – 12999	2
13000 – 14999	6
15000 – 16999	14
17000 – 18999	9
19000 – 20999	4
21000 – 22999	2
23000 – 24999	0
25000 – 26999	2
	39

    There is no absolute definition of an outlier. The section on boxplots suggests that any score more than $(1.5)(IQR)$ from the median is an outlier.

$IQR = Q_3 - Q_1 = x_{30} - x_{10} = 18580-15077 = 3503$

$16792 \pm 2(3503)$

16792 ± 7006

9786 to 23798

By this definition, the 25399 and 25989 may be considered outliers.

c. original claim: $\mu < 5$ hours = 18000 seconds

H_o: $\mu = 18000$

H_1: $\mu < 18000$

$\alpha = .05$

C.R. $t < -t_{38,.05} = -1.686$

calculations:

$t_{\bar{x}} = (\bar{x} - \mu)/s_{\bar{x}}$

$= (17198.3 - 18000)/(3107.2/\sqrt{39})$

$= -801.7/497.6$

$= -1.611$

$.05 < \text{P-value} < .10$

conclusion:

Do not reject H_o; there is not sufficient evidence to conclude that $\mu < 5$ hours.

d. Let the females be group 1.

original claim: $\mu_1-\mu_2 \neq 0$

$\bar{x}_1-\bar{x}_2 = 17198.3 - 15415.2 = 1783.1$

H_o: $\mu_1-\mu_2 = 0$

H_1: $\mu_1-\mu_2 \neq 0$

$\alpha = .05$

C.R. $t < -t_{38,.025} = -2.024$

$t > t_{38,.025} = 2.024$

calculations:

$t_{\bar{x}_1-\bar{x}_2} = (\bar{x}_1-\bar{x}_2 - \mu_{\bar{x}_1-\bar{x}_2})/s_{\bar{x}_1-\bar{x}_2}$

$= (1783.1 - 0)/\sqrt{(3107)^2/39+(3037)^2/111}$

$= 1783.1/572.0$

$= 3.101$

$\text{P-value} = 2 \cdot P(t_{38}>3.101) < .01$

conclusion:

Reject H_o; there is sufficient evidence to conclude that $\mu_1-\mu_2 \neq 0$ (in fact, that $\mu_1-\mu_2 < 0$).

e. Let the females be group 1.

$\hat{p}_1 = x_1/n_1 = 39/150 = .26$

$\hat{p}_2 = x_2/n_2 = 111/150 = .74$

The methods of this chapter cannot be used to make inferences about $p_1 - p_2$ because the data is not from two independent samples. The "n_2" above is really the same 150 that was n_1. Whatever the value of $\hat{p}_1$ is, the value of $\hat{p}_2$ will have to be $1-\hat{p}_1$. To test whether p_1 and p_2 are equal (which means each must be .5), use only the data for one of the genders – for example, test H_o: $p_1 = .50$.

Chapter 9

Correlation and Regression

9-2 Correlation

1. a. From Table A-6, CV = ±.707; therefore r = .993 indicates a significant (positive) linear correlation.
   b. The proportion of the variation in weight that can be explained in terms of the variation in chest size is $r^2 = (.993)^2 = .986$, or 98.6%.

3. a. From Table A-6, CV = ±.444; therefore r = -.133 does not indicate a significant linear correlation.
   b. The proportion of the variation in Super Bowl points that can be explained in terms of the variation in the DJIA high values is $r^2 = (-.133)^2 = .017$, or 1.7%.

NOTE: In addition to the value of n, calculation of r requires five sums: Σx, Σy, Σx^2, Σy^2 and Σxy. The next problem shows the chart prepared to find these sums. As the sums can usually be found conveniently using a calculator and without constructing the chart, subsequent problems typically give only the values of the sums and do not show a chart.

Also, calculation of r also involves three subcalculations.
(1) $n(\Sigma xy) - (\Sigma x)(\Sigma y)$ determines the sign of r. If large values of x are associated with large values of y, it will be positive. If large values of x are associated with small values of y, it will be negative. If not, a mistake has been made.
(2) $n(\Sigma x^2) - (\Sigma x)^2$ cannot be negative. If it is, a mistake has been made.
(3) $n(\Sigma y^2) - (\Sigma y)^2$ cannot be negative. If it is, a mistake has been made.

Finally, r must be between -1 and 1 inclusive. If not, a mistake has been made. If this or any of the previous mistakes occurs, stop immediately and find the error; continuing will be a fruitless waste of effort.

5.
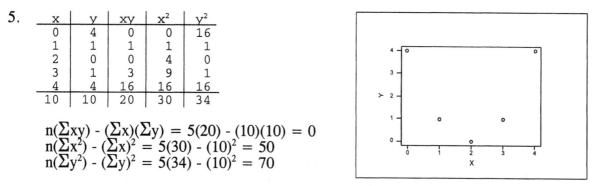

x	y	xy	x^2	y^2
0	4	0	0	16
1	1	1	1	1
2	0	0	4	0
3	1	3	9	1
4	4	16	16	16
10	10	20	30	34

$n(\Sigma xy) - (\Sigma x)(\Sigma y) = 5(20) - (10)(10) = 0$
$n(\Sigma x^2) - (\Sigma x)^2 = 5(30) - (10)^2 = 50$
$n(\Sigma y^2) - (\Sigma y)^2 = 5(34) - (10)^2 = 70$

a. According to the scatter diagram, there appears to be a significant "U-shaped" pattern, but no <u>linear</u> relationship between x and y. Expect a value for r close to 0.
b. $r = [n(\Sigma xy) - (\Sigma x)(\Sigma y)]/[\sqrt{n(\Sigma x^2) - (\Sigma x)^2} \cdot \sqrt{n(\Sigma y^2) - (\Sigma y)^2}]$
   $= 0/[\sqrt{50} \cdot \sqrt{70}]$
   $= 0$
   From Table A-6, assuming $\alpha = .05$, CV = ±.878; therefore r = 0 does not indicate a significant linear correlation. This agrees with the interpretation of the scatter diagram in part (a).

7. The following table and summary statistics apply to all parts of this exercise.
x: 1 1 1 2 2 2 3 3 3 10
y: 1 2 3 1 2 3 1 2 3 10
using all the points: n=10 $\Sigma x=28$ $\Sigma y=28$ $\Sigma xy=136$ $\Sigma x^2=142$ $\Sigma y^2=142$
without the outlier: n=9 $\Sigma x=18$ $\Sigma y=18$ $\Sigma xy=36$ $\Sigma x^2=42$ $\Sigma y^2=42$
a. There appears to be a strong positive linear correlation, with r close to 1.
b. $n(\Sigma xy) - (\Sigma x)(\Sigma y) = 10(136) - (28)(28) = 576$
$n(\Sigma x^2) - (\Sigma x)^2 = 10(142) - (28)^2 = 636$
$n(\Sigma y^2) - (\Sigma y)^2 = 10(142) - (28)^2 = 636$
$r = [n(\Sigma xy) - (\Sigma x)(\Sigma y)]/[\sqrt{n(\Sigma x^2) - (\Sigma x)^2} \cdot \sqrt{n(\Sigma y^2) - (\Sigma y)^2}]$
   $= 576/[\sqrt{636} \cdot \sqrt{636}]$
   $= .906$
From Table A-6, assuming $\alpha=.05$, CV $= \pm.632$; therefore r $= .906$ indicates a
significant (positive) linear correlation. This agrees with the interpretation of the scatter
diagram in part (a).
c. There appears to be no linear correlation, with r close to 0.
$n(\Sigma xy) - (\Sigma x)(\Sigma y) = 9(36) - (18)(18) = 0$
$n(\Sigma x^2) - (\Sigma x)^2 = 9(42) - (18)^2 = 54$
$n(\Sigma y^2) - (\Sigma y)^2 = 9(42) - (18)^2 = 54$
$r = [n(\Sigma xy) - (\Sigma x)(\Sigma y)]/[\sqrt{n(\Sigma x^2) - (\Sigma x)^2} \cdot \sqrt{n(\Sigma y^2) - (\Sigma y)^2}]$
   $= 0/[\sqrt{54} \cdot \sqrt{54}]$
   $= 0$
From Table A-6, assuming $\alpha=.05$, CV $= \pm.666$; therefore r $= 0$ does not indicate a
significant linear correlation. This agrees with the interpretation of the scatter diagram.
d. The effect of a single pair of values can be dramatic, changing the conclusion entirely.

NOTE: In each of exercises 8-14, the first variable listed is designated x, and the second
variable listed is designated y. In correlation problems, the designation of x and y is arbitrary –
so long as a person remains consistent after making the designation. For use in the next section,
the following summary statistics should be saved for each exercise: n, Σx, Σy, Σx^2, Σy^2, Σxy.

9. a. n = 8
Σx = 188.2
Σy = 52.0
Σx^2 = 11850.04
Σy^2 = 402.90
Σxy = 1141.22

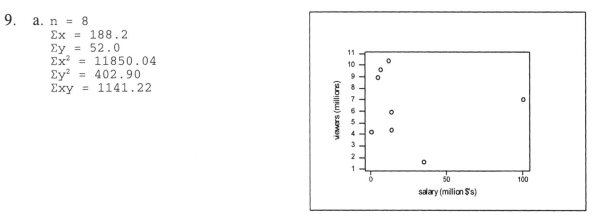

b. $n(\Sigma xy) - (\Sigma x)(\Sigma y) = 8(1141.22) - (188.2)(52.0) = -656.64$
$n(\Sigma x^2) - (\Sigma x)^2 = 8(11850.04) - (188.2)^2 = 59381.08$
$n(\Sigma y^2) - (\Sigma y)^2 = 8(402.90) - (52.0)^2 = 519.20$
$r = [n(\Sigma xy) - (\Sigma x)(\Sigma y)]/[\sqrt{n(\Sigma x^2) - (\Sigma x)^2} \cdot \sqrt{n(\Sigma y^2) - (\Sigma y)^2}]$
   $= -656.64/[\sqrt{59381.08} \cdot \sqrt{519.20}]$
   $= -.118$

c. H_o: $\rho = 0$
   H_1: $\rho \neq 0$
   $\alpha = .05$
   C.R. r < -.707   OR   C.R. t < $-t_{6,.025}$ = -2.447
        r > .707   t > $t_{6,.025}$ = 2.447

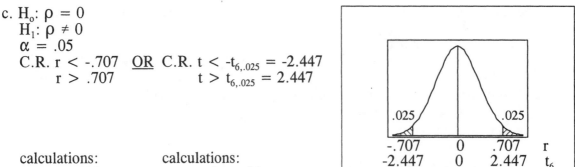

calculations: calculations:
   r = -.118 $t_r = (r - \mu_r)/s_r$
                                 $= (-.118 - 0)/\sqrt{(1-(-.118)^2)/6}$
conclusion: $= -.118/.4054 = -.292$

   Do not reject H_o; there is not sufficient evidence to reject the claim that $\rho = 0$.

No; there is no significant correlation between salary and number of viewers.
Susan Lucci (1/4.2 = $0.24 per viewer) has the lowest cost per viewer, and Kelsey
Grammar (35.2/1.6 = $22.00 per viewer) has the highest cost per viewer.

11. a. n = 14
        Σx = 1875
        Σy = 1241
        Σx^2 = 252179
        Σy^2 = 111459
        Σxy = 167023

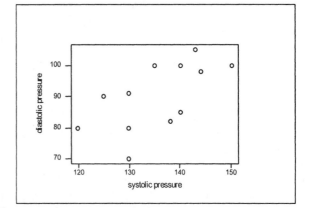

b. $n(\Sigma xy) - (\Sigma x)(\Sigma y) = 14(167023) - (1875)(1241) = 11447$
   $n(\Sigma x^2) - (\Sigma x)^2 = 14(252179) - (1875)^2 = 14881$
   $n(\Sigma y^2) - (\Sigma y)^2 = 14(111459) - (1241)^2 = 20345$
   $r = [n(\Sigma xy) - (\Sigma x)(\Sigma y)]/[\sqrt{n(\Sigma x^2) - (\Sigma x)^2} \cdot \sqrt{n(\Sigma y^2) - (\Sigma y)^2}]$
   $= 11447/[\sqrt{14881} \cdot \sqrt{20345}]$
   $= .658$

c. H_o: $\rho = 0$
   H_1: $\rho \neq 0$
   $\alpha = .05$
   C.R. r < -.532   OR   C.R. t < $-t_{12,.025}$ = -2.179
        r > .532   t > $t_{12,.025}$ = 2.179

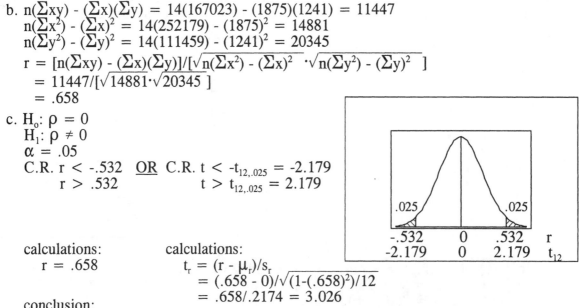

calculations: calculations:
   r = .658 $t_r = (r - \mu_r)/s_r$
                                 $= (.658 - 0)/\sqrt{(1-(.658)^2)/12}$
                                 $= .658/.2174 = 3.026$
conclusion:

   Reject H_o; there is sufficient evidence to reject the claim that $\rho = 0$ and to conclude
   that $\rho \neq 0$ (in fact, $\rho > 0$).

Yes; there is a correlation between systolic and diastolic pressure.
Yes; since the measurements were on the same patient, one could study the variability
in a person's blood pressure – or in the variability of medical student readings?

13.   a. n = 12
      Σx = 175
      Σy = 2102.46
      Σx² = 5155
      Σy² = 601709.8240
      Σxy = 37111.51

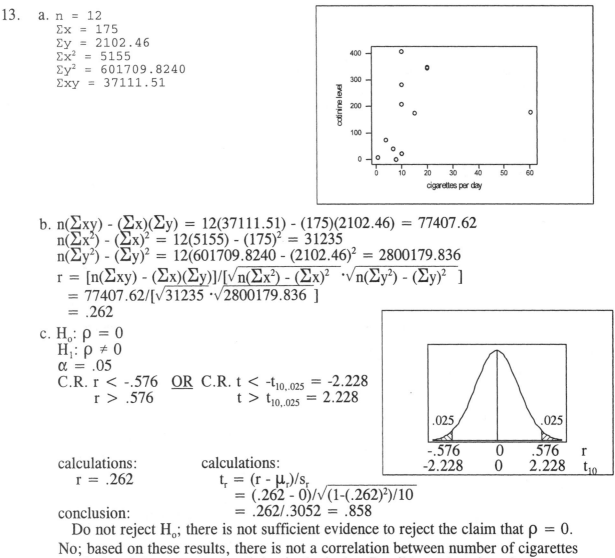

   b. $n(\Sigma xy) - (\Sigma x)(\Sigma y) = 12(37111.51) - (175)(2102.46) = 77407.62$
      $n(\Sigma x^2) - (\Sigma x)^2 = 12(5155) - (175)^2 = 31235$
      $n(\Sigma y^2) - (\Sigma y)^2 = 12(601709.8240 - (2102.46)^2 = 2800179.836$
      $r = [n(\Sigma xy) - (\Sigma x)(\Sigma y)]/[\sqrt{n(\Sigma x^2) - (\Sigma x)^2} \cdot \sqrt{n(\Sigma y^2) - (\Sigma y)^2}\ ]$
      $= 77407.62/[\sqrt{31235} \cdot \sqrt{2800179.836}\ ]$
      $= .262$

   c. $H_o: \rho = 0$
      $H_1: \rho \neq 0$
      $\alpha = .05$
      C.R. $r < -.576$ OR   C.R. $t < -t_{10,.025} = -2.228$
           $r > .576$ $t > t_{10,.025} = 2.228$

   calculations: calculations:
      $r = .262$ $t_r = (r - \mu_r)/s_r$
                                 $= (.262 - 0)/\sqrt{(1-(.262)^2)/10}$
   conclusion: $= .262/.3052 = .858$

   Do not reject H_o; there is not sufficient evidence to reject the claim that $\rho = 0$.

   No; based on these results, there is not a correlation between number of cigarettes
   smoked and the amount of cotinine in the body. Possible explanations for not
   identifying the expected correlation are (1) the subjects may not have accurately self-
   reported their number of cigarettes smoked, (2) the survey did not consider whether or
   not the cigarettes smoked had filters (which would affect the amounts nicotine reaching
   the body), and (3) the results could be affected by exposure to second-hand smoke.

15.   a. n = 16
      Σx = .35
      Σy = 60.2
      Σx² = .0143
      Σy² = 227.24
      Σxy = 1.342

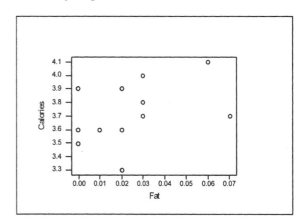

b. $n(\Sigma xy) - (\Sigma x)(\Sigma y) = 16(1.342) - (.35)(60.2) = .402$
$n(\Sigma x^2) - (\Sigma x)^2 = 16(.0143) - (.35)^2 = .1063$
$n(\Sigma y^2) - (\Sigma y)^2 = 16(227.24) - (60.2)^2 = 11.80$
$r = [n(\Sigma xy) - (\Sigma x)(\Sigma y)]/[\sqrt{n(\Sigma x^2) - (\Sigma x)^2} \cdot \sqrt{n(\Sigma y^2) - (\Sigma y)^2}]$
$= .402/[\sqrt{.1063} \cdot \sqrt{11.80}]$
$= .359$

c. H_o: $\rho = 0$
H_1: $\rho \neq 0$
$\alpha = .05$
C.R. $r < -.497$ OR   C.R. $t < -t_{14,.025} = -2.145$
$r > .497$ $t > t_{14,.025} = 2.145$

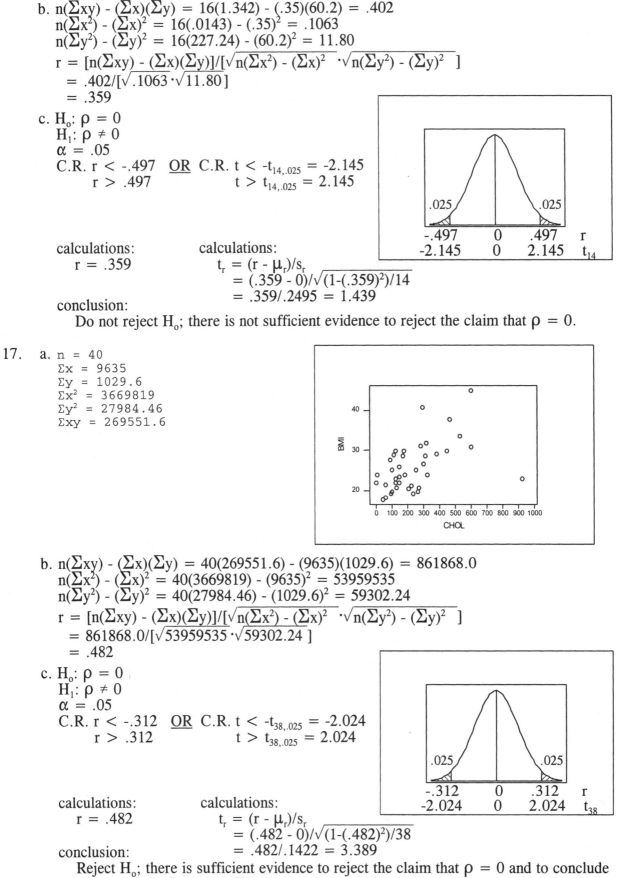

calculations: calculations:
$r = .359$ $t_r = (r - \mu_r)/s_r$
$= (.359 - 0)/\sqrt{(1-(.359)^2)/14}$
$= .359/.2495 = 1.439$
conclusion:
Do not reject H_o; there is not sufficient evidence to reject the claim that $\rho = 0$.

17.   a. n = 40
$\Sigma x = 9635$
$\Sigma y = 1029.6$
$\Sigma x^2 = 3669819$
$\Sigma y^2 = 27984.46$
$\Sigma xy = 269551.6$

b. $n(\Sigma xy) - (\Sigma x)(\Sigma y) = 40(269551.6) - (9635)(1029.6) = 861868.0$
$n(\Sigma x^2) - (\Sigma x)^2 = 40(3669819) - (9635)^2 = 53959535$
$n(\Sigma y^2) - (\Sigma y)^2 = 40(27984.46) - (1029.6)^2 = 59302.24$
$r = [n(\Sigma xy) - (\Sigma x)(\Sigma y)]/[\sqrt{n(\Sigma x^2) - (\Sigma x)^2} \cdot \sqrt{n(\Sigma y^2) - (\Sigma y)^2}]$
$= 861868.0/[\sqrt{53959535} \cdot \sqrt{59302.24}]$
$= .482$

c. H_o: $\rho = 0$
H_1: $\rho \neq 0$
$\alpha = .05$
C.R. $r < -.312$ OR   C.R. $t < -t_{38,.025} = -2.024$
$r > .312$ $t > t_{38,.025} = 2.024$

calculations: calculations:
$r = .482$ $t_r = (r - \mu_r)/s_r$
$= (.482 - 0)/\sqrt{(1-(.482)^2)/38}$
conclusion: $= .482/.1422 = 3.389$
Reject H_o; there is sufficient evidence to reject the claim that $\rho = 0$ and to conclude
that $\rho \neq 0$ (in fact, $\rho > 0$).

19.   A. Selling price-list price relationship?

a. n = 50
   Σx = 8917
   Σy = 8517.0
   Σx² = 1899037
   Σy² = 1710311.50
   Σxy = 1801273.0

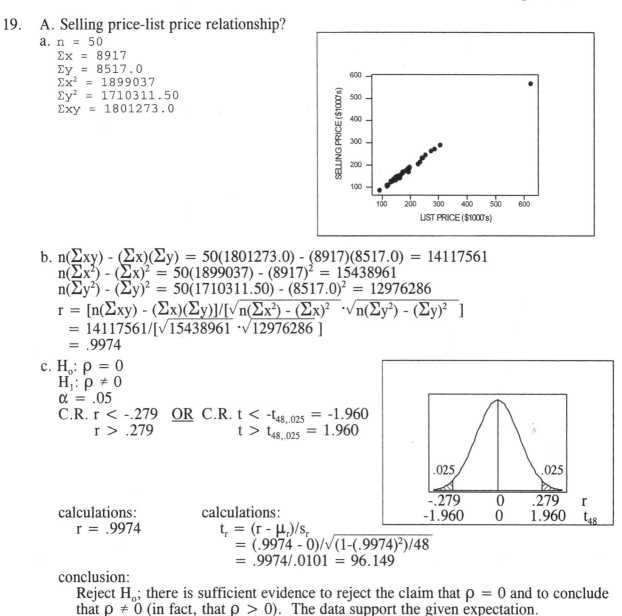

b. n(Σxy) - (Σx)(Σy) = 50(1801273.0) - (8917)(8517.0) = 14117561
   n(Σx²) - (Σx)² = 50(1899037) - (8917)² = 15438961
   n(Σy²) - (Σy)² = 50(1710311.50) - (8517.0)² = 12976286
   r = [n(Σxy) - (Σx)(Σy)]/[√‾n(Σx²) - (Σx)²‾ ·√‾n(Σy²) - (Σy)²‾]
      = 14117561/[√‾15438961‾ ·√‾12976286‾]
      = .9974

c. H₀: ρ = 0
   H₁: ρ ≠ 0
   α = .05
   C.R. r < -.279   OR C.R. t < -t₄₈,.₀₂₅ = -1.960
        r > .279   t > t₄₈,.₀₂₅ = 1.960

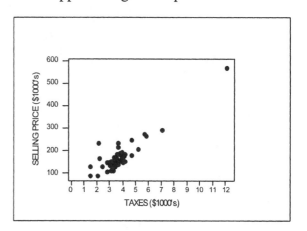

calculations: calculations:
   r = .9974   t_r = (r - μ_r)/s_r
                             = (.9974 - 0)/√‾(1-(.9974)²)/48‾
                             = .9974/.0101 = 96.149

conclusion:
   Reject H₀; there is sufficient evidence to reject the claim that ρ = 0 and to conclude
   that ρ ≠ 0 (in fact, that ρ > 0). The data support the given expectation.

B. Selling price-taxes relationship?

a. n = 50
   Σx = 185305
   Σy = 8517.0
   Σx² = 809069344
   Σy² = 1710311.50
   Σxy = 36631916.5

b. n(Σxy) - (Σx)(Σy) = 50(36631916.5) - (185305)(8517.0) = 253353140
   n(Σx²) - (Σx)² = 50(809069344) - (185305)² = 6115524175
   n(Σy²) - (Σy)² = 50(1710311.50) - (8517.0)² = 12976286

$$r = [n(\Sigma xy) - (\Sigma x)(\Sigma y)]/[\sqrt{n(\Sigma x^2) - (\Sigma x)^2} \cdot \sqrt{n(\Sigma y^2) - (\Sigma y)^2}\ ]$$
$$= 253353140/[\sqrt{6115524175} \cdot \sqrt{12976286}\ ]$$
$$= .8994$$

c. H_o: $\rho = 0$
   H_1: $\rho \neq 0$
   $\alpha = .05$
   C.R. r < -.279   OR   C.R. t < $-t_{48,.025}$ = -1.960
        r > .279   t > $t_{48,.025}$ = 1.960

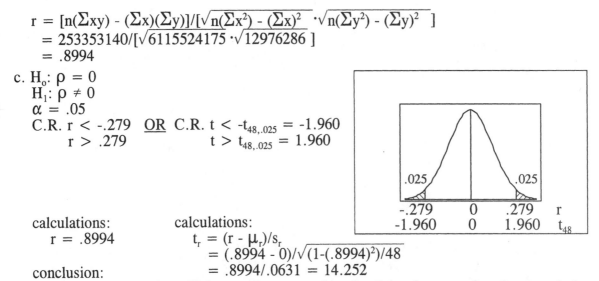

calculations: calculations:
   r = .8994 $t_r = (r - \mu_r)/s_r$
                                 $= (.8994 - 0)/\sqrt{(1-(.8994)^2)/48}$
conclusion: $= .8994/.0631 = 14.252$

Reject H_o; there is sufficient evidence to reject the claim that $\rho = 0$ and to conclude that $\rho \neq 0$ (in fact, that $\rho > 0$).

Yes; the tax bill appears to be based on the value of the house – since the selling price is assumed to represent the true value of the house, and there is a strong correlation between selling price and the amount of the taxes. The tax assessments appear to be valid.

21.   A. Actual-fiveday relationship?
   a. n = 31
      Σx = 1069
      Σy = 1082
      Σx^2 = 37381
      Σy^2 = 38352
      Σxy = 37628

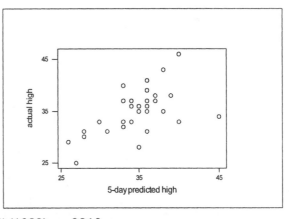

   b. $n(\Sigma xy) - (\Sigma x)(\Sigma y) = 31(37628) - (1069)(1082) = 9810$
      $n(\Sigma x^2) - (\Sigma x)^2 = 31(37381) - (1069)^2 = 16050$
      $n(\Sigma y^2) - (\Sigma y)^2 = 31(38352) - (1082)^2 = 18188$
      $$r = [n(\Sigma xy) - (\Sigma x)(\Sigma y)]/[\sqrt{n(\Sigma x^2) - (\Sigma x)^2} \cdot \sqrt{n(\Sigma y^2) - (\Sigma y)^2}\ ]$$
      $$= 9810/[\sqrt{16050} \cdot \sqrt{18188}\ ]$$
      $$= .574$$

   c. H_o: $\rho = 0$
      H_1: $\rho \neq 0$
      $\alpha = .05$
      C.R. r < -.361   OR   C.R. t < $-t_{29,.025}$ = -2.045
           r > .361   t > $t_{29,.025}$ = 2.045

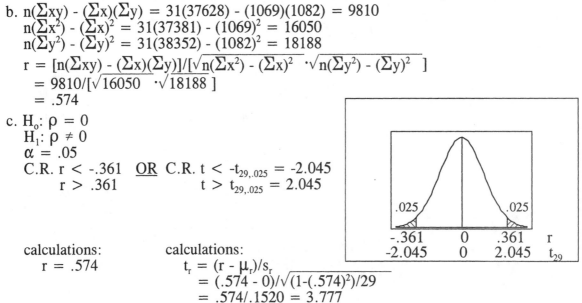

calculations: calculations:
   r = .574 $t_r = (r - \mu_r)/s_r$
                                 $= (.574 - 0)/\sqrt{(1-(.574)^2)/29}$
                                 $= .574/.1520 = 3.777$

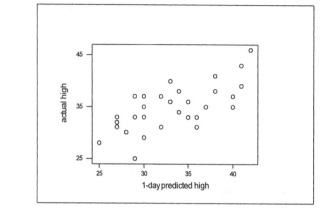

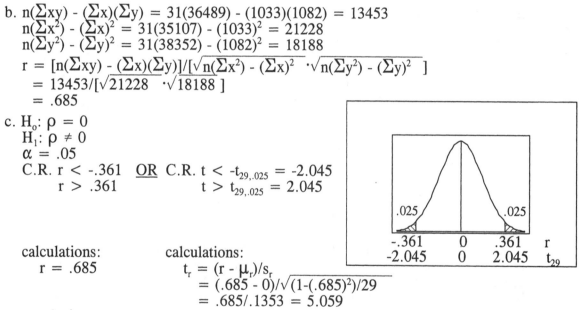

conclusion:
> Reject H_o; there is sufficient evidence to reject the claim that $\rho = 0$ and to conclude that $\rho \neq 0$ (in fact, that $\rho > 0$).

Yes; there is a linear correlation between the 5-day forecast and the actual temperature. No; a correlation means only that there is a relationship between the values, not that they agree with each other. If the 5-day forecasts were always exactly 20 degrees too cold, for example, there would be a perfected correlation but not agreement.

B. Actual-oneday relationship?

a. n = 31
Σx = 1033
Σy = 1082
Σx² = 35107
Σy² = 38352
Σxy = 36489

b. $n(\Sigma xy) - (\Sigma x)(\Sigma y) = 31(36489) - (1033)(1082) = 13453$
$n(\Sigma x^2) - (\Sigma x)^2 = 31(35107) - (1033)^2 = 21228$
$n(\Sigma y^2) - (\Sigma y)^2 = 31(38352) - (1082)^2 = 18188$
$r = [n(\Sigma xy) - (\Sigma x)(\Sigma y)]/[\sqrt{n(\Sigma x^2) - (\Sigma x)^2} \cdot \sqrt{n(\Sigma y^2) - (\Sigma y)^2}]$
$= 13453/[\sqrt{21228} \cdot \sqrt{18188}]$
$= .685$

c. $H_o: \rho = 0$
$H_1: \rho \neq 0$
$\alpha = .05$
C.R. r < -.361   OR   C.R. t < $-t_{29,.025}$ = -2.045
    r > .361   t > $t_{29,.025}$ = 2.045

calculations: calculations:
r = .685 $t_r = (r - \mu_r)/s_r$
                       $= (.685 - 0)/\sqrt{(1-(.685)^2)/29}$
                       $= .685/.1353 = 5.059$

conclusion:
> Reject H_o; there is sufficient evidence to reject the claim that $\rho = 0$ and to conclude that $\rho \neq 0$ (in fact, that $\rho > 0$).

Yes; there is a linear correlation between the 1-day forecast and the actual temperature. No; a correlation means only that there is a relationship between the values, not that they agree with each other. If the 1-day forecasts were always exactly 20 degrees too cold, for example, there would be a perfected correlation but not agreement.

C. One would expect the 1-day forecasts to have a higher correlation with the actual temperatures than the 5-day forecasts would, and the results support this expectation. But even a very high correlation means only that there is a relationship between the predicted and the actual temperatures, not that they agree with each other. If the predictions were always exactly 20 degrees too cold, for example, there would be a perfected correlation but not agreement.

23.   A. Interval-duration relationship?

    a. $n = 50$
       $\Sigma x = 10832$
       $\Sigma y = 4033$
       $\Sigma x^2 = 2513280$
       $\Sigma y^2 = 332331$
       $\Sigma xy = 903488$

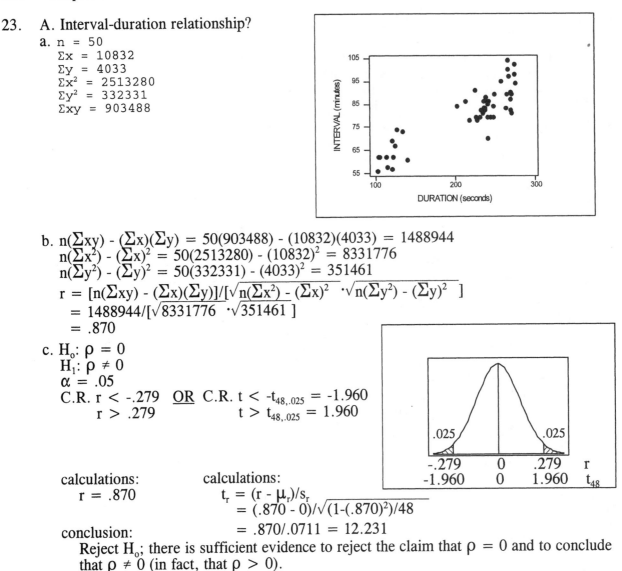

    b. $n(\Sigma xy) - (\Sigma x)(\Sigma y) = 50(903488) - (10832)(4033) = 1488944$
      $n(\Sigma x^2) - (\Sigma x)^2 = 50(2513280) - (10832)^2 = 8331776$
      $n(\Sigma y^2) - (\Sigma y)^2 = 50(332331) - (4033)^2 = 351461$

      $r = [n(\Sigma xy) - (\Sigma x)(\Sigma y)]/[\sqrt{n(\Sigma x^2) - (\Sigma x)^2} \cdot \sqrt{n(\Sigma y^2) - (\Sigma y)^2}\,]$
        $= 1488944/[\sqrt{8331776} \cdot \sqrt{351461}\,]$
        $= .870$

    c. H_o: $\rho = 0$
      H_1: $\rho \neq 0$
      $\alpha = .05$
      C.R. $r < -.279$ <u>OR</u> C.R. $t < -t_{48,.025} = -1.960$
           $r > .279$ $t > t_{48,.025} = 1.960$

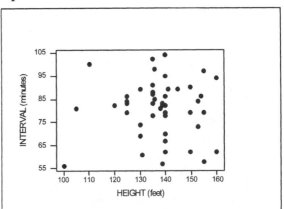

calculations: calculations:
  $r = .870$ $t_r = (r - \mu_r)/s_r$
                    $= (.870 - 0)/\sqrt{(1-(.870)^2)/48}$
conclusion: $= .870/.0711 = 12.231$

    Reject H_o; there is sufficient evidence to reject the claim that $\rho = 0$ and to conclude
    that $\rho \neq 0$ (in fact, that $\rho > 0$).
Yes; there is a significant positive linear correlation, suggesting that the interval after an
eruption is related to the duration of the eruption. NOTE: The longer the duration of an
eruption, the more pressure has been released and the longer it will take the geyser to
build back up for another eruption. In fact, the park rangers use the duration of one
eruption to predict the time of the next eruption.

    B. Interval-height relationship?

    a. $n = 50$
       $\Sigma x = 6904$
       $\Sigma y = 4033$
       $\Sigma x^2 = 961150$
       $\Sigma y^2 = 332331$
       $\Sigma xy = 556804$

b. $n(\Sigma xy) - (\Sigma x)(\Sigma y) = 50(556804) - (6904)(4033) = -3632$
$n(\Sigma x^2) - (\Sigma x)^2 = 50(961150) - (6904)^2 = 392284$
$n(\Sigma y^2) - (\Sigma y)^2 = 50(332331) - (4033)^2 = 351461$
$r = [n(\Sigma xy) - (\Sigma x)(\Sigma y)]/[\sqrt{n(\Sigma x^2) - (\Sigma x)^2} \cdot \sqrt{n(\Sigma y^2) - (\Sigma y)^2}\,]$
$= -3632/[\sqrt{392284} \cdot \sqrt{351461}\,]$
$= -.00978$

c. $H_o: \rho = 0$
$H_1: \rho \neq 0$
$\alpha = .05$
C.R. $r < -.279$ OR C.R. $t < -t_{48,.025} = -1.960$
$r > .279$ $t > t_{48,.025} = 1.960$

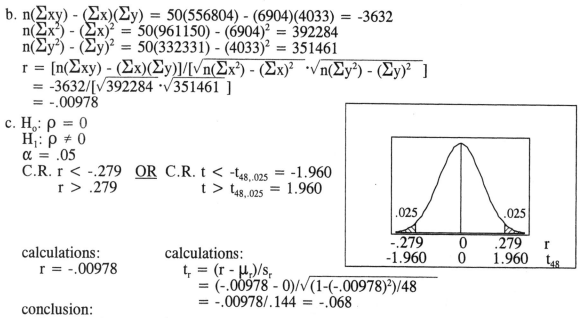

.025   .025
-.279   0   .279   r
-1.960   0   1.960 t_{48}

calculations: calculations:
$r = -.00978$ $t_r = (r - \mu_r)/s_r$
$= (-.00978 - 0)/\sqrt{(1-(-.00978)^2)/48}$
$= -.00978/.144 = -.068$

conclusion:
Do not reject H_o; there is not sufficient evidence to reject the claim that $\rho = 0$.

No; there is not a significant linear correlation, suggesting that the interval after an eruption is not so related to the height of the eruption.

C. Duration is the more relevant predictor of the interval until the next eruption -- because it has a significant correlation with interval, while height does not.

25. A linear correlation coefficient very close to zero indicates <u>no</u> significant linear correlation and no tendencies can be inferred.

27. A linear correlation coefficient very close to zero indicates no significant <u>linear</u> correlation, but there may some other type of relationship between the variables.

29. There are n = 13 data points.
From the scatterplot, it appears that the correlation is about +.75.
$H_o: \rho = 0$
$H_1: \rho \neq 0$
$\alpha = .05$
C.R. $r < -.553$ OR   C.R. $t < -t_{11,.025} = -2.201$
$r > .553$ $t > t_{11,.025} = 2.201$

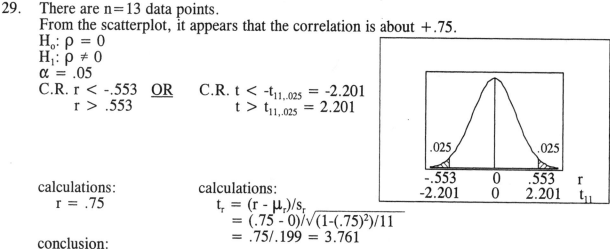

.025   .025
-.553   0   .553   r
-2.201   0   2.201 t_{11}

calculations: calculations:
$r = .75$ $t_r = (r - \mu_r)/s_r$
$= (.75 - 0)/\sqrt{(1-(.75)^2)/11}$
$= .75/.199 = 3.761$

conclusion:
Reject H_o; there is sufficient evidence to reject the claim that $\rho = 0$ and to conclude that $\rho \neq 0$ (in fact, that $\rho > 0$).

Yes; based on this result there is a significant positive linear correlation between the mortality rate and the rate of infections acquired in intensive care units.

31. a. For $\pm t_{48,.025} = \pm 1.960$,
        the critical values are $r = \pm 1.960/\sqrt{(\pm 1.960)^2 + 48} = \pm .272$.

     b. For $\pm t_{73,.05} = \pm 1.645$,
        the critical values are $r = \pm 1.645/\sqrt{(\pm 1.645)^2 + 73} = \pm .189$.

     c. For $-t_{18,.05} = -1.734$,
        the critical value is $r = -1.734/\sqrt{(-1.734)^2 + 18} = -.378$.

     d. For $t_{8,.05} = 1.860$,
        the critical value is $r = 1.860/\sqrt{(1.860)^2 + 8} = .549$.

     e. For $t_{10,.01} = 2.764$,
        the critical value is $r = 2.764/\sqrt{(2.764)^2 + 10} = .658$.

33. For $r = .600$, $(1+r)/(1-r) = (1.6)/(.4) = 4$. Following the procedure outlined in the text,

     Step a. $z_{.025} = 1.960$

     Step b. $w_L = \frac{1}{2} \cdot \ln(4) - 1.960/\sqrt{47} = .407$
             $w_R = \frac{1}{2} \cdot \ln(4) + 1.960/\sqrt{47} = .979$

     Step c. $(e^{.814} - 1)/(e^{.814} + 1) < \rho < (e^{1.958} - 1)/(e^{1.958} + 1)$
             $(1.258)/(3.258) < \rho < (6.086)/(8.086)$
             $.386 < \rho < .753$

NOTE: While the distribution of r is not normal, $\tanh^{-1}(r)$ [i.e., the inverse hyperbolic tangent of r] follows a normal distribution with $\mu_r = \tanh^{-1}(\rho)$ and $\sigma_r = 1/\sqrt{n-3}$. The steps above are equivalent to finding $\tanh^{-1}(r)$, constructing a confidence interval for $\tanh^{-1}(\rho)$, and then applying the hyperbolic tangent function to the endpoints to produce a confidence interval for ρ.

9-3 Regression

1. a. For n=20, CV=$\pm .444$; since r=.987 > .444, use the regression line for predictions.
        $\hat{y} = 6.00 + 4.00x$
        $\hat{y}_{3.00} = 6.00 + 4.00(3.00) = 18.00$
     b. For n=20, CV=$\pm .444$; since r=.052 < .444, use the mean for predictions.
        $\hat{y} = \bar{y}$
        $\hat{y}_{3.00} = \bar{y} = 5.00$

3. For n=8, CV=$\pm .707$; since r=.993 > .707, use the regression line for predictions.
        $\hat{y} = -187 + 11.3x$
        $\hat{y}_{52} = -187 + 11.3(52) = 401$

NOTE: For exercises 5-24, the exact summary statistics (i.e., without any rounding) are given on the right. While the intermediate calculations on the left are presented rounded to various degrees of accuracy, the entire unrounded values were preserved in the calculator until the end.

5. $\bar{x} = 2.00$ $n = 5$
    $\bar{y} = 2.00$ $\Sigma x = 10$
    $b_1 = [n(\Sigma xy) - (\Sigma x)(\Sigma y)]/[n(\Sigma x^2) - (\Sigma x)^2]$ $\Sigma y = 10$
        $= [5(20) - (10)(10)]/[5(30) - (10)^2]$ $\Sigma x^2 = 30$
        $= 0/50$ $\Sigma y^2 = 344$
        $= 0.00$ $\Sigma xy = 20$
    $b_0 = \bar{y} - b_1\bar{x}$
        $= 2.00 - (0.00)(2.00) = 2.00$
    $\hat{y} = b_0 + b_1x$
        $= 2.00 + 0.00x$

7. a. $\bar{x} = 2.80$
   $\bar{y} = 2.80$
   $b_1 = [n(\Sigma xy) - (\Sigma x)(\Sigma y)]/[n(\Sigma x^2) - (\Sigma x)^2]$
   $= 576/636$
   $= .906$
   $b_0 = \bar{y} - b_1\bar{x}$
   $= 2.80 - .906(2.80) = .264$
   $\hat{y} = b_0 + b_1 x$
   $= .264 + .906x$

   $n = 10$
   $\Sigma x = 28$
   $\Sigma y = 28$
   $\Sigma x^2 = 142$
   $\Sigma y^2 = 142$
   $\Sigma xy = 136$

   b. $\bar{x} = 2.00$
   $\bar{y} = 2.00$
   $b_1 = [n(\Sigma xy) - (\Sigma x)(\Sigma y)]/[n(\Sigma x^2) - (\Sigma x)^2]$
   $= 0/54$
   $= .000$
   $b_0 = \bar{y} - b_1\bar{x}$
   $= 2.00 - .000(2.00) = 2.00$
   $\hat{y} = b_0 + b_1 x$
   $= 2.00 + 0.00x$

   $n = 9$
   $\Sigma x = 18$
   $\Sigma y = 18$
   $\Sigma x^2 = 42$
   $\Sigma y^2 = 42$
   $\Sigma xy = 136$

   c. The two regression lines are very different, illustrating that one point can affect the regression dramatically.

NOTE: In the exercises that follow, this manual uses the full accuracy of b_0 and b_1 when calculating $\hat{y}$. Using only the rounded values as stated in the equation produces slightly different answers. More detail for calculating the numerator and denominator of b_1 is given in the corresponding exercise in the previous section.

9. $\bar{x} = 23.525$
   $\bar{y} = 6.500$
   $b_1 = [n(\Sigma xy) - (\Sigma x)(\Sigma y)]/[n(\Sigma x^2) - (\Sigma x)^2]$
   $= -656.64/59381.08 = -.0111$
   $b_0 = \bar{y} - b_1\bar{x}$
   $= 6.5000 - (-.0111)(23.525) = 6.76$
   $\hat{y} = b_0 + b_1 x$
   $= 6.76 - .0111x$
   $\hat{y}_{16} = \bar{y} = 6.5$ [no significant correlation]
   In this instance, the predicted value is far from actual value.

   $n = 8$
   $\Sigma x = 188.2$
   $\Sigma y = 52.0$
   $\Sigma x^2 = 11850.04$
   $\Sigma y^2 = 402.90$
   $\Sigma xy = 1141.22$

11. $\bar{x} = 133.93$
    $\bar{y} = 88.64$
    $b_1 = [n(\Sigma xy) - (\Sigma x)(\Sigma y)]/[n(\Sigma x^2) - (\Sigma x)^2]$
    $= 11447/14881 = .769$
    $b_0 = \bar{y} - b_1\bar{x}$
    $= 88.64 - (.769)(133.93) = -14.4$
    $\hat{y} = b_0 + b_1 x$
    $= -14.4 + .769x$
    $\hat{y}_{122} = -14.4 + .769(112) = 79$

    $n = 14$
    $\Sigma x = 1875$
    $\Sigma y = 1241$
    $\Sigma x^2 = 252179$
    $\Sigma y^2 = 111459$
    $\Sigma xy = 167023$

13. $\bar{x} = 14.583$
    $\bar{y} = 175.205$
    $b_1 = [n(\Sigma xy) - (\Sigma x)(\Sigma y)]/[n(\Sigma x^2) - (\Sigma x)^2]$
    $= 77407.62/31235 = 2.478$
    $b_0 = \bar{y} - b_1\bar{x}$
    $= 175.205 - (2.478)(14.583) = 139.1$
    $\hat{y} = b_0 + b_1 x$
    $= 139.1 + 2.478x$
    $\hat{y}_{40} = \bar{y} = 175.2$ [no significant correlation]

    $n = 12$
    $\Sigma x = 175$
    $\Sigma y = 2102.46$
    $\Sigma x^2 = 5155$
    $\Sigma y^2 = 601709.8240$
    $\Sigma xy = 37111.51$

15. $\bar{x} = .0219$
$\bar{y} = 3.7625$
$b_1 = [n(\Sigma xy) - (\Sigma x)(\Sigma y)]/[n(\Sigma x^2) - (\Sigma x)^2]$
$\quad = .402/.1063 = 3.78$
$b_0 = \bar{y} - b_1\bar{x}$
$\quad = 3.7625 - (3.78)(.0219) = 3.68$
$\hat{y} = b_0 + b_1x$
$\quad = 3.68 + 3.78x$
$\hat{y}_{.05} = \bar{y} = 3.76$ [no significant correlation]

$n = 16$
$\Sigma x = .35$
$\Sigma y = 60.2$
$\Sigma x^2 = .0143$
$\Sigma y^2 = 227.24$
$\Sigma xy = 1.342$

17. $\bar{x} = 240.875$
$\bar{y} = 25.740$
$b_1 = [n(\Sigma xy) - (\Sigma x)(\Sigma y)]/[n(\Sigma x^2) - (\Sigma x)^2]$
$\quad = 861868.8/53959535 = .0160$
$b_0 = \bar{y} - b_1\bar{x}$
$\quad = 25.740 - (.0160)(240.875) = 21.9$
$\hat{y} = b_0 + b_1x$
$\quad = 21.9 + .0160x$
$\hat{y}_{500} = 21.9 + .0160(500) = 29.9$

$n = 40$
$\Sigma x = 9635$
$\Sigma y = 1029.6$
$\Sigma x^2 = 3669819$
$\Sigma y^2 = 27984.46$
$\Sigma xy = 269551.6$

19. a. selling price/asking price relationship
$\bar{x} = 178.34$
$\bar{y} = 170.34$
$b_1 = [n(\Sigma xy) - (\Sigma x)(\Sigma y)]/[n(\Sigma x^2) - (\Sigma x)^2]$
$\quad = 14117561/15438961 = .914$
$b_0 = \bar{y} - b_1\bar{x}$
$\quad = 170.34 - (.914)(178.34) = 7.26$
$\hat{y} = b_0 + b_1x$
$\quad = 7.26 + .914x$
$\hat{y}_{200} = 7.26 + .914(200) = 190.1$ [\$190,100]

$n = 50$
$\Sigma x = 8917$
$\Sigma y = 8517.0$
$\Sigma x^2 = 1899037$
$\Sigma y^2 = 1710311.50$
$\Sigma xy = 1801273$

b. NOTE: These x-y values are reversed from what they were in the corresponding exercise in section 9.2.
selling price/taxes relationship
$\bar{x} = 3706.10$
$\bar{y} = 170.34$
$b_1 = [n(\Sigma xy) - (\Sigma x)(\Sigma y)]/[n(\Sigma x^2) - (\Sigma x)^2]$
$\quad = 253353140/12976286 = 19.5$
$b_0 = \bar{y} - b_1\bar{x}$
$\quad = 3706.10 - (19.5)(170.34) = 380.3$
$\hat{y} = b_0 + b_1x$
$\quad = 380.3 + 19.5x$
$\hat{y}_{400} = 380.3 + 19.5(400) = 8190$

$n = 50$
$\Sigma x = 8517.0$
$\Sigma y = 185305$
$\Sigma x^2 = 1710311.50$
$\Sigma y^2 = 809069344$
$\Sigma xy = 36631916.5$

21. a. actual/five-day relationship
$\bar{x} = 34.48$
$\bar{y} = 34.90$
$b_1 = [n(\Sigma xy) - (\Sigma x)(\Sigma y)]/[n(\Sigma x^2) - (\Sigma x)^2]$
$\quad = 9810/16050 = .611$
$b_0 = \bar{y} - b_1\bar{x}$
$\quad = 34.90 - (.611)(34.48) = 13.8$
$\hat{y} = b_0 + b_1x$
$\quad = 13.8 + .611x$
$\hat{y}_{28} = 13.8 + .611(28) = 31$

$n = 31$
$\Sigma x = 1069$
$\Sigma y = 1082$
$\Sigma x^2 = 37381$
$\Sigma y^2 = 38352$
$\Sigma xy = 37628$

b. actual/one-day relationship

$\bar{x} = 33.32$ $n = 31$

$\bar{y} = 34.90$ $\Sigma x = 1033$

$b_1 = [n(\Sigma xy) - (\Sigma x)(\Sigma y)]/[n(\Sigma x^2) - (\Sigma x)^2]$ $\Sigma y = 1082$

$\quad = 13453/21228 = .634$ $\Sigma x^2 = 35107$

$b_o = \bar{y} - b_1\bar{x}$ $\Sigma y^2 = 38352$

$\quad = 34.90 - (.634)(33.32) = 13.8$ $\Sigma xy = 36489$

$\hat{y} = b_o + b_1 x$

$\quad = 13.8 + .634x$

$\hat{y}_{28} = 13.8 + .634(28) = 32$

c. The predicted value in part (b) is better, because the correlation was higher.

23. a. interval/duration relationship

$\bar{x} = 216.64$ $n = 50$

$\bar{y} = 80.66$ $\Sigma x = 10832$

$b_1 = [n(\Sigma xy) - (\Sigma x)(\Sigma y)]/[n(\Sigma x^2) - (\Sigma x)^2]$ $\Sigma y = 4033$

$\quad = 1488944/8331776 = .179$ $\Sigma x^2 = 2513280$

$b_o = \bar{y} - b_1\bar{x}$ $\Sigma y^2 = 332331$

$\quad = 80.66 - (.179)(216.64) = 41.9$ $\Sigma xy = 903488$

$\hat{y} = b_o + b_1 x$

$\quad = 41.9 + .179x$

$\hat{y}_{210} = 41.9 + .179(210) = 79.5$ minutes

b. interval/height relationship

$\bar{x} = 138.08$ $n = 50$

$\bar{y} = 80.66$ $\Sigma x = 6904$

$b_1 = [n(\Sigma xy) - (\Sigma x)(\Sigma y)]/[n(\Sigma x^2) - (\Sigma x)^2]$ $\Sigma y = 4033$

$\quad = -3632/392284 = -.00926$ $\Sigma x^2 = 961150$

$b_o = \bar{y} - b_1\bar{x}$ $\Sigma y^2 = 332331$

$\quad = 80.66 - (-.00926)(138.08) = 81.9$ $\Sigma xy = 556804$

$\hat{y} = b_o + b_1 x$

$\quad = 81.9 - .00926x$

$\hat{y}_{275} = \bar{y} = 80.7$ minutes [no significant correlation]

c. The predicted time in part (a) is better, since interval and duration are significantly correlated. Since interval and height are not significantly correlated, the predicted time in part (b) did not even use the height data.

25. Yes; the point is an outlier, since it is far from the other data points – 120 is far from the other boat values, and 160 is far from the other manatee death values.

No; the point is not an influential one, since it will not change the regression line by very much – the original regression line predicts $\hat{y} = -113 + 2.27(120) = 159.4$, and so the new point is consistent with the others.

27. original data   original data divided by 1000

$n=5$ $n = 5$

$\Sigma x = 4,234,178$ $\Sigma x = 4,234.178$

$\Sigma y = 576$ $\Sigma y = 576$

$\Sigma x^2 = 3,595,324,583,102$ $\Sigma x^2 = 3,595,324.583102$

$\Sigma y^2 = 67552$ $\Sigma y^2 = 67552$

$\Sigma xy = 491,173,342$ $\Sigma xy = 491,173.342$

<u>original data</u>

$\overline{x} = 846835.6$

$\overline{y} = 115.2$

$n\Sigma xy - (\Sigma x)(\Sigma y) = 16,980,182$

$n\Sigma x^2 - (\Sigma x)^2 = 48,459,579,826$

$b_1 = 16,980,182/48,459,579,826$

$\quad = .0003504$

$b_0 = \overline{y} - b_1\overline{x}$

$\quad = 115.2 - .0003504(846835.6)$

$\quad = -181.53$

$\hat{y} = b_0 + b_1 x$

$\quad = -181.53 + .0003504x$

<u>original data divided by 1000</u>

$\overline{x} = 846.8356$

$\overline{y} = 115.2$

$n\Sigma xy - (\Sigma x)(\Sigma y) = 16,980.182$

$n\Sigma x^2 - (\Sigma x)^2 = 48,459.579826$

$b_1 = 16,980.182/48,459.579826$

$\quad = .3504$

$b_0 = \overline{y} - b_1\overline{x}$

$\quad = 115.2 - .3504(846.8356)$

$\quad = -181.53$

$\hat{y} = b_0 + b_1 x$

$\quad = -181.53 + .3504x$

Dividing each x by 1000 multiplies b_1, the coefficient of x in the regression equation, by 1000; multiplying the x coefficient by 1000 and dividing x by 1000 will "cancel out" and all predictions remain the same.

Dividing each y by 1000 divides both b_1 and b_0 by 1000; consistent with the new "units" for y, all predictions will also turn out divided by 1000.

29. •original data

x	y
2.0	12.0
2.5	18.7
4.2	53.0
10.0	225.0

$n = 4$

$\Sigma x = 18.7$

$\Sigma y = 308.7$

$\Sigma x^2 = 127.89$

$\Sigma y^2 = 53927.69$

$\Sigma xy = 2543.35$

$b_1 = [n(\Sigma xy) - (\Sigma x)(\Sigma y)]/[n(\Sigma x^2) - (\Sigma x)^2] = 4400.71/161.87 = 27.2$

$b_0 = \overline{y} - b_1\overline{x} = (308.7/4) - (27.2)(18.7/4) = -49.9$

$\hat{y} = b_0 + b_1 x$

$\quad = -49.9 + 27.2x$

$r = [n(\Sigma xy) - (\Sigma x)(\Sigma y)]/[\sqrt{n(\Sigma x^2) - (\Sigma x)^2} \cdot \sqrt{n(\Sigma y^2) - (\Sigma y)^2}]$

$\quad = 4400.71/[\sqrt{161.87} \cdot \sqrt{120415.07}]$

$\quad = .9968$

•using ln(x) for x

x	y
.693	12.0
.916	18.7
1.435	53.0
2.303	225.0

$n = 4$

$\Sigma x = 5.347$

$\Sigma y = 308.7$

$\Sigma x^2 = 8.681$

$\Sigma y^2 = 53927.69$

$\Sigma xy = 619.594$

$b_1 = [n(\Sigma xy) - (\Sigma x)(\Sigma y)]/[n(\Sigma x^2) - (\Sigma x)^2] = 827.7220/6.134071 = 134.9$

$b_0 = \overline{y} - b_1\overline{x} = (308.7/4) - (134.9)(5.347/4) = 103.2$

$\hat{y} = b_0 + b_1 \cdot ln(x)$

$\quad = -103.2 + 134.9 \cdot ln(x)$ [since the "x" is really ln(x)]

$r = [n(\Sigma xy) - (\Sigma x)(\Sigma y)]/[\sqrt{n(\Sigma x^2) - (\Sigma x)^2} \cdot \sqrt{n(\Sigma y^2) - (\Sigma y)^2}]$

$\quad = 827.7220/[\sqrt{6.134071} \cdot \sqrt{120415.07}]$

$\quad = .9631$

Based on the value of the associated correlations (.9968 > .9631), the equation using the original data seems to fit the data better than the equation using ln(x) instead of x.

NOTE: Both x and y (perhaps, especially y) seem to grow exponentially. A wiser choice for a transformation might be to use both ln(x) for x and ln(y) for y [or, perhaps, only ln(y) for y].

31. Refer to the table below at the left, where $\hat{y}=-112.71+2.2741x$ and the residual is $y-\hat{y}$. The residual plot is given below at the right, with x on the horizontal axis and the residuals on the vertical axis.

year	x	y	$\hat{y}$	residual
1991	68	53	41.9	11.1
1992	68	38	41.9	-3.9
1993	67	35	39.7	-4.7
1994	70	49	46.5	2.5
1995	71	42	48.8	-6.8
1996	73	60	53.3	6.7
1997	76	54	60.1	-6.1
1998	81	67	71.5	-4.5
1999	83	82	76.0	6.0
2000	84	78	78.3	-0.3
	741	558	558.0	0.0

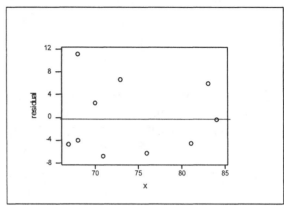

9-4 Variation and Prediction Intervals

1. The coefficient of determination is $r^2 = (.8)^2 = .64$.
   The portion of the total variation explained by the regression line is $r^2 = .64 = 64\%$

3. The coefficient of determination is $r^2 = (-.503)^2 = .253$.
   The portion of the total variation explained by the regression line is $r^2 = .253 = 25.3\%$.

5. Since $r^2 = .924$, $r = +.961$ (positive because $b_1 = 12.5$ is positive).
   From Table A-6, the critical values necessary for significance are $\pm.361$.
   Since $.961 > .361$, we conclude there is a significant positive linear correlation between the tar and nicotine contents of cigarettes.

7. Use 1.2599, from the "fit" value on the Minitab display. This also agrees with the regression equation: $\hat{y} = .154 + .0651x$
   $$\hat{y}_{17} = .154 + .0651(17) = 1.26$$

NOTE: The following summary statistics apply to exercises 9-12 and 13-16. They may be used in the chapter formulas to calculate any values needed in the process of working the problems.

exercise 9	exercise 10	exercise 11	exercise 12
n = 9	n = 14	n = 14	n = 12
$\Sigma x = 632.0$	$\Sigma x = 1875$	$\Sigma x = 66.4$	$\Sigma x = 51.4$
$\Sigma y = 1089$	$\Sigma y = 1241$	$\Sigma y = 669.1$	$\Sigma y = 969.0$
$\Sigma x^2 = 44399.50$	$\Sigma x^2 = 252179$	$\Sigma x^2 = 444.54$	$\Sigma x^2 = 220.42$
$\Sigma y^2 = 132223$	$\Sigma y^2 = 111459$	$\Sigma y^2 = 37365.71$	$\Sigma y^2 = 78487.82$
$\Sigma xy = 76546.0$	$\Sigma xy = 167023$	$\Sigma xy = 3865.67$	$\Sigma xy = 4144.61$
see also 9.2-3, #10	see also 9.2-3,#11	see also 9.2-3,#14	not used previously

9. The predicted values were calculated using the regression line $\hat{y} = -151.6977 + 3.883382x$.

x	y	$\hat{y}$	$\overline{y}$	$\hat{y}-\overline{y}$	$(\hat{y}-\overline{y})^2$	$y-\hat{y}$	$(y-\hat{y})^2$	$y-\overline{y}$	$(y-\overline{y})^2$
71.0	125	124.02	121	3.0204	9.123	0.97958	0.960	4	16
70.5	119	122.08	121	1.0787	1.164	-3.07874	9.479	-2	4
71.0	128	124.02	121	3.0204	9.123	3.97958	15.837	7	49
72.0	128	127.90	121	6.9038	47.663	0.09619	0.009	7	49
70.0	119	120.14	121	-0.8630	0.745	-1.13702	1.293	-2	4
70.0	127	120.14	121	-0.8630	0.745	6.86298	47.100	6	36
66.5	105	106.55	121	-14.4548	208.941	-1.54520	2.388	-16	256
70.0	123	120.14	121	-0.8630	0.745	2.86298	8.197	2	4
71.0	115	124.02	121	3.0204	9.123	-9.02042	81.368	-6	36
632.0	1089	1089.00	1089	0.0001	287.371	-0.00006	166.630	0	454

a. The explained variation is $\Sigma(\hat{y}-\bar{y})^2 = 287.37$
b. The unexplained variation is $\Sigma(y-\hat{y})^2 = 166.63$
c. The total variation is $\Sigma(y-\bar{y})^2 = 454$
d. $r^2 = \Sigma(\hat{y}-\bar{y})^2/\Sigma(y-\bar{y})^2 = 287.37/454 = .6330$
e. $s_e^2 = \Sigma(y-\hat{y})^2/(n-2) = 166.63/7 = 23.8043$
   $s_e = 4.8790$

NOTE: A table such as the one in the preceding problem organizes the work and provides all the values needed to discuss variation. In such a table, the following must always be true (except for minor discrepancies due to rounding) and can be used as a check before proceeding.
* $\Sigma y = \Sigma\hat{y} = \Sigma\bar{y}$ * $\Sigma(\hat{y}-\bar{y}) = \Sigma(y-\hat{y}) = \Sigma(y-\bar{y}) = 0$ * $\Sigma(y-\bar{y})^2 + \Sigma(y-\hat{y})^2 = \Sigma(y-\bar{y})^2$

11. The predicted values were calculated using the regression line $\hat{y} = 22.46293 + 5.340648x$.

x	y	$\hat{y}$	$\bar{y}$	$\hat{y}-\bar{y}$	$(\hat{y}-\bar{y})^2$	$y-\hat{y}$	$(y-\hat{y})^2$	$y-\bar{y}$	$(y-\bar{y})^2$
1.8	21.0	32.076	47.793	-15.7169	247.02	-11.0761	122.68	-26.793	717.86
1.9	33.5	32.610	47.793	-15.1828	230.52	0.8898	0.79	-14.293	204.29
1.8	24.6	32.076	47.793	-15.7169	247.02	-7.4761	55.89	-23.193	537.92
2.4	40.7	35.280	47.793	-12.5125	156.56	5.4195	29.37	-7.093	50.31
5.1	73.2	49.700	47.793	1.9072	3.64	23.4998	552.24	25.407	645.52
3.1	24.9	39.019	47.793	-8.7741	76.98	-14.1189	199.34	-22.893	524.09
5.5	40.4	51.836	47.793	4.0435	16.35	-11.4365	130.79	-7.393	54.66
5.1	45.3	49.700	47.793	1.9072	3.64	-4.4002	19.36	-2.493	6.22
8.3	53.5	66.790	47.793	18.9973	360.90	-13.2903	176.63	5.707	32.57
13.7	93.8	95.630	47.793	47.8368	2288.36	-1.8298	3.35	46.007	2116.64
5.3	64.0	50.768	47.793	2.9754	8.85	13.2316	175.08	16.207	262.67
4.9	62.7	48.632	47.793	0.8391	0.70	14.0679	197.91	14.907	222.22
3.7	47.2	42.223	47.793	-5.5697	31.02	4.9767	24.77	-0.593	0.35
3.8	44.3	42.757	47.793	-5.0356	25.36	1.5426	2.38	-3.493	12.20
66.4	669.1	669.100	669.102	-0.0019	3696.93	-0.0001	1690.58	-0.002	5387.51

a. The explained variation is $\Sigma(\hat{y}-\bar{y})^2 = 3696.93$
b. The unexplained variation is $\Sigma(y-\hat{y})^2 = 1690.58$
c. The total variation is $\Sigma(y-\bar{y})^2 = 5387.51$
d. $r^2 = \Sigma(\hat{y}-\bar{y})^2/\Sigma(y-\bar{y})^2 = 3696.93/5387.51 = .6862$
e. $s_e^2 = \Sigma(y-\hat{y})^2/(n-2) = 1690.58/12 = 140.8817$
   $s_e = 11.8694$

13. a. $\hat{y} = -151.6977 + 3.883382x$.
    $\hat{y}_{69} = -151.6977 + 3.883382(69) = 116.29$
   b. preliminary calculations
      $n = 9$
      $\Sigma x = 632.0$ $\bar{x} = (\Sigma x)/n = 632.0/9 = 70.222$
      $\Sigma x^2 = 44399.50$ $n\Sigma x^2-(\Sigma x)^2 = 9(44399.50) - (632.0)^2 = 171.50$
      $\hat{y} \pm t_{n-2,\alpha/2}s_e\sqrt{1 + 1/n + n(x_o-\bar{x})^2/[n\Sigma x^2-(\Sigma x)^2]}$
      $\hat{y}_{69} \pm t_{7,.025}(4.8790)\sqrt{1 + 1/9 + 9(69-70.222)^2/[171.50]}$
      $116.29 \pm (2.365)(4.8790)\sqrt{1.18948}$
      116.29 ± 12.58
      $103.7 < y_{69} < 128.9$

15. a. $\hat{y} = 22.46293 + 5.340648x$
    $\hat{y}_{4.0} = 22.46293 + 5.340648(4.0) = 43.83$
   b. preliminary calculations
      $n = 14$
      $\Sigma x = 66.4$ $\bar{x} = 66.4/14 = 4.743$
      $\Sigma x^2 = 444.54$ $n\Sigma x^2-(\Sigma x)^2 = 14(444.54) - (66.4)^2 = 1814.60$

$$\hat{y} \pm t_{n-2,\alpha/2}s_e\sqrt{1 + 1/n + n(x_o-\bar{x})^2/[n\Sigma x^2-(\Sigma x)^2]}$$

$$\hat{y}_{4.0} \pm t_{12,.005}(11.8694)\sqrt{1 + 1/14 + 14(4.0-4.743)^2/[1814.60]}$$

$$43.83 \pm (3.055)(11.8694)\sqrt{1.07569}$$

$$43.83 \pm 37.61$$

$$6.2 < y_{4.0} < 81.4$$

Exercises 17-20 refer to the chapter problem of Table 9-1. They use the following, which are calculated and/or discussed at various places in the text,

    $n = 10$

    $\Sigma x = 741$ $\hat{y} = -112.7098976 + 2.274087687x$

    $\Sigma x^2 = 55289$ $s_e = 6.6123487$

and the values obtained below.

    $\bar{x} = (\Sigma x)/n = 741/10 = 74.1$

    $n\Sigma x^2-(\Sigma x)^2 = 10(55289) - (741)^2 = 3809$

17. $\hat{y}_{85} = -112.7098976 + 2.274087687(85) = 80.58$

    $\hat{y} \pm t_{n-2,\alpha/2}s_e\sqrt{1 + 1/n + n(x_o-\bar{x})^2/[n\Sigma x^2-(\Sigma x)^2]}$

    $\hat{y}_{85} \pm t_{8,.005}(6.6123487)\sqrt{1 + 1/10 + 10(85-74.1)^2/[3809]}$

    $80.58 \pm (3.355)(6.6123487)\sqrt{1.41192}$

    80.58 ± 26.36

    $54.2 < y_{85} < 106.9$

19. $\hat{y}_{90} = -112.7098976 + 2.274087687(90) = 91.96$

    $\hat{y} \pm t_{n-2,\alpha/2}s_e\sqrt{1 + 1/n + n(x_o-\bar{x})^2/[n\Sigma x^2-(\Sigma x)^2]}$

    $\hat{y}_{90} \pm t_{8,.025}(6.6123487)\sqrt{1 + 1/10 + 10(90-74.1)^2/[3809]}$

    $91.96 \pm (2.306)(6.6123487)\sqrt{1.76372}$

    91.96 ± 20.25

    $71.7 < y_{90} < 112.2$

21. This exercise uses the following values from the chapter problem of Table 9-1, which are calculated and/or discussed at various places in the text,

    $n = 10$ $\Sigma x = 741$ $b_o = -112.7098976$ $s_e = 6.6123487$

                  $\Sigma x^2 = 55289$ $b_1 = 2.274087687$

and the values obtained below.

    $\bar{x} = (\Sigma x)/n = 741/10 = 74.1$

    $\Sigma x^2-(\Sigma x)^2/n = 55289 - (741)^2/10 = 380.9$

  a. $b_o \pm t_{n-2,\alpha/2}s_e\sqrt{1/n + \bar{x}^2/[\Sigma x^2-(\Sigma x)^2/n]}$

    $-112.7098976 \pm t_{8,.025}(6.6123487)\sqrt{1/10 + (74.1)^2/[380.9]}$

    $-112.7098976 \pm (2.306)(6.6123487)\sqrt{14.5153}$

    $-112.7098976 \pm 58.093686$

    $-170.8 < \beta_o < 54.6$

  b. $b_1 \pm t_{n-2,\alpha/2}s_e/\sqrt{\Sigma x^2-(\Sigma x)^2/n}$

    $2.274087687 \pm t_{8,.025}(6.6123487)/\sqrt{380.9}$

    $2.274087687 \pm (2.306)(6.6123487)/\sqrt{380.9}$

    $2.274087687 \pm .781285$

    $1.5 < \beta_1 < 3.1$

23. a. Since $s_e^2 = \Sigma(y-\hat{y})^2/(n-2)$,

    $(n-2)\cdot s_e^2 = \Sigma(y-\hat{y})^2$.

    And so (explained variation) $= (n-2)\cdot s_e^2$.

b. Let EXPV = explained variation = $\Sigma(\hat{y}-\bar{y})^2$.
   Let UNXV = unexplained variation = $\Sigma(y-\hat{y})^2$.
   Let TOTV = total variation = $\Sigma(y-\bar{y})^2$.

$$r^2 = (EXPV)/(TOTV)$$
$$(TOTV)\cdot r^2 = (EXPV)$$
$$(EXPV + UNXV)\cdot r^2 = (EXPV)$$
$$(EXPV)\cdot r^2 + (UNXV)\cdot r^2 = (EXPV)$$
$$(UNXV)\cdot r^2 = (EXPV) - (EXPV)\cdot r^2$$
$$(UNXV)\cdot r^2 = (EXPV)\cdot(1-r^2)$$
$$(UNXV)\cdot r^2/(1-r^2) = (EXPV)$$

And so (explained variation) = $[r^2/(1-r^2)]\cdot$(unexplained variation)

c. If $r^2 = .900$, than $r = \pm.949$.
   Since the regression line has a negative slope (i.e., $b_1 = -2$), choose the negative root.
   And so the linear correlation coefficient is -.949.

9-5 Multiple Regression

1. WEIGHT = -271.71 -.870(HEADLENGTH) +.554(LENGTH) +12.153(CHEST)
   $\hat{y} = -271.71 - .870x_1 + .554x_2 + 12.153x_3$

3. Yes, because the adjusted R^2 is .924. Although the P-value is less than .05, that alone in multiple regression problems does not necessarily indicate the regression is of practical significance. Increasing the number of x variables, like increasing the sample size in the previous chapters, can create statistical significance that is not of practical value. The adjusted R^2 takes into account the number of x variables.

5. Use HWY, the highway fuel consumption, because it has the highest adjusted R^2 [.853] among all the regressions with a single predictor variable.

7. Use HWY and WT, highway fuel consumption and weight, because it had the highest adjusted R^2 [.861] among all the regressions given. A reasonable argument could be made for using HWY alone, since its adjusted R^2 [.853] is only slightly less – i.e., the addition of a second predictor variable does not substantially improve the regression.

9. a. HEIGHT = 21.56 + .6899(MOTHER)
      $\hat{y} = 21.56 + .6899x_1$
      adjusted $R^2 = .347$
   b. HEIGHT = 45.72 + .2926(FATHER)
      $\hat{y} = 45.72 + .2926x2$
      adjusted $R^2 = .050$
   c. HEIGHT = 9.80 + .6580(MOTHER) + .2004(FATHER)
      $\hat{y} = 9.80 + .6580x_1 - .2004x_2$
      adjusted $R^2 = .366$
   d. The regression in part (c) is the best, because it has the highest adjusted R^2. The regression in part (a) is almost as good, however, as adding a second variable improves the regression very little.
   e. No, since the adjusted R^2 of the best equation is only .366.

11. a. CALORIES = 3.680 + 3.782(FAT)
    $\hat{y} = 3.680 + 3.782x_1$
    adjusted R^2 = .067
    b. CALORIES = 3.464 + 1.012(SUGAR)
    $\hat{y} = 3.464 + 1.012x_2$
    adjusted R^2 = .556
    c. CALORIES = 3.403 + 3.198(FAT) + .982(SUGAR)
    $\hat{y} = 3.403 + 3.198x_1 - .982x_2$
    adjusted R^2 = .628
    d. The regression in part (c) is the best, because it has the highest adjusted R^2.
    e. Yes, since the adjusted R^2 of the best equation .628.

13. To find the best variables for a regression to predict NICOTINE, consider the correlations. Arranged in one table, the correlations are as follows.

    ```
                 TAR        NICOTINE  CO
    TAR        1.000
    NICOTINE    .961       1.000
    CO          .934        .863      1.000
    ```
    The variable having the highest correlation [.961] with NICOTINE is TAR. The positive correlation indicates cigarettes with more nicotine tend to contain more tar. The variable having the second highest correlation [.863] with NICOTINE is CO. But since the correlation [.934] between TAR and CO is so high, those two variables can be accurately predicted from each other and give duplicate information. They should not both be included. The best multiple regression equation to predict NICOTINE, therefore, is probably the simple linear regression
    NICOTINE = .1540 + .0651(TAR)
    R^2 = 92.4% = .924
    adjusted R^2 = 92.1% = .921
    overall P-value = .000

15. To find the best variables for a regression to predict the selling price SELL, consider the correlations. Arranged in one table, the correlations are as follows.

    ```
            SELL   LIST   AREA   ROOM   BEDR   BATH   AGEH   ACRE   TAXH
    SELL   1.000
    LIST    .997  1.000
    AREA    .879   .892  1.000
    ROOM    .560   .571   .751  1.000
    BEDR    .335   .320   .476   .657  1.000
    BATH    .640   .640   .668   .555   .458  1.000
    AGEH   -.147  -.130   .125   .371   .141  -.023  1.000
    ACRE    .169   .167   .177   .282   .037   .301   .304  1.000
    TAXH    .899   .907   .810   .513   .314   .583  -.189   .060  1.000
    ```
    The variable having the highest correlation [.997] with SELL is LIST. This positive correlation indicates that the houses with the higher selling prices tend to be those with the high listing prices. This would be the best single variable for predicting selling price. The best additional predictor variable to add would be one having a low correlation with LIST (so that it is not merely duplicating the information provided by LIST). This suggests that AGEH or ACRE would be appropriate variables to consider next. But since the seller and/or realtor presumably considered each of the other variables when determining LIST, there is a sense in which the information of the other variables is already contained in LIST. The best multiple regression equation with SELL as the dependent variable, therefore, is probably

SELL = 7.26 + .914(LIST)
    R^2 = 99.5% = .995
    adjusted R^2 = 99.5% = .995
    overall P-value = .000

The adjusted R^2 and overall P-value indicate that this is a suitable regression equation for predicting selling price. Since the adjusted R^2 is so close to 1.000, these is little room for improvement and no additional variable could be of any value.

9-6 Modeling

1. The graph appears to be that of a quadratic function.
    Try a regression of the form $y = ax^2 + bx + c$.
    This produces $y = 2x^2 - 12x + 18$ with adjusted R^2 = 100.0%, a perfect fit.

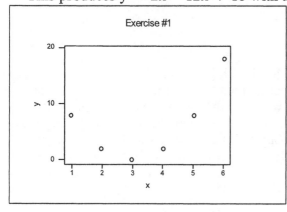

3. The graph appears to be that of an exponential function.
    The equation has the form $y = a \cdot b^x$
        $\ln y = \ln (a \cdot b^x)$ [taking the natural log of both sides]
            $= \ln a + x \cdot (\ln b)$
    Try a regression of the form $\ln y = c + dx$
    This produces $\ln y = -.00009 + 1.1x$ with adjusted $R^2 \approx 100.0\%$, a "perfect" fit.
        solving for the original parameters: $\ln a = -.0009$ $\ln b = 1.1$
                                $a = e^{-.00009}$ $b = e^{1.1}$
                                $= 1.00$ $= 3.00$
    which yields the exponential equation: $y = (1.00) \cdot (3.00)^x$

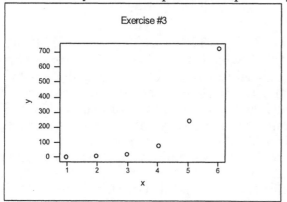

5. Code the years as 1,2,3,... for 1980,1981,1982,...
   The graph appears to be either that of a straight line or that of a quadratic function.
   Try a regression of the form $y = ax + b$.
      This produces $y = 2.6455x + 14.329$ with adjusted $R^2 = 79.2\%$.
   Try a regression of the form $y = ax^2 + bx + c$.
      This produces $y = .05167x^2 + 1.509x + 18.686$ with adjusted $R^2 = 79.0\%$.
   The linear model appears to be the better model.
      $\hat{y} = 2.6455x + 14.329$
      $\hat{y}_{22} = 2.6455(22) + 14.329 = 73$, which is considerably lower than the actual value 82.

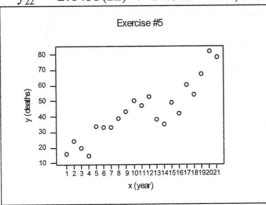

7. Code the years as 1,2,3,... for 1990,1991,1992,...
   The graph appears to be either that of a straight line or that of a quadratic function.
   Try a regression of the form $y = ax + b$.
      This produces $y = 56.982x + 340.382$ with adjusted $R^2 = 99.5\%$.
   Try a regression of the form $y = ax^2 + bx + c$.
      This produces $y = 1.2145x^2 + 42.408x + 371.958$ with adjusted $R^2 = 99.9\%$.
   The quadratic model is the better model.
      $\hat{y} = 1.2145x^2 + 42.408x + 371.958$
      $\hat{y}_{16} = 1.2145(16)^2 + 42.408(16) + 371.958 = 1361$

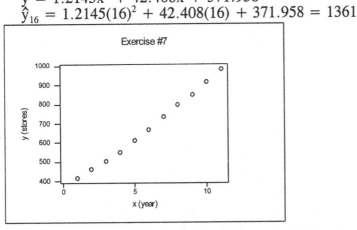

9. a. Moore's law is modeled by an exponential function.
      If y has value "a" at year $x=1$ and doubles every 1 year,
         $y = a{\cdot}2^{x-1} = a{\cdot}2^{(x-1)/1}$.
      If y has value "a" at year $x=1$ and doubles every 1.5 years,
         $y = a{\cdot}2^{(x-1)/1.5}$.
      NOTE: $y = a{\cdot}2^{(x-1)/1.5}$
            $= a{\cdot}2^{(2/3)(x-1)}$
            $= a{\cdot}2^{(2/3)x} {\cdot}2^{-2/3}$
            $= [a/2^{2/3}]{\cdot}[(2^{2/3})^x]$
            $= (a/2^{2/3}){\cdot}(1.587)^x$

b. The exponential function has the form $y = a \cdot b^x$
$$\ln y = \ln (a \cdot b^x) \text{ [taking the natural log of both sides]}$$
$$= \ln a + x \cdot (\ln b)$$
   Try a regression of the form $\ln y = c + dx$
   This produces $\ln y = .8417 + .311794x$ with adjusted $R^2 \approx 99.3\%$.
      solving for the original parameters: $\ln a = .8417 \qquad \ln b = .311794$
$$a = e^{.8417} \qquad b = e^{.311794}$$
$$= 2.3203 \qquad = 1.36587$$
   which yields the exponential equation: $y = (2.3203) \cdot (1.36587)^x$
   With adjusted $R^2 = .993$, the exponential model appears to be the best model.

c. No; considering the base of x in the exponential model, Moore's law appears to be off. Part (a) gives the base for a quantity that doubles every 18 months as 1.587; part (b) gives an actual base of $1.366 = 2^{.45}$, which corresponds to doubling every $1/.45$ years (or about every 27 months).

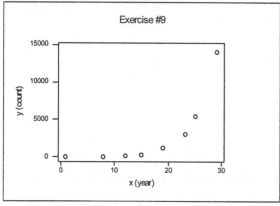

Exercise #9

11. Using the formula provided for $\hat{y}$ produces the following table.

x	y	$\bar{y}$	$\hat{y}$	$(y-\bar{y})^2$	$(y-\hat{y})^2$
1	5	101.27	9.63	9268.4	21.420
2	10	101.27	14.57	8330.7	20.927
3	17	101.27	21.94	7101.9	24.403
4	31	101.27	32.76	4938.3	3.088
5	50	101.27	48.33	2628.9	2.797
6	76	101.27	70.10	638.7	34.799
7	106	101.27	99.35	22.3	44.217
8	132	101.27	136.59	944.2	21.092
9	179	101.27	180.91	6041.5	3.633
10	227	101.27	229.57	15807.3	6.626
11	281	101.27	278.54	32301.9	6.051
66	1114			88024.2	189.052

a. $\Sigma(y-\hat{y})^2 = 189.1$
b. $R^2 = 1 - \Sigma(y-\hat{y})^2/\Sigma(y-\bar{y})^2 = 1 - 189.1/88024.2 = .9979$
c. For the quadratic model, $R^2 = .9991688$ is given in th text.
   Solving $R^2 = 1 - \Sigma(y-\hat{y})^2/\Sigma(y-\bar{y})^2$ gives $\Sigma(y-\hat{y})^2 = (1 - R^2) \cdot \Sigma(y-\bar{y})^2$
$$= (.0008312) \cdot (88024.2)$$
$$= 73.2$$

The quadratic model is better because its $\Sigma(y-\hat{y})^2$ is lower and its R^2 is higher.

Review Exercises

1. $n = 8$
   $\Sigma x = 276.6$
   $\Sigma y = 1.66$
   $\Sigma x^2 = 10680.48$
   $\Sigma y^2 = .35220$
   $\Sigma xy = 57.191$

   $n(\Sigma xy) - (\Sigma x)(\Sigma y) = 8(57.191) - (276.6)(1.66) = -1.628$
   $n(\Sigma x^2) - (\Sigma x)^2 = 8(10680.48) - (276.6)^2 = 8936.28$
   $n(\Sigma y^2) - (\Sigma y)^2 = 8(.3522) - (1.66)^2 = .0620$
   $r = [n(\Sigma xy) - (\Sigma x)(\Sigma y)]/[\sqrt{n(\Sigma x^2) - (\Sigma x)^2} \cdot \sqrt{n(\Sigma y^2) - (\Sigma y)^2}]$
   $= -1.628/[\sqrt{8936.28} \cdot \sqrt{.0620}]$
   $= -.069$

   $H_o: \rho = 0$
   $H_1: \rho \neq 0$
   $\alpha = .05$
   C.R. $r < -.707$ OR
   $\quad\quad r > .707$

   C.R. $t < -t_{6,.025} = -2.447$
   $\quad\quad t > t_{6,.025} = 2.447$

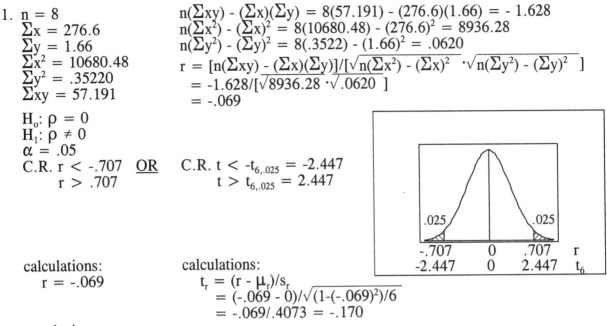

   calculations:
   $r = -.069$

   calculations:
   $t_r = (r - \mu_r)/s_r$
   $= (-.069 - 0)/\sqrt{(1-(-.069)^2)/6}$
   $= -.069/.4073 = -.170$

   conclusion:
   Do not reject H_o; there is not sufficient evidence to reject the claim that $\rho = 0$.

   No; there is no significant linear correlation between age and BAC.

2. a. $n = 6$
   $\Sigma x = 441.84$
   $\Sigma y = 63.58$
   $\Sigma x^2 = 36754.1408$
   $\Sigma y^2 = 809.5364$
   $\Sigma xy = 5308.7436$

   $n(\Sigma xy) - (\Sigma x)(\Sigma y) = 6(5308.7436)-(441.84)(63.58) = 3760.2744$
   $n(\Sigma x^2) - (\Sigma x)^2 = 6(36754.1408) - (441.84)^2 = 25302.2592$
   $n(\Sigma y^2) - (\Sigma y)^2 = 6(809.5364) - (63.58)^2 = 414.8020$
   $r = [n(\Sigma xy) - (\Sigma x)(\Sigma y)]/[\sqrt{n(\Sigma x^2) - (\Sigma x)^2} \cdot \sqrt{n(\Sigma y^2) - (\Sigma y)^2}]$
   $= 3760.2744/[\sqrt{25302.2592} \cdot \sqrt{414.8020}]$
   $= .828$

   $H_o: \rho = 0$
   $H_1: \rho \neq 0$
   $\alpha = .05$
   C.R. $r < -.811$ OR C.R. $t < -t_{4,.025} = -2.776$
   $\quad\quad r > .811$ $\quad\quad\quad\quad\quad t > t_{4,.025} = 2.776$

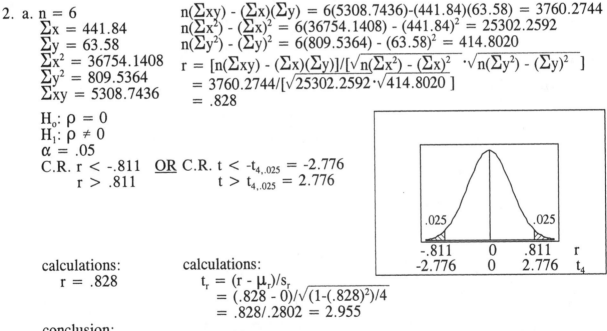

   calculations:
   $r = .828$

   calculations:
   $t_r = (r - \mu_r)/s_r$
   $= (.828 - 0)/\sqrt{(1-(.828)^2)/4}$
   $= .828/.2802 = 2.955$

   conclusion:
   Reject H_o; there is sufficient evidence to reject the claim that $\rho = 0$ and conclude that $\rho \neq 0$ (in fact, that $\rho > 0$).

b. $\bar{x} = 441.84/6 = 73.64$
$\bar{y} = 63.58/6 = 10.597$
$b_1 = [n(\Sigma xy) - (\Sigma x)(\Sigma y)]/[n(\Sigma x^2) - (\Sigma x)^2]$
$\quad = 3760.2744/25302.2592 = .1486$
$b_o = \bar{y} - b_1\bar{x}$
$\quad = 10.597 - (.1486)(73.64) = -.3473$
$\hat{y} = b_o + b_1x$
$\quad = -.347 + .149x$

The tips left are slightly less than 15% of the bill – to follow the established pattern, leave 35¢ less than 14.9% of the bill.

3. Let x be the price and y be the consumption.

$n = 10$ $n(\Sigma xy) - (\Sigma x)(\Sigma y) = 10(4.74330) - (13.66)(3.468)$
$\Sigma x = 13.66$ $= .06012$
$\Sigma y = 3.468$ $n(\Sigma x^2) - (\Sigma x)^2 = 10(18.6694) - (13.66)^2$
$\Sigma x^2 = 18.6694$ $= .0984$
$\Sigma y^2 = 1.234888$ $n(\Sigma y^2) - (\Sigma y)^2 = 10(1.234888) - (3.468)^2$
$\Sigma xy = 4.74330$ $= .321856$

$r = [n(\Sigma xy) - (\Sigma x)(\Sigma y)]/[\sqrt{n(\Sigma x^2) - (\Sigma x)^2} \cdot \sqrt{n(\Sigma y^2) - (\Sigma y)^2}]$
$\quad = [.06012]/[\sqrt{.0984} \cdot \sqrt{.321856}]$
$\quad = .338$

a. $H_o: \rho = 0$
$H_1: \rho \neq 0$
$\alpha = .05$
C.R. $r < -.632$ __OR__   C.R. $t < -t_{8,.025} = -2.306$
      $r > .632$ $t > t_{8,.025} = 2.306$

	.025		.025	
	-.632	0	.632	r
	-2.306	0	2.306	t_8

calculations: calculations:
   $r = .338$ $t_r = (r - \mu_r)/s_r$
                     $= (.338 - 0)/\sqrt{(1-(.338)^2)/8}$
                     $= .338/.333 = 1.015$

conclusion:
    Do not reject H_o; there is not sufficient evidence to reject the claim that $\rho = 0$.

b. $r^2 = (.338)^2 = .114 = 11.4\%$
c. $b_1 = [n(\Sigma xy) - (\Sigma x)(\Sigma y)]/[n(\Sigma x^2) - (\Sigma x)^2]$
$\quad = .06012/.0984 = .611$
$b_o = \bar{y} - b_1\bar{x}$
$\quad = (3.468/10) - (.611)(13.66/10) = -.488$
$\hat{y} = b_o + b_1x$
$\quad = -.488 + .611x$

d. $\hat{y} = -.488 + .611x$
$\hat{y}_{1.38} = \bar{y} = .3468$ pints per capita per week [no significant correlation]

4. Let x be the income and y be the consumption.

$n = 10$ $n(\Sigma xy) - (\Sigma x)(\Sigma y) = 10(1230.996) - (3548)(3.468)$
$\Sigma x = 3548$ $= 5.496$
$\Sigma y = 3.468$ $n(\Sigma x^2) - (\Sigma x)^2 = 10(1259524) - (3548)^2$
$\Sigma x^2 = 1259524$ $= 6936$
$\Sigma y^2 = 1.234888$ $n(\Sigma y^2) - (\Sigma y)^2 = 10(1.234888) - (3.468)^2$
$\Sigma xy = 1230.996$ $= .321856$

$r = [n(\Sigma xy) - (\Sigma x)(\Sigma y)]/[\sqrt{n(\Sigma x^2) - (\Sigma x)^2} \cdot \sqrt{n(\Sigma y^2) - (\Sigma y)^2}]$

$= [5.496]/[\sqrt{6936} \cdot \sqrt{.321856}]$

$= .116$

a. H_o: $\rho = 0$

H_1: $\rho \neq 0$

$\alpha = .05$

C.R. r < -.632   OR   C.R. t < $-t_{8,.025}$ = -2.306

r > .632   t > $t_{8,.025}$ = 2.306

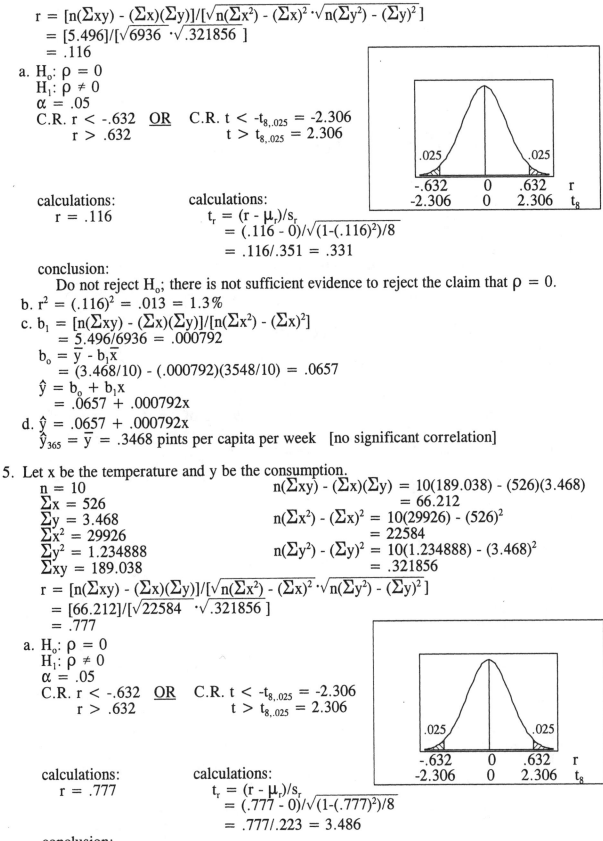

calculations: calculations:

r = .116 $t_r = (r - \mu_r)/s_r$

$= (.116 - 0)/\sqrt{(1-(.116)^2)/8}$

$= .116/.351 = .331$

conclusion:

Do not reject H_o; there is not sufficient evidence to reject the claim that $\rho = 0$.

b. $r^2 = (.116)^2 = .013 = 1.3\%$

c. $b_1 = [n(\Sigma xy) - (\Sigma x)(\Sigma y)]/[n(\Sigma x^2) - (\Sigma x)^2]$

$= 5.496/6936 = .000792$

$b_o = \bar{y} - b_1\bar{x}$

$= (3.468/10) - (.000792)(3548/10) = .0657$

$\hat{y} = b_o + b_1 x$

$= .0657 + .000792x$

d. $\hat{y} = .0657 + .000792x$

$\hat{y}_{365} = \bar{y} = .3468$ pints per capita per week   [no significant correlation]

5. Let x be the temperature and y be the consumption.

n = 10 $n(\Sigma xy) - (\Sigma x)(\Sigma y) = 10(189.038) - (526)(3.468)$

$\Sigma x = 526$ $= 66.212$

$\Sigma y = 3.468$ $n(\Sigma x^2) - (\Sigma x)^2 = 10(29926) - (526)^2$

$\Sigma x^2 = 29926$ $= 22584$

$\Sigma y^2 = 1.234888$ $n(\Sigma y^2) - (\Sigma y)^2 = 10(1.234888) - (3.468)^2$

$\Sigma xy = 189.038$ $= .321856$

$r = [n(\Sigma xy) - (\Sigma x)(\Sigma y)]/[\sqrt{n(\Sigma x^2) - (\Sigma x)^2} \cdot \sqrt{n(\Sigma y^2) - (\Sigma y)^2}]$

$= [66.212]/[\sqrt{22584} \cdot \sqrt{.321856}]$

$= .777$

a. H_o: $\rho = 0$

H_1: $\rho \neq 0$

$\alpha = .05$

C.R. r < -.632   OR   C.R. t < $-t_{8,.025}$ = -2.306

r > .632   t > $t_{8,.025}$ = 2.306

calculations: calculations:

r = .777 $t_r = (r - \mu_r)/s_r$

$= (.777 - 0)/\sqrt{(1-(.777)^2)/8}$

$= .777/.223 = 3.486$

conclusion:

Reject H_o; there is sufficient evidence to reject the claim that $\rho = 0$ and to conclude that $\rho \neq 0$ (in fact, that $\rho > 0$).

b. $r^2 = (.777)^2 = .604 = 60.4\%$

c. $b_1 = [n(\Sigma xy) - (\Sigma x)(\Sigma y)]/[n(\Sigma x^2) - (\Sigma x)^2]$
    $= 66.212/22584 = .00293$

   $b_0 = \bar{y} - b_1\bar{x}$
    $= (3.468/10) - (.00293)(526/10) = .193$

   $\hat{y} = b_0 + b_1 x$
    $= .193 + .00293x$

d. $\hat{y} = .193 + .00293x$
   $\hat{y}_{32} = .193 + .00293(32) = .2864$ pints per capita per week

6. Using Minitab and the given notation,
   $\hat{y} = -.053 + .747x_1 - .00220x_2 + .00303x_3$
   $R^2 = 72.6\% = .726$
   adjusted $R^2 = 58.9\% = .589$
   overall P-value $= .040$
   Yes; since the overall P-value of .040 is less than .05, the regression equation can be used
   to predict ice cream consumption. Individually, x_1 and x_2 were not significantly related to
   consumption; only x_3 could be used to predict consumption. This suggests that variables x_1
   and x_2 might not be making a worthwhile contribution to the regression, and that the
   equation from exercise #5 (using x_3 alone) might actually be a better and more efficient
   predictive equation.
   To check this, calculate the adjusted R^2 for exercise #5.
      adjusted $R^2 = 1 - \{(n-1)/[n-(k+1)]\}\cdot(1-R^2)$
         $= 1 - \{9/[10-(2)]\}\cdot(1-R^2) = 1 - (9/8)\cdot(1-.777^2) = .554$
   Since the adjusted R^2 of .589 is higher than the adjusted R^2 of .554 from exercise #3,
   however, there is a well-defined statistical sense in which this multiple regression equation
   is the best of the equations considered in these exercises.

Cumulative Review

1. Let x be the F-K Grade Level score and y be the Flesch Reading Ease score.
   $n = 12$ $n(\Sigma xy) - (\Sigma x)(\Sigma y) = 12(6540.86) - (101.2)(793.8)$
   $\Sigma x = 101.2$ $= -1842.24$
   $\Sigma y = 793.8$ $n(\Sigma x^2) - (\Sigma x)^2 = 12(897.82) - (101.2)^2$
   $\Sigma x^2 = 897.82$ $= 532.40$
   $\Sigma y^2 = 53189.10$ $n(\Sigma y^2) - (\Sigma y)^2 = 12(53189.10) - (793.8)^2$
   $\Sigma xy = 6540.86$ $= 8150.76$

   $r = [n(\Sigma xy) - (\Sigma x)(\Sigma y)]/[\sqrt{n(\Sigma x^2) - (\Sigma x)^2}\cdot\sqrt{n(\Sigma y^2) - (\Sigma y)^2}]$
    $= [-1842.24]/[\sqrt{532.40}\cdot\sqrt{8150.76}]$
    $= -.884$

a. $H_0: \rho = 0$
   $H_1: \rho \neq 0$
   $\alpha = .05$
   C.R. $r < -.576$ OR   C.R. $t < -t_{10,.025} = -2.228$
       $r > .576$ $t > t_{10,.025} = 2.228$

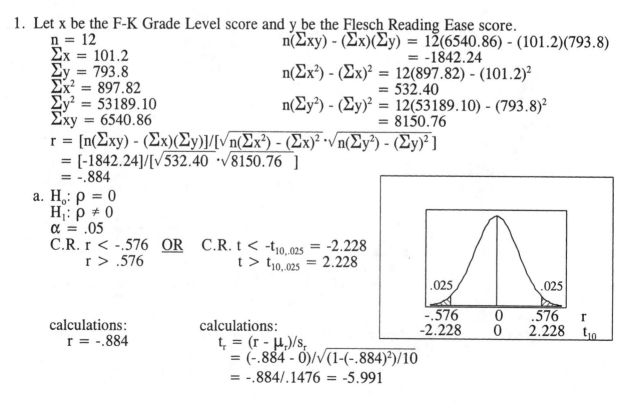

   calculations: calculations:
      $r = -.884$ $t_r = (r - \mu_r)/s_r$
                               $= (-.884 - 0)/\sqrt{(1-(-.884)^2)/10}$
                               $= -.884/.1476 = -5.991$

conclusion:
  Reject H_o; there is sufficient evidence to reject the claim that $\rho = 0$ and to conclude that $\rho \neq 0$ (in fact, that $\rho < 0$).

b. $b_1 = [n(\Sigma xy) - (\Sigma x)(\Sigma y)]/[n(\Sigma x^2) - (\Sigma x)^2]$
  $= -1842.24/532.40 = -3.460$
  $b_o = \bar{y} - b_1\bar{x}$
  $= (793.80/12) - (-3.460)(101.20/12) = 95.33$
  $\hat{y} = b_o + b_1 x$
  $= 95.33 - 3.460x$

c. Yes; in general it's always possible to test wether two parameter values are equal. No; since the two measures do not use the same approach or the same units to the problem, it would not make sense to test wether the values are numerically equal.

d. First find the mean and standard deviation of y, the Flesch Reading Ease scores.
  $\bar{y} = (\Sigma y)/n = 793.8/12 = 66.15$
  $s^2 = [n(\Sigma y^2) - (\Sigma y)^2]/[n(n-1)]$
  $= [12(53189.10) - (793.8)^2]/[12(11)]$
  $= (8150.76)/132 = 61.748$
  $s = 7.858$
  $\bar{x} \pm t_{11,.025} \cdot s/\sqrt{n}$
  $66.15 \pm 2.201 \cdot (7.858)/\sqrt{12}$
  66.15 ± 4.99
  $61.16 < \mu < 71.14$

2. Let x and y be as given.
  $n = 12$ $n(\Sigma xy) - (\Sigma x)(\Sigma y) = 12(122836) - (1189)(1234)$
  $\Sigma x = 1189$ $= 6806$
  $\Sigma y = 1234$ $n(\Sigma x^2) - (\Sigma x)^2 = 12(118599) - (1189)^2$
  $\Sigma x^2 = 118599$ $= 9467$
  $\Sigma y^2 = 127724$ $n(\Sigma y^2) - (\Sigma y)^2 = 12(127724) - (1234)^2$
  $\Sigma xy = 122836$ $= 9932$
  $r = [n(\Sigma xy) - (\Sigma x)(\Sigma y)]/[\sqrt{n(\Sigma x^2) - (\Sigma x)^2} \cdot \sqrt{n(\Sigma y^2) - (\Sigma y)^2}]$
  $= [6806]/[\sqrt{9467} \sqrt{9932}]$
  $= .702$

a. $\bar{x} = 1189/12 = 99.1$
  $s_x = \sqrt{9467/(12 \cdot 11)} = 8.47$

b. $\bar{y} = 1234/12 = 102.8$
  $s_y = \sqrt{9932/(12 \cdot 11)} = 8.67$

c. No; there does not appear to be a difference between the means of the two populations. In exploring the relationship between the IQ's of twins, such a two sample approach is not appropriate because it completely ignores the pairings of the scores.

d. Combing the scores produces the following summary statistics.
  summary statistics: $n = 24$, $\Sigma x = 2423$, $\Sigma x^2 = 246,323$, $\bar{x} = 100.958$, $s = 8.5997$
  original claim: $\mu \neq 100$
  $H_o: \mu = 100$
  $H_1: \mu \neq 100$
  $\alpha = .05$
  C.R. $t < -t_{23,.025} = -2.069$
  $\qquad t > t_{23,.025} = 2.069$
  calculations:
  $t_{\bar{x}} = (\bar{x} - \mu)/s_{\bar{x}}$
  $= (100.958 - 100)/(8.5997/\sqrt{24})$
  $= .958/1.755 = .546$
  P-value $= 2 \cdot P(t_{23} > .546) > .20$

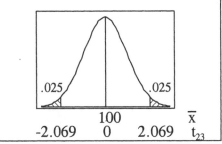

conclusion:
  Do not reject H_o; there is not sufficient evidence to conclude that $\mu \neq 120$.

e. Correlation would be appropriate for answering the question "is there a (linear) relationship?"

H_o: $\rho = 0$

H_1: $\rho \neq 0$

$\alpha = .05$ [assumed]

C.R. r < -.576   OR   C.R. t < -$t_{10,.025}$ = -2.228
       r > .576   t > $t_{10,.025}$ = 2.228

calculations: calculations:
  r = .702 $t_r = (r - \mu_r)/s_r$
                              $= (.702 - 0)/\sqrt{(1-(.702)^2)/10}$
                              $= .702/.225 = 3.116$

conclusion:
  Reject H_o; there is sufficient evidence to reject the claim that $\rho = 0$ and to conclude that $\rho \neq 0$ (in fact, $\rho > 0$).

  Ordinarily, the conclusion would be that about $R^2 = (.702)^2 = .493 = 49.3\%$ of the variation in x can be explained in terms of y (and vice-versa). In this context that means that about 49.3% of the variation among the IQ's in one group can be explained in terms of the IQ's of their twins (i.e, in terms of heredity). In simplest terms, it seems intelligence is about ½ due to heredity and ½ due to environment.

NOTE: Exercise #2 contains at least two interesting subtleties. (1) Correlation addresses only whether there is a relationship between the IQ's, and not whether the IQ's are close to each other. If each older twin had an IQ 20 points higher than the corresponding younger twin, there would be a perfect correlation – but the twins would not be similar in IQ at all. Beware of the misconception that correlation implies similarity. (2) Within each pair, the older twin was designated x. That was an arbitrary decision to produce an objective rule. The x-y designation within pairs could just as properly have been made alphabetically, randomly, or by another rule. But the rule affects the results. If it happens to designate all the twins with the higher IQ as x, the correlation rises to r = .810. If it happens to designate the twins with the higher IQ as x in the first six pairs and as y in the last six, the correlation falls to r = .633.

One technique which addresses both of the above issues (i.e., it tests for similarity and does not depend upon an x-y designation at all) involves comparing the variation within pairs to the overall variation in IQ's. If there is significantly less variability between the twins than there is variability in the general population (or between non-identical twins raised apart), then there is a significant relationship (i.e., in the sense of similarity) between the IQ's of identical twins.

Chapter 10

Multinomial Experiments and Contingency Tables

10-2 Multinomial Experiments: Goodness-of-Fit

1. a. H_o: $p_1 = p_2 = p_3 = p_4 = .25$
    b. Since $\Sigma O = (5+6+8+13) = 32$, the hypothesis of equally likely categories indicates $E_i = .25(32) = 8$ for $i=1,2,3,4$.
    c. $\chi^2 = \Sigma[(O-E)^2/E] = (5-8)^2/8 + (6-8)^2/8 + (8-8)^2/8 + (13-8)^2/8$
    $= 9/8 + 4/8 + 0/8 + 25/8$
    $= 38/8 = 4.750$
    d. $\chi^2_{3,.05} = 7.815$
    e. There is not enough evidence to reject the claim that the four categories are equally likely.

3. a. $\chi^2_{37,.10} = 51.085$
    b. Since $\chi^2_{37,.10} = 51.085 > 38.232 > 29.051 = \chi^2_{37,.90}$, $.10 < $ P-value $ < .90$.
    c. There is not enough evidence to reject the claim that the 38 results are equally likely.

NOTES FOR THE REMAINING EXERCISES:
(1) In multinomial problems, always verify that $\Sigma E = \Sigma O$ before proceeding. If these sums are not equal, then an error has been made and further calculations have no meaning.
(2) As in the previous uses of the chi-squared distribution, the accompanying illustrations follow the "usual" shape – even though that shape is not correct for df=1 or df=2.

5. H_o: $p_1 = p_2 = p_3 = ... = p_6 = 1/6$
    H_1: at least one of the proportions is different from 1/6
    $\alpha = .05$
    C.R. $\chi^2 > \chi^2_{5,.05} = 11.071$
    calculations:

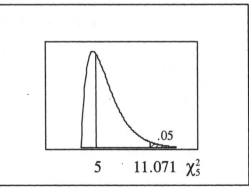

outcome	O	E	$(O-E)^2/E$
1	27	33.33	1.2033
2	31	33.33	.1633
3	42	33.33	2.2533
4	40	33.33	1.3333
5	28	33.33	.8533
6	32	33.33	.0533
	200	200.00	5.8600

$\chi^2 = \Sigma[(O-E)^2/E]$
$= 5.860$

conclusion:
Do not reject H_o; there is not sufficient evidence to conclude that at least one of the proportions is different from 1/6.
No, this particular loaded die did not behave noticeably differently from a fair die.

7. H_o: $p_{Sun} = p_{Mon} = p_{Tue} = \ldots = p_{Sat} = 1/7$
   H_1: at least one of the proportions is different from 1/7
   $\alpha = .05$
   C.R. $\chi^2 > \chi^2_{6,.05} = 12.592$
   calculations:

day	O	E	$(O-E)^2/E$
Sun	31	25.714	1.0865
Mon	20	25.714	1.2698
Tue	20	25.714	1.2698
Wed	22	25.714	.5365
Thu	22	25.714	.5365
Fri	29	25.714	.4198
Sat	36	25.714	4.1143
	180	180.000	9.2332

$\chi^2 = \Sigma[(O-E)^2/E]$
$\quad = 9.233$

conclusion:
   Do not reject H_o; there is not sufficient evidence to conclude that at least one of the proportions is different from 1/7.

9. H_o: $p_{Mon} = p_{Tue} = p_{Wed} = p_{Thu} = p_{Fri} = 1/5$
   H_1: at least one of the proportions is different from 1/5
   $\alpha = .05$ [assumed]
   C.R. $\chi^2 > \chi^2_{4,.05} = 9.488$
   calculations:

day	O	E	$(O-E)^2/E$
Mon	31	29.4	.087
Tue	42	29.4	5.400
Wed	18	29.4	4.420
Thu	25	29.4	.659
Fri	31	29.4	.087
	147	147.0	10.653

$\chi^2 = \Sigma[(O-E)^2/E]$
$\quad = 10.653$

conclusion:
   Reject H_o; there is sufficient evidence to conclude that at least one of the proportions is different from 1/5.
The accidents seem to occur less frequently on Wednesday, and with increasing frequency toward the start and end of the week. This could be due to Monday workers still thinking about the past weekend and Friday workers thinking about the coming weekend. The only exception to this pattern is Tuesday.

11. H_o: $p_1 = p_2 = p_3 = \ldots = p_8 = 1/8$
    H_1: at least one of the proportions is different from 1/8
    $\alpha = .05$ [assumed]
    C.R. $\chi^2 > \chi^2_{7,.05} = 14.067$
    calculations:

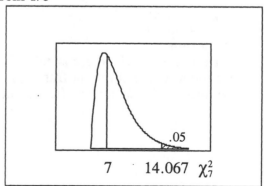

start	O	E	$(O-E)^2/E$
1	29	18	6.722
2	19	18	.056
3	18	18	.000
4	25	18	2.722
5	17	18	.056
6	10	18	3.556
7	15	18	.500
8	11	18	2.722
	144	144	16.333

$\chi^2 = \Sigma[(O-E)^2/E]$
$\quad = 16.333$

conclusion:
>   Reject H_o; there is sufficient evidence to conclude that at least one of the proportions is different from 1/8 – i.e., that the probabilities of winning in the different starting positions are not all the same.

13. H_o: $p_0 = p_1 = p_2 = ... = p_9 = 1/10$
> H_1: at least one of the proportions is different from 1/10
> $\alpha = .05$
>   C.R. $\chi^2 > \chi^2_{9,.05} = 16.919$
>   calculations:

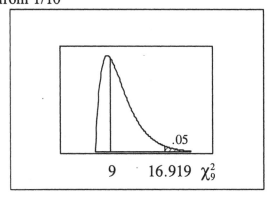

digit	O	E	$(O-E)^2/E$
0	18	16	0.2500
1	12	16	1.0000
2	14	16	0.2500
3	9	16	3.0625
4	17	16	0.0625
5	20	16	1.0000
6	21	16	1.5625
7	26	16	6.2500
8	7	16	5.0625
9	16	16	0.0000
	160	160	18.5000

$$\chi^2 = \Sigma[(O-E)^2/E]$$
$$= 18.500$$

conclusion:
>   Reject H_o; there is sufficient evidence to conclude that at least one of the proportions is different from 1/10 – i.e., that the digits are not equally likely.

If the .01 level of significance is used, the critical region is $\chi^2 > \chi^2_{9,.01} = 21.666$ and the decision changes. While we can be 95% confident that the digits are not being equally selected, we cannot be 99% confident that such is the case. If the digits are not being equally selected, (1) something is wrong with the device/method used to make the selections and (2) a person could use that knowledge to increase his chances of winning.

15. H_o: $p_{Bro} = .3$, $p_{Yel} = .2$, $p_{Red} = .2$, $p_{Ora} = .1$, $p_{Gre} = .1$, $p_{Blu} = .1$
> H_1: at least one of the proportions is not as claimed
> $\alpha = .05$
>   C.R. $\chi^2 > \chi^2_{5,.05} = 11.071$
>   calculations:

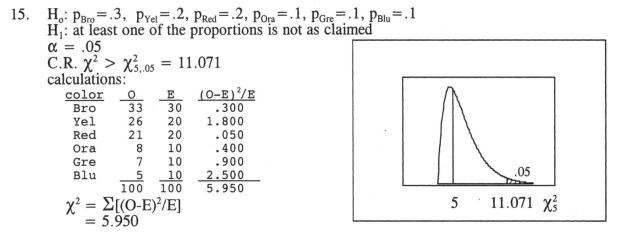

color	O	E	$(O-E)^2/E$
Bro	33	30	.300
Yel	26	20	1.800
Red	21	20	.050
Ora	8	10	.400
Gre	7	10	.900
Blu	5	10	2.500
	100	100	5.950

$$\chi^2 = \Sigma[(O-E)^2/E]$$
$$= 5.950$$

conclusion:
>   Do not reject H_o; there is not sufficient evidence to conclude that at least one of the proportions is not as claimed.

17. H_0: $p_0 = p_1 = p_2 = \ldots = p_9 = 1/10$
       H_1: at least one of the proportions is different from 1/10
       $\alpha = .05$
       C.R. $\chi^2 > \chi^2_{9,.05} = 16.919$

calculations:

digit	O	E	$(O-E)^2/E$
0	8	10	.400
1	8	10	.400
2	12	10	.400
3	11	10	.100
4	10	10	.000
5	8	10	.400
6	9	10	.100
7	8	10	.400
8	12	10	.400
9	14	10	1.600
	100	100	4.200

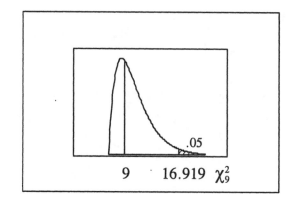

$\chi^2 = \Sigma[(O-E)^2/E]$
$\quad = 18.500$

conclusion:
    Do not reject H_0; there is not sufficient evidence to conclude that at least one of the
    proportions is different from 1/10.

19. H_0: the leading digit proportions conform to Benford's law.
       H_1: at least one of the proportions is different from those specified by Benford's law.
       $\alpha = .05$
       C.R. $\chi^2 > \chi^2_{8,.05} = 15.507$
calculations:

digit	O	p	E	$(O-E)^2/E$
1	72	.301	60.2	2.3130
2	23	.176	35.2	4.2284
3	26	.125	25.0	.0400
4	20	.097	19.4	.0186
5	21	.079	15.8	1.7114
6	18	.067	13.4	1.5791
7	8	.058	11.6	1.1172
8	8	.051	10.2	.4745
9	4	.046	9.2	2.9391
	200	1.000	200.0	14.4213

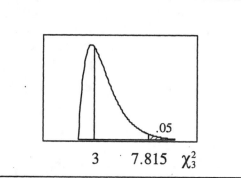

$\chi^2 = \Sigma[(O-E)^2/E]$
$\quad = 14.421$

conclusion:
    Do not reject H_0; there is not sufficient evidence to conclude that at least one of the
    proportions is different those specified by Benford's law.

21. H_0: $p_{LF} = p_{RF} = p_{LR} = p_{RR} = .25$
       H_1: at least one of the proportions is different from .25
       $\alpha = .05$
       C.R. $\chi^2 > \chi^2_{3,.05} = 7.815$
calculations:

tire	O	E	$(O-E)^2/E$
LF	11	23.5	6.649
RF	15	23.5	3.074
LR	8	23.5	10.223
RR	60	23.5	56.691
	94	94.0	76.638

$\chi^2 = \Sigma[(O-E)^2/E]$
$\quad = 76.638$

conclusion:
Reject H_o; there is sufficient evidence to conclude that at least one of the proportions is different from .25.
In general, an "outlier" is a point that is different from the normal pattern. Since that (i.e., "difference from the normal pattern") is essentially what the goodness-of-fit test is designed to detect, an outlier will tend to increase the probability of rejecting H_o.

23. NOTE: Both outcomes having the same expected frequency is equivalent to $p_1 = p_2 = .5$.

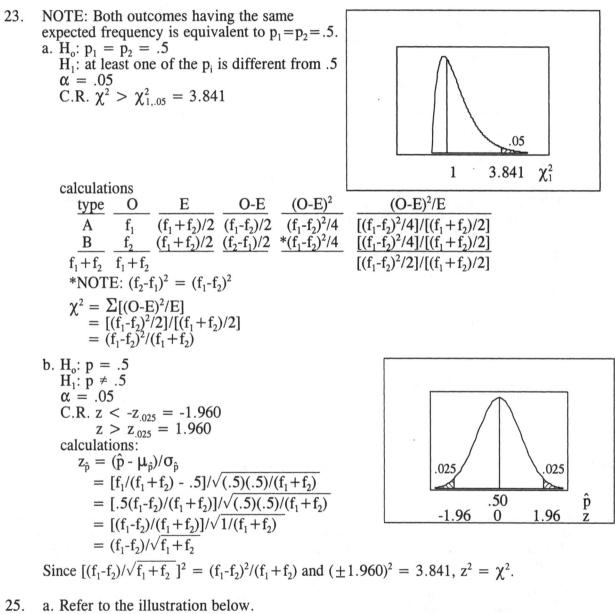

a. $H_o: p_1 = p_2 = .5$
$H_1:$ at least one of the p_i is different from .5
$\alpha = .05$
C.R. $\chi^2 > \chi^2_{1,.05} = 3.841$

calculations

type	O	E	O-E	$(O-E)^2$	$(O-E)^2/E$
A	f_1	$(f_1+f_2)/2$	$(f_1-f_2)/2$	$(f_1-f_2)^2/4$	$[(f_1-f_2)^2/4]/[(f_1+f_2)/2]$
B	f_2	$(f_1+f_2)/2$	$(f_2-f_1)/2$	*$(f_1-f_2)^2/4$	$[(f_1-f_2)^2/4]/[(f_1+f_2)/2]$
	f_1+f_2	f_1+f_2			$[(f_1-f_2)^2/2]/[(f_1+f_2)/2]$

*NOTE: $(f_2-f_1)^2 = (f_1-f_2)^2$

$\chi^2 = \Sigma[(O-E)^2/E]$
$= [(f_1-f_2)^2/2]/[(f_1+f_2)/2]$
$= (f_1-f_2)^2/(f_1+f_2)$

b. $H_o: p = .5$
$H_1: p \neq .5$
$\alpha = .05$
C.R. $z < -z_{.025} = -1.960$
$z > z_{.025} = 1.960$
calculations:
$z_{\hat{p}} = (\hat{p} - \mu_{\hat{p}})/\sigma_{\hat{p}}$
$= [f_1/(f_1+f_2) - .5]/\sqrt{(.5)(.5)/(f_1+f_2)}$
$= [.5(f_1-f_2)/(f_1+f_2)]/\sqrt{(.5)(.5)/(f_1+f_2)}$
$= [(f_1-f_2)/(f_1+f_2)]/\sqrt{1/(f_1+f_2)}$
$= (f_1-f_2)/\sqrt{f_1+f_2}$

Since $[(f_1-f_2)/\sqrt{f_1+f_2}]^2 = (f_1-f_2)^2/(f_1+f_2)$ and $(\pm1.960)^2 = 3.841$, $z^2 = \chi^2$.

25. a. Refer to the illustration below.
$P(x<79.5) = P(z<-1.37)$
$= .0853$
$P(79.5<x<95.5) = P(z<-.30) - P(z<-1.37)$
$= .3821 - .0853 = .2968$
$P(95.5<x<110.5) = P(z<.70) - P(z<-.30)$
$= .7580 - .3821 = .3759$
$P(110.5<x<120.5) = P(z<1.37) - P(z<.70)$
$= .9147 - .7850 = .1567$
$P(x>120.5) = 1 - P(z<1.37)$
$= 1 - .9147 = .0853$

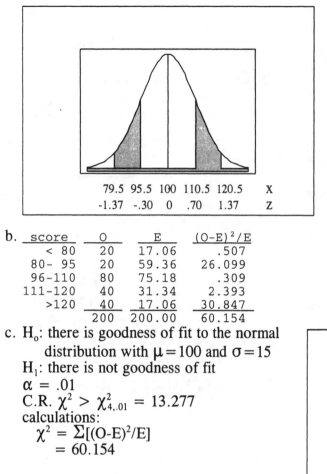

b.

score	O	E	$(O-E)^2/E$
< 80	20	17.06	.507
80- 95	20	59.36	26.099
96-110	80	75.18	.309
111-120	40	31.34	2.393
>120	40	17.06	30.847
	200	200.00	60.154

c. H_o: there is goodness of fit to the normal
   distribution with $\mu = 100$ and $\sigma = 15$
   H_1: there is not goodness of fit
   $\alpha = .01$
   C.R. $\chi^2 > \chi^2_{4,.01} = 13.277$
   calculations:
   $$\chi^2 = \Sigma[(O-E)^2/E]$$
   $$= 60.154$$

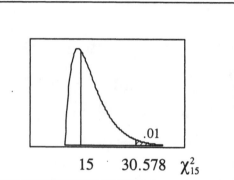

conclusion:
   Reject H_o; there is sufficient evidence to conclude that the observed frequencies do
   not fit a normal distribution with $\mu = 100$ and $\sigma = 15$.

10-3 Contingency Tables: Independence and Homogeneity

NOTE: For each row and each column it must be true that $\Sigma O = \Sigma E$. After the marginal row
and column totals are calculated, both the row totals and the column totals must sum to produce
the same grand total. If either of the preceding is not true, then an error has been made and
further calculations have no meaning. In addition, the following are true for all χ^2 contingency
table analyses in this manual.
* The E values for each cell are given in parentheses below the O values.
* The addends used to calculate the χ^2 test statistic follow the physical arrangement of the cells
    in the original contingency table. This practice makes it easier to monitor the large
    number of intermediate steps involved and helps to prevent errors caused by missing or
    double-counting cells.
* The accompanying chi-square illustrations follow the "usual" shape, even though that shape is
    not correct for df=1 or df=2.

1. H_0: ethnicity and being stopped are independent*
    H_1: ethnicity and being stopped are related*
    $\alpha = .05$
    C.R. $\chi^2 > \chi^2_{1,.05} = 3.841$
    calculations:

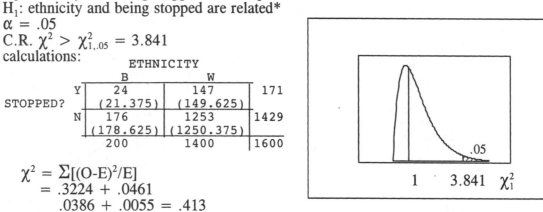

ETHNICITY

		B	W	
STOPPED?	Y	24 (21.375)	147 (149.625)	171
	N	176 (178.625)	1253 (1250.375)	1429
		200	1400	1600

$$\chi^2 = \Sigma[(O-E)^2/E]$$
$$= .3224 + .0461$$
$$.0386 + .0055 = .413$$

conclusion:
   Do not reject H_0; there is not sufficient evidence to reject the claim that ethnicity and being stopped are independent.
   NOTE: The P-value from Minitab can also be used to make the decision. Since .521 > .05, do not reject H_0.
No, we cannot conclude that racial profiling is being used.
*NOTE: While the table claims to summarize "results for randomly selected drivers stopped by police," it appears that fixed numbers (200 black, 1400 white) of the ethnic groups were asked whether or not they had been stopped by the police. If so, the test should use "H_0: the proportion of drivers stopped is the same for both ethnic groups" to test homogeneity and not independence.

3. H_0: response is independent of company status
    H_1: response is related to company status
    $\alpha = .05$
    C.R. $\chi^2 > \chi^2_{1,.05} = 3.841$
    calculations:

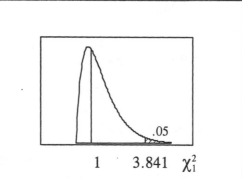

OPINION

		Y	N	
STATUS	W	192 (181.60)	244 (254.40)	436
	W	40 (50.40)	81 (70.60)	121
		232	325	557

$$\chi^2 = \Sigma[(O-E)^2/E]$$
$$= .5954 + .4250$$
$$2.1454 + 1.5316 = 4.697$$

conclusion:
   Reject H_0; there is sufficient evidence to reject the claim that response is independent of company status and to conclude that the two variables are related.
Yes, the decision changes if the .01 level is used – since 4.697 > $\chi^2_{1,.01} = 6.635$.
No, workers and bosses no not appear to agree on this issue.

5. H_o: there is homogeneity of proportions across gender of interviewer
    H_1: there is not homogeneity of proportions across gender of interviewer
    $\alpha = .01$
    C.R. $\chi^2 > \chi^2_{1,.01} = 6.635$
    calculations:

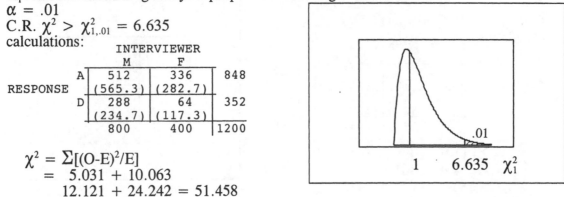

		INTERVIEWER		
		M	F	
	A	512	336	848
RESPONSE		(565.3)	(282.7)	
	D	288	64	352
		(234.7)	(117.3)	
		800	400	1200

$\chi^2 = \Sigma[(O-E)^2/E]$
$= \ \ 5.031 + 10.063$
$\ \ \ 12.121 + 24.242 = 51.458$

conclusion:

Reject H_o; there is sufficient evidence to reject the claim that the proportion of agree/disagree responses are homogeneous across the gender of the interviewer.

7. The lack of precision in the reported results prevents this exercise from having a well-defined unique solution. Any $482 \le x \le 491$, for example, produces the $x/1014 = 48\%$ males as stated. The exercise is worked using the closest integers that yield the stated percents, and the possible extreme solutions are given in a NOTE at the end.
    H_o: gender and fear are flying are independent
    H_1: gender and fear of flying are related
    $\alpha = .05$
    C.R. $\chi^2 > \chi^2_{1,.05} = 3.841$
    calculations:

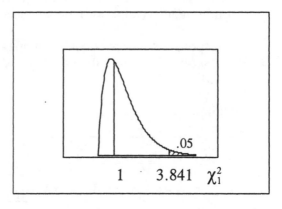

		GENDER		
		M	F	
	Y	58	174	232
FEAR?		(111.42)	(120.58)	
	N	429	353	782
		(375.58)	(406.42)	
		487	527	1014

$\chi^2 = \Sigma[(O-E)^2/E]$
$= 25.615 + 23.671$
$\ \ \ 7.599 + \ \ 7.023 = 63.908$

conclusion:

Reject H_o; there is sufficient evidence to reject the claim that gender and fear of flying are independent and to conclude that the two variables are related.

NOTE: The following spread over more than 10 in the calculated χ^2 is possible from the following frequencies which also meet all the criteria in the statement of the exercise.

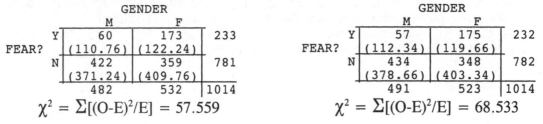

		GENDER		
		M	F	
	Y	60	173	233
FEAR?		(110.76)	(122.24)	
	N	422	359	781
		(371.24)	(409.76)	
		482	532	1014

$\chi^2 = \Sigma[(O-E)^2/E] = 57.559$

		GENDER		
		M	F	
	Y	57	175	232
FEAR?		(112.34)	(119.66)	
	N	434	348	782
		(378.66)	(403.34)	
		491	523	1014

$\chi^2 = \Sigma[(O-E)^2/E] = 68.533$

9. H_0: success in stopping to smoke is independent of the method used
    H_1: success in stopping to smoke is related to the method used.
    $\alpha = .05$
    C.R. $\chi^2 > \chi^2_{2,.05} = 5.991$
    calculations:

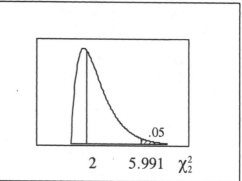

	METHOD			
	GUM	PATCH	INHALER	
N	191	263	95	549
STOP?	(198.34)	(253.87)	(96.79)	
Y	59	57	27	143
	(51.66)	(66.13)	(25.21)	
	250	320	122	692

$\chi^2 = \Sigma[(O-E)^2/E]$
$= 0.271 + 0.328 + 0.033$
$\quad 1.042 + 1.260 + 0.127 = 3.062$

conclusion:
    Do not reject H_0; there is not sufficient evidence to reject the claim that success in stopping to smoke is independent of the method used.

11. H_0: death by homicide and occupation are independent
    H_1: death by homicide and occupation are related
    $\alpha = .05$ [assumed]
    C.R. $\chi^2 > \chi^2_{3,.05} = 7.815$
    calculations:

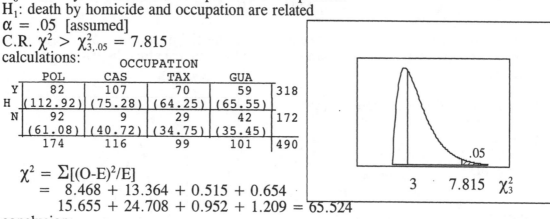

	OCCUPATION				
	POL	CAS	TAX	GUA	
Y	82	107	70	59	318
H	(112.92)	(75.28)	(64.25)	(65.55)	
N	92	9	29	42	172
	(61.08)	(40.72)	(34.75)	(35.45)	
	174	116	99	101	490

$\chi^2 = \Sigma[(O-E)^2/E]$
$= 8.468 + 13.364 + 0.515 + 0.654$
$\quad 15.655 + 24.708 + 0.952 + 1.209 = 65.524$

conclusion:
    Reject H_0; there is sufficient evidence to reject the claim that death by homicide and occupation are independent.
    Yes, the occupation cashier appears to be significantly more prone to death by homicide.

13. Refer to the following table.

	AGE						
	18-21	22-29	30-39	40-49	50-59	60+	
Y	73	255	245	136	138	202	1049
CO-OP	(73.13)	(239.4)	(242.0)	(132.3)	(143.6)	(218.5)	
N	11	20	33	16	27	49	156
	(10.87)	(35.6)	(36.0)	(19.7)	(21.4)	(33.5)	
	84	275	278	152	165	251	1205

H_0: age and cooperation are independent
H_1: age and cooperation are related
$\alpha = .01$
C.R. $\chi^2 > \chi^2_{5,.01} = 15.086$

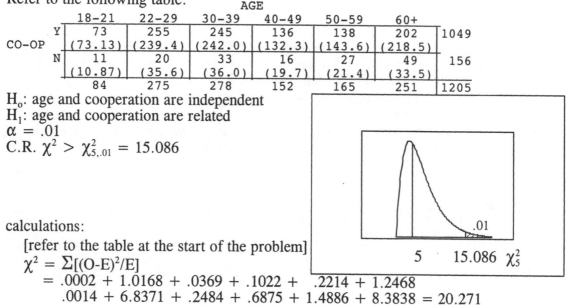

calculations:
    [refer to the table at the start of the problem]
    $\chi^2 = \Sigma[(O-E)^2/E]$
    $= .0002 + 1.0168 + .0369 + .1022 + .2214 + 1.2468$
    $\quad .0014 + 6.8371 + .2484 + .6875 + 1.4886 + 8.3838 = 20.271$

conclusion:
   Reject H_o; there is sufficient evidence to conclude that a person's age and level of cooperation are related.
Yes, those aged 60+ appear to be particularly more uncooperative than the others.

15. H_o: type of crime and criminal/victim connection are independent
H_1: type of crime and criminal/victim connection are related
$\alpha = .05$
C.R. $\chi^2 > \chi^2_{2,.05} = 5.991$
calculations:

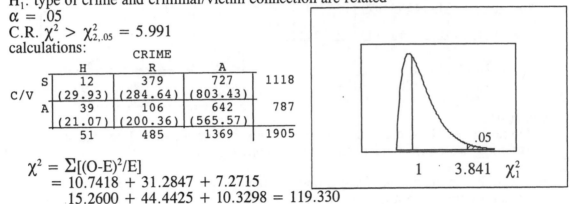

		CRIME H	R	A	
C/V	S	12 (29.93)	379 (284.64)	727 (803.43)	1118
	A	39 (21.07)	106 (200.36)	642 (565.57)	787
		51	485	1369	1905

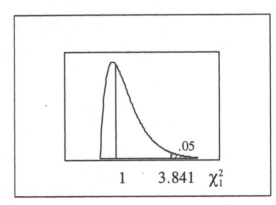

$\chi^2 = \Sigma[(O-E)^2/E]$
  $= 10.7418 + 31.2847 + 7.2715$
     $15.2600 + 44.4425 + 10.3298 = 119.330$
conclusion:
   Reject H_o; there is sufficient evidence to reject the claim that the type of crime and the criminal/victim connection are related and to conclude that the two variables are related.

17. H_o: sentence and plea are independent
H_1: sentence and plea are related
$\alpha = .05$
C.R. $\chi^2 > \chi^2_{1,.05} = 3.841$
calculations:

		PLEA G	NG	
SENTENCE	P	392 (418.48)	58 (31.52)	450
	NP	564 (537.52)	14 (40.48)	578
		956	72	1028

$\chi^2 = \Sigma[(O-E)^2/E]$
  $= 1.6759 + 22.2518$
     $1.3047 + 17.3241 = 42.557$
conclusion:
   Reject H_o; there is sufficient evidence to reject the claim that sentence and plea are independent and to conclude that a person's sentence and his original plea are related.
Yes; assuming that those who are really guilty will indeed be convicted with a trial, these results suggest that a guilty plea should be encouraged. But the study reported only those who plead not guilty and were convicted in trials. Suppose there were also guilty 50 persons who plead not guilty and were acquitted. Including them in the no prison category changes the conclusion entirely and yields the following.

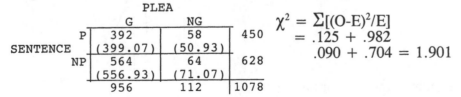

		PLEA G	NG	
SENTENCE	P	392 (399.07)	58 (50.93)	450
	NP	564 (556.93)	64 (71.07)	628
		956	112	1078

$\chi^2 = \Sigma[(O-E)^2/E]$
  $= .125 + .982$
     $.090 + .704 = 1.901$

19. H_o: getting a headache and the amount pf drug used are independent
    H_1: getting a headache and the amount of drug used are related
    $\alpha = .05$
    C.R. $\chi^2 > \chi^2_{3,.05} = 7.815$
    calculations:

		DOSE (mg)				
A		0	10	20/40	80	
C Y		19	47	8	6	80
H		(16.10)	(51.45)	(6.86)	(5.60)	
E N		251	816	107	88	203
?		(253.90)	(811.55)	(108.14)	(88.40)	
		270	863	115	94	1342

$$\chi^2 = \Sigma[(O-E)^2/E]$$
$$= .524 + .384 + .191 + .028$$
$$.033 + .024 + .012 + .002 = 1.199$$

conclusion:
   Do not reject H_o; there is not sufficient evidence to conclude that size of home advantage and the sport are related.

21. H_o: ethnicity and being stopped are independent
    H_1: ethnicity and being stopped are related
    $\alpha = .05$
    C.R. $\chi^2 > \chi^2_{1,.05} = 3.841$
    calculations:

		ETHNICITY		
		B	W	
	Y	24	147	171
STOPPED?		(21.375)	(149.625)	
	N	176	1253	1429
		(178.625)	(1250.375)	
		200	1400	1600

$$\chi^2 = \Sigma[(|O-E|-.5)^2/E]$$
$$= .2113 + .0302$$
$$.0253 + .0036 = .270$$

conclusion:
   Do not reject H_o; there is not sufficient evidence to reject the claim that ethnicity and being stopped are independent.
No, we cannot conclude that racial profiling is being used.
Without the correction for continuity [see the solution for exercise #1 for the details] the calculated test statistic is .413. Since $(|O-E|-.5)^2 < (O-E)^2$ whenever $|O-E| > .25$, Yates' correction generally lowers the calculated test statistic.

Review Exercises

1. H_o: $p_{Mon} = p_{Tue} = p_{Wed} = p_{Thu} = p_{Fri} = 1/5$
    H_1: at least one of the proportions is different from 1/5
    $\alpha = .05$
    C.R. $\chi^2 > \chi^2_{4,.05} = 9.488$
    calculations:

day	O	E	$(O-E)^2/E$
Mon	98	75	7.053
Tue	68	75	.653
Wed	89	75	2.613
Thu	64	75	1.613
Fri	56	75	4.813
	375	375	16.747

$\chi^2 = \Sigma[(O-E)^2/E]$
$\quad = 16.747$

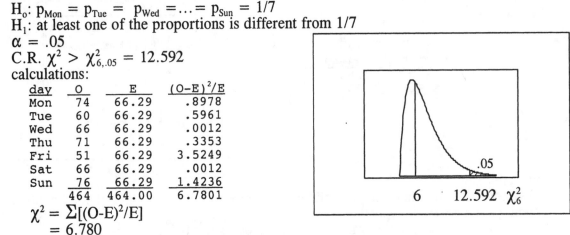

conclusion:
    Reject H_o; there is sufficient evidence to conclude that at least one of the proportions is different from 1/5.
Since the calls are not uniformly distributed over the days of the business week, it would not make sense to have the staffing level the same each day – unless the company staffed to handle the higher numbers of calls and had other work available to do during slack times.

2. H_o: $p_{Mon} = p_{Tue} = p_{Wed} = \ldots = p_{Sun} = 1/7$
    H_1: at least one of the proportions is different from 1/7
    $\alpha = .05$
    C.R. $\chi^2 > \chi^2_{6,.05} = 12.592$
    calculations:

day	O	E	$(O-E)^2/E$
Mon	74	66.29	.8978
Tue	60	66.29	.5961
Wed	66	66.29	.0012
Thu	71	66.29	.3353
Fri	51	66.29	3.5249
Sat	66	66.29	.0012
Sun	76	66.29	1.4236
	464	464.00	6.7801

$\chi^2 = \Sigma[(O-E)^2/E]$
$\quad = 6.780$

conclusion:
    Do not reject H_o; there is not sufficient evidence to reject the claim that the gunfire death rates are the same for the different days of the week.
No, there is not support for the theory theat more gunfire deaths occur on weekends.

3. Refer to the following table and calculations.

		arson	rape	violence	CRIME stealing	coining	fraud	
USE	Y	50 (49.11)	88 (79.21)	155 (139.93)	379 (358.55)	18 (16.90)	63 (109.31)	753
	N	43 (43.89)	62 (70.79)	110 (125.07)	300 (320.45)	14 (15.10)	144 (97.69)	673
		93	150	265	679	32	207	1426

$\chi^2 = \Sigma[(O-E)^2/E]$
$\quad = .016 + 0.976 + 1.622 + 1.167 + .072 + 19.617$
$\quad\quad .018 + 1.092 + 1.815 + 1.306 + .080 + 21.949 = 49.731$

H_o: alcohol use and type of crime are independent
H_1: alcohol use and type of crime are related
α = .05 [assumed]
C.R. $\chi^2 > \chi^2_{5,.05}$ = 11.071
calculations:

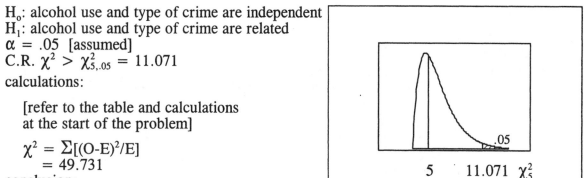

[refer to the table and calculations
at the start of the problem]

$\chi^2 = \Sigma[(O\text{-}E)^2/E]$
  = 49.731
conclusion:

Reject H_o; there is sufficient evidence to conclude that alcohol use and the type of crime committed are related.

Fraud seems to be different from the other crimes in that it is more likely to be committed by someone who abstains from alcohol.

4. H_o: success in stopping to smoke is independent of the method used
    H_1: success in stopping to smoke is related to the method used
    α = .05
    C.R. $\chi^2 > \chi^2_{1,.05}$ = 3.841
    calculations:

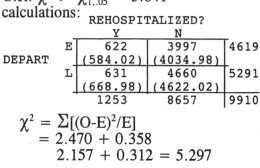

REHOSPITALIZED?

		Y	N	
DEPART	E	622 (584.02)	3997 (4034.98)	4619
	L	631 (668.98)	4660 (4622.02)	5291
		1253	8657	9910

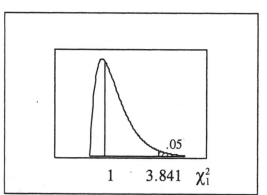

$\chi^2 = \Sigma[(O\text{-}E)^2/E]$
  = 2.470 + 0.358
    2.157 + 0.312 = 5.297

conclusion:

Reject H_o; there is sufficient evidence to reject the claim that whether a newborn is rehospitalized is independent of how soon it was discharged and to conclude that the two variables are related.

Yes, the conclusion changes if the level of significance becomes .01 [CV = 6.635]. This means we can be 95% confidence such a relationship exists, but not 99% confident.

Cumulative Review Exercises

1. scores in order: 66 75 77 80 82 84 89 94; n = 8 Σx = 647 Σx^2 = 52847
    $\overline{x}$ = 647/8 = 80.9
    $\tilde{x}$ = (80 + 82)/2 = 81.0
    R = 94 - 66 = 28
    $s^2 = [8(52847) - (647)^2]/[8(7)]$ = 74.41
    s = 8.6
    five-number summary:
       x_1 = 66
       P_{25} = (75 + 77)/2 = 76.0 [(.25)8 = 2, a whole number, use $x_{2.5}$]
       $P_{50} = \tilde{x}$ = 81.0
       P_{75} = (84 + 89)/2 = 86.5 [(.75)8 = 6, a whole number, use $x_{6.5}$]
       x_8 = 94

2. Consider the table at the right.
    a. $P(C) = 176/647 = .272$
    b. $P(M) = 303/647 = .468$
    c. $P(M \text{ or } C) = P(M) + P(C) - P(M \text{ and } C)$
    $$= (303/647) + (176/647) - (82/647)$$
    $$= 397/647 = .614$$
    d. $P(F_1 \text{ and } F_2) = P(F_1) \cdot P(F_2 | F_1)$
    $$= (344/647) \cdot (343/646) = .282$$

	A	B	C	D	
M	66	80	82	75	303
F	77	89	94	84	344
	143	169	176	159	647

3. H_o: gender and selection are independent
    H_1: gender and selection are related
    $\alpha = .05$ [assumed]
    C.R. $\chi^2 > \chi^2_{3,.05} = 7.815$
    calculations:

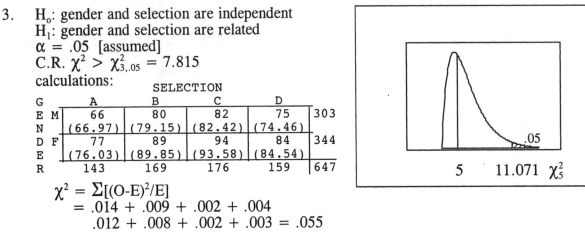

SELECTION

G		A	B	C	D	
E	M	66	80	82	75	303
N		(66.97)	(79.15)	(82.42)	(74.46)	
D	F	77	89	94	84	344
E		(76.03)	(89.85)	(93.58)	(84.54)	
R		143	169	176	159	647

5   11.071 χ^2_5

$$\chi^2 = \Sigma[(O-E)^2/E]$$
$$= .014 + .009 + .002 + .004$$
$$.012 + .008 + .002 + .003 = .055$$

conclusion:
    Do not reject H_o; there is not sufficient evidence to reject the claim that gender and selection are independent.

4. $n = 4$ $\Sigma xy = 26210$ $n(\Sigma xy) - (\Sigma x)(\Sigma y) = 4(26210) - (303)(344) = 608$
    $\Sigma x = 303$ $\Sigma x^2 = 23105$ $n(\Sigma x^2) - (\Sigma x)^2 = 4(23105) - (303)^2 = 611$
    $\Sigma y = 344$ $\Sigma y^2 = 29742$ $n(\Sigma y^2) - (\Sigma y)^2 = 4(29742) - (344)^2 = 632$
    $r = [n(\Sigma xy) - (\Sigma x)(\Sigma y)]/[\sqrt{n(\Sigma x^2) - (\Sigma x)^2} \cdot \sqrt{n(\Sigma y^2) - (\Sigma y)^2}]$
    $$= 608/[\sqrt{611} \cdot \sqrt{632}]$$
    $$= .978$$

    original claim: $\rho \neq 0$
    H_o: $\rho = 0$
    H_1: $\rho \neq 0$
    $\alpha = .05$ [assumed]
    C.R. $r < -.950$ OR   C.R. $t < -t_{2,.025} = -4.303$
         $r > .950$ $t > t_{2,.025} = 4.303$

.025 / .025

-.950   0   .950   r
-4.303   0   4.303 t_2

    calculations: calculations:
      $r = .978$ $t_r = (r - \mu_r)/s_r$
    $$= (.978 - 0)/\sqrt{(1-(.978)^2)/2}$$
    $$= .978/.146$$
    $$= 6.696$$

conclusion:
    Reject H_o; there is sufficient evidence to conclude that $\rho \neq 0$ (in fact, $\rho > 0$).

5. $d = y - x$: 11 9 12 9

$n = 4$ $\qquad \Sigma d = 41$ $\qquad \Sigma d^2 = 427$

$\qquad\qquad\quad \bar{d} = 10.25$ $\qquad s_d = 1.50$

original claim: $\mu_d > 0$

H_o: $\mu_d = 0$

H_1: $\mu_d > 0$

$\alpha = .05$ [assumed]

C.R. $t > t_{3,.05} = 2.353$

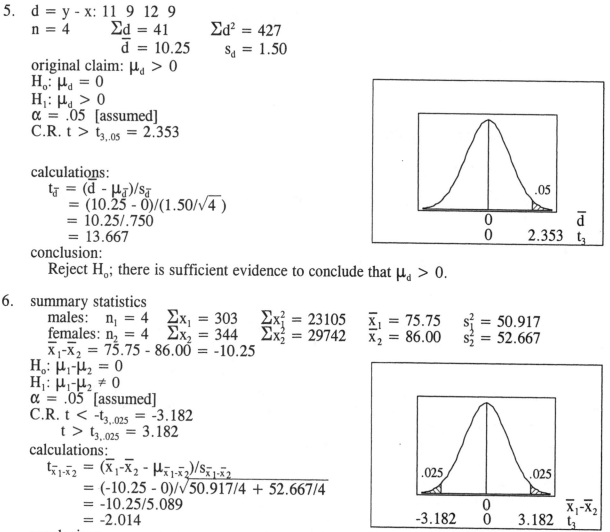

calculations:

$\quad t_{\bar{d}} = (\bar{d} - \mu_{\bar{d}})/s_{\bar{d}}$

$\qquad = (10.25 - 0)/(1.50/\sqrt{4}\,)$

$\qquad = 10.25/.750$

$\qquad = 13.667$

conclusion:

Reject H_o; there is sufficient evidence to conclude that $\mu_d > 0$.

6. summary statistics

males: $n_1 = 4$ $\quad \Sigma x_1 = 303$ $\quad \Sigma x_1^2 = 23105$ $\quad \bar{x}_1 = 75.75$ $\quad s_1^2 = 50.917$

females: $n_2 = 4$ $\quad \Sigma x_2 = 344$ $\quad \Sigma x_2^2 = 29742$ $\quad \bar{x}_2 = 86.00$ $\quad s_2^2 = 52.667$

$\bar{x}_1 - \bar{x}_2 = 75.75 - 86.00 = -10.25$

H_o: $\mu_1 - \mu_2 = 0$

H_1: $\mu_1 - \mu_2 \neq 0$

$\alpha = .05$ [assumed]

C.R. $t < -t_{3,.025} = -3.182$

$\qquad t > t_{3,.025} = 3.182$

calculations:

$\quad t_{\bar{x}_1 - \bar{x}_2} = (\bar{x}_1 - \bar{x}_2 - \mu_{\bar{x}_1 - \bar{x}_2})/s_{\bar{x}_1 - \bar{x}_2}$

$\qquad = (-10.25 - 0)/\sqrt{50.917/4 + 52.667/4}$

$\qquad = -10.25/5.089$

$\qquad = -2.014$

conclusion:

Do not reject H_o; there is not sufficient evidence to reject the claim that $\mu_1 - \mu_2 = 0$.

Chapter 11

Analysis of Variance

11-2 One-Way ANOVA

1. a. H_o: $\mu_1 = \mu_2 = \mu_3$
    b. H_1: at least one mean is different
    c. $F_{33}^2 = 8.98$
    d. $F_{33,.05}^2 = 3.3158$ [closest entry in Table A-5]
    e. P-value = .001
    f. Since $8.98 > 3.3158$, reject H_o and conclude the three mean F-K Grade Level scores are not all the same.

3. a. H_o: $\mu_1 = \mu_2 = \mu_3$
    b. H_1: at least one mean is different
    c. $F_{108}^2 = .1887$
    d. $F_{108,.05}^2 = 3.0804$ [from Excel]
    e. P-value = .8283
    f. No; since $.1887 < 3.0804$, there is not sufficient evidence to support the claim that the means for three different age groups are not all the same.

NOTE: This section is calculation-oriented. Do not get so involved with the formulas that you miss concepts. This manual arranges the calculations to promote both computational efficiency and understanding of the underlying principles. The following notation is used in this section.

k = the number of groups

n_i = the number of scores in group i (where i = 1,2,...,k)

$\bar{x}_i$ = the mean of group i

s_i^2 = the variance of group i

$\bar{\bar{x}}$ = the overall mean of all the scores in all the groups

$\quad = \Sigma n_i \bar{x}_i / \Sigma n_i$ = the (weighted) mean of the group means

$\quad = \Sigma \bar{x}_i / k$ = simplified form when each group has equal size n

s_B^2 = the variance between the groups

$\quad = \Sigma n_i (\bar{x}_i - \bar{\bar{x}})^2 / (k-1)$

$\quad = n\Sigma (\bar{x}_i - \bar{\bar{x}})^2 / (k-1) = ns_{\bar{x}}^2$ = simplified form when each group has equal size n

s_p^2 = the variance within the groups

$\quad = \Sigma df_i s_i^2 / \Sigma df_i$ = the (weighted) mean of the group variances

$\qquad\qquad$ = the two-sample formula for s_p^2 generalized to k samples

$\quad = \Sigma s_i^2 / k$ = simplified form when each group has equal size n

numerator df = k-1

denominator df = Σdf_i

$\qquad\qquad$ = k(n-1) = simplified form when each group has equal size n

$F = s_B^2 / s_p^2$ = (variance between groups)/(variance within groups)

5. Since each group has equal size n, the simplified forms can be used.
The following preliminary values are identified and/or calculated.

	subcompact	compact	midsize	full size
n	5	5	5	5
Σx	3344	2779	2434	2689
Σx^2	2470638	1577659	1297312	1541765
$\overline{x}$	668.8	555.8	486.8	537.8
s^2	58542.7	8272.8	28110.2	23905.2

$k = 4$ $\overline{\overline{x}} = \Sigma\overline{x}_i/k$
$n = 5$ $= 562.3$
$s_{\overline{x}}^2 = \Sigma(\overline{x}_i-\overline{\overline{x}})^2/(k-1)$ $s_p^2 = \Sigma s_i^2/k$
     $= 5895$ $= 29707.725$

$H_o: \mu_1 = \mu_2 = \mu_3 = \mu_4$
H_1: at least one mean is different
$\alpha = .05$
C.R. $F > F_{16,.05}^3 = 3.2389$
calculations:
     $F = ns_{\overline{x}}^2/s_p^2$
       $= 5(5895)/29797.725$
       $= .9922$

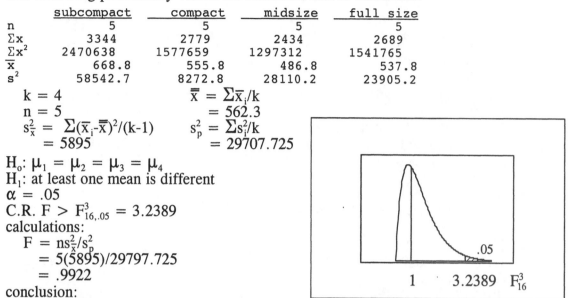

conclusion:
Do not reject H_o; there is not sufficient evidence to conclude that at least one mean is different.
No, these data do not suggest that larger cars are safer.

7. Since each group has equal size n, the simplified forms can be used.
The following preliminary values are identified and/or calculated.

	4000 BC	1850 BC	150 AD
n	9	9	9
Σx	1194	1210	1243
Σx^2	158544	162768	171853
$\overline{x}$	132.67	134.44	138.11
s^2	17.500	11.278	22.611

$k = 3$ $\overline{\overline{x}} = \Sigma\overline{x}_i/k$
$n = 9$ $= 135.07$
$s_{\overline{x}}^2 = \Sigma(\overline{x}_i-\overline{\overline{x}})^2/(k-1)$ $s_p^2 = \Sigma s_i^2/k$
     $= 15.416/2$ $= 51.389/3$
     $= 7.708$ $= 17.130$

$H_o: \mu_1 = \mu_2 = \mu_3$
H_1: at least one mean is different
$\alpha = .05$
C.R. $F > F_{24,.05}^2 = 3.4028$
calculations:
     $F = ns_{\overline{x}}^2/s_p^2$
       $= 9(7.708)/17.130$
       $= 4.0498$

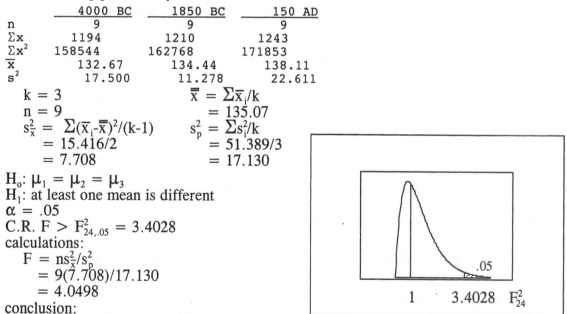

conclusion:
Reject H_o; there is sufficient evidence to conclude that at least one mean is different.

9. The following preliminary values are identified and/or calculated.

	R	O	Y	Br	Bl	G	total
n	21	8	26	33	5	7	100
Σx	19.104	7.401	23.849	30.123	4.507	6.846	91.470
Σx^2	17.934278	6.862429	21.904543	27.546793	4.075729	6.018734	83.802506
$\overline{x}$	.90971	.92513	.91272	.91282	.90140	.92657	.91470
s^2	.000755	.002226	.001144	.001562	.003280	.001499	

$\overline{\overline{x}} = \Sigma n_i \overline{x}_i / \Sigma n_i$
    $= [21(.90971) + 8(.92513) + 26(.91272) + 33(.91282) + 5(.90140) + 7(.92657)]/100$
    $= 91.470/100$
    $= .91470$ [NOTE: $\overline{\overline{x}}$ must always agree with the $\overline{x}$ in the "total" column.]

$\Sigma n_i(\overline{x}_i - \overline{\overline{x}})^2 = 21(.90971-.91470)^2 + 8(.92513-.91470)^2 + 26(.91272-.91470)^2 +$
        $33(.91282-.91470)^2 + 5(.90140-.91470)^2 + 7(.92657-.91470)^2$
        $= .00355$

$\Sigma df_i s_i^2 = 20(.000755) + 7(.002226) + 25(.001144) +$
        $32(.001562) + 4(.003280) + 6(.001499)$
        $= .13135$

$s_B^2 = \Sigma n_i(\overline{x}_i - \overline{\overline{x}})^2/(k-1)$
    $= .00355/5 = .00071$

$s_p^2 = \Sigma df_i s_i^2 / \Sigma df_i$
    $= .13135/94 = .001398$

$H_o: \mu_R = \mu_O = \mu_Y = \mu_{Br} = \mu_{Bl} = \mu_G$
$H_1:$ at least one mean is different
$\alpha = .05$
C.R. $F > F_{94,.05}^5 = 2.2899$
calculations:
    $F = s_B^2/s_p^2$
        $= .00071/.001398$
        $= .5081$

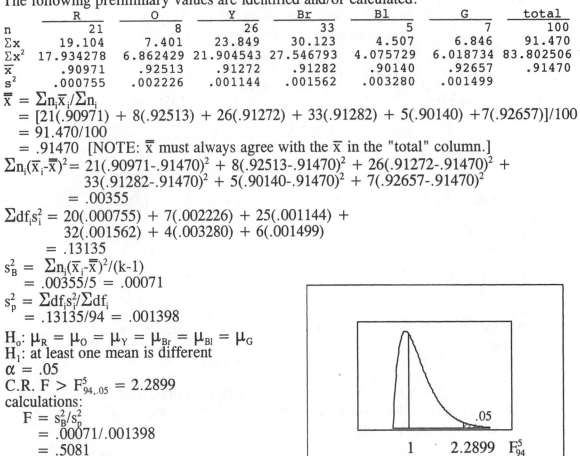

conclusion:
    Do not reject H_o; there is not sufficient evidence to conclude that at least one mean is different.

No. If the intent is to make the different colors have the same mean, there is no evidence that this is not being accomplished. Corrective action is not required.

NOTE: An ANOVA table may be completed as follows:

source	SS	df	MS	F
Trt	.00355	5	.00071	.51
Error	.13135	94	.00140	
Total	.13490	99		

$F = MS_{Trt}/MS_{Err}$
  $= .00071/.00140$
  $= .51$

(1) Enter $SS_{Trt} = \Sigma n_i(\overline{x}_i - \overline{\overline{x}})^2$ and $SS_{Err} = \Sigma df_i s_i^2$ values from the preliminary calculations.
(2) Enter $df_{Trt} = k-1$ and $df_{Err} = \Sigma df_i = \Sigma(n_i - 1) = \Sigma n_i - k$.
(3) Add the SS and df columns to find SS_{Tot} and df_{Tot}. [The df_{Tot} must equal $\Sigma n_i - 1$.]
(4) Calculate $MS_{Trt} = SS_{Trt}/df_{Trt}$ and $MS_{Err} = SS_{Err}/df_{Err}$.
(5) Calculate $F = MS_{Trt}/MS_{Err}$.
As a final check, calculate s^2 (i.e., the variance of all the scores in one large group) two different ways as indicated below. If these answers agree, the problem is probably correct.
    * from the "total" column in the table for the preliminary calculations:
        $s^2 = [n\Sigma x^2 - (\Sigma x)^2]/[n(n-1)]$
            $= [100(83.802506) - (91.470)^2]/[100(99)]$
            $= 13.4897/9900 = .001363$
    * from the "total" row of the ANOVA table
        $s^2 = SS_{Tot}/df_{Tot}$
            $= .13490/99 = .001363$

11. The following preliminary values are identified and/or calculated. ["shelf 3" = shelf 3/4]

	shelf 1	shelf 2	shelf 3	total
n	6	6	4	16
Σx	1.19	2.73	.80	4.72
Σx^2	.3853	1.2445	.1846	1.8144
$\overline{x}$	.1983	.4550	.2000	.2950
s^2	.029857	.000470	.008200	

$\overline{\overline{x}} = \Sigma n_i \overline{x}_i / \Sigma n_i$
$= [6(.1983) + 6(.4550) + 4(.2000)]/16 = .2950$
$\Sigma n_i(\overline{x}_i - \overline{\overline{x}})^2 = 6(.1983 - .2950)^2 + 6(.4550 - .2950)^2 + 4(.2000 - .295)^2 = .24577$
$\Sigma df_i s_i^2 = 5(.029857) + 5(.000470) + 3(.008200) = .17624$
$s_B^2 = \Sigma n_i(\overline{x}_i - \overline{\overline{x}})^2/(k-1)$
$= .24577/2 = .12288$
$s_p^2 = \Sigma df_i s_i^2 / \Sigma df_i$
$= .17624/13 = .01356$

$H_o: \mu_1 = \mu_2 = \mu_3$
$H_1:$ at least one mean is different
$\alpha = .05$
C.R. $F > F_{13,.05}^2 = 3.8056$
calculations:
$\quad F = s_B^2 / s_p^2$
$\quad\quad = .12288/.01356$
$\quad\quad = 9.0645$

conclusion:
Reject H_o; there is sufficient evidence to conclude that at least one mean is different.
It appears that the cereals are not placed on the shelves at random, but that those with higher sugar amounts are on shelf 2 – at children's eye-level.

13. a. $_5C_2 = 5!/2!3! = 10$
    b. P(no type I error in one t test) = .95
       assuming independence, P(no Type I error in ten t tests) = $(.95)^{10}$ = .599
    c. P(no type I error in one F test) = .95
    d. the analysis of variance test

11-3 Two-Way ANOVA

NOTE: The formulas and principles in this section are logical extensions of the previous ones.
$\quad SS_{Row} = \Sigma n_i(\overline{x}_i - \overline{\overline{x}})^2$ for i=1,2,3... [for each row]
$\quad SS_{Col} = \Sigma n_j(\overline{x}_j - \overline{\overline{x}})^2$ for j=1,2,3... [for each column]
$\quad SS_{Tot} = \Sigma(x - \overline{\overline{x}})^2$ [for all the x's]
When there is only one observation per cell the unexplained variation is
$\quad\quad SS_{Err} = SS_{Tot} - SS_{Row} - SS_{Col}$
    and there is not enough data to measure interaction.
When there is more than one observation per cell the unexplained variation (i.e., failure of items in the same cell to respond the same) is
$\quad\quad SS_{Err} = \Sigma(x - \overline{x}_{ij})^2 = \Sigma df_{ij} s_{ij}^2$ [for each cell – i,e., for each i,j (row,col) combination]
    and the interaction sum of squares is
$\quad\quad SS_{Int} = SS_{Tot} - SS_{Row} - SS_{Col} - SS_{Err}.$
Since the data will be analyzed from statistical software packages, however, the above formulas need not be used by hand.

1. The term "two-way" refers to the fact that the data can be analyzed from two different perspectives. There are two factors acting simultaneously that may or may not affect the response, and this technique allows for simultaneous analyses of both factors.
    The term "analysis of variance" refers to the fact that we are making judgments about means by analyzing variances. If the variance between the means is larger than the unexplained natural variation in the problem, then the means are declared to be different.

3. If there is a significant interaction, no general statements can be made about either of the factors – differences in the levels of factor A may be significant at one level of factor B, but not at another level of factor B.

5. H_o: there is no site-age interaction effect
    H_1: there is a site-age interaction effect
    $\alpha = .05$
    C.R. $F > F_{18,.05}^{4} = 2.9277$
    calculations:
    $\quad F = MS_{INT}/MS_E$
    $\quad\quad = 4.43/3.44$
    $\quad\quad = 1.28$
    $\quad$P-value = .313 [from Minitab]

    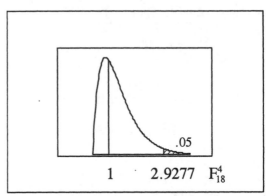

    conclusion:
    $\quad$Do not reject H_o; there is not sufficient evidence to conclude that there is any interaction between a falcon's site and its age (in determining its DDT level).

7. H_o: $\mu_1 = \mu_2 = \mu_3$ [there is no age effect]
    H_1: at least one μ_i is different
    $\alpha = .05$
    C.R. $F > F_{18,.05}^{2} = 3.5546$
    calculations:
    $\quad F = MS_{AGE}/MS_E$
    $\quad\quad = 860.59/3.44$
    $\quad\quad = 249.85$
    $\quad$P-value = .000 [from Minitab]

    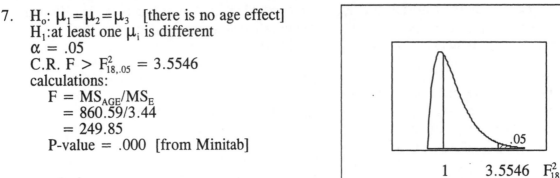

    conclusion:
    $\quad$Reject H_o; there is sufficient evidence to conclude that age has an effect on the amount of DDT.

9. H_o: $\mu_M = \mu_F$ [there is no gender effect]
    H_1: the gender means are different
    $\alpha = .05$
    C.R. $F > F_{36,.05}^{1} = 4.0847$
    calculations:
    $\quad F = MS_{GENDER}/MS_E$
    $\quad\quad = 52635/10465$
    $\quad\quad = 5.03$
    $\quad$P-value = .031 [from Minitab]

    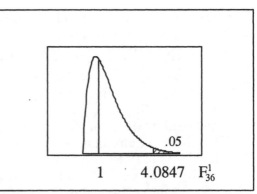

    conclusion:
    $\quad$Reject H_o; there is sufficient evidence to conclude that gender affects SAT score.

11. $H_0: \mu_1 = \mu_2 = \dots = \mu_{24}$ [there is no subject effect]
H_1: at least one μ_i is different
$\alpha = .05$
C.R. $F > F^{23}_{69,.05} = 1.7001$
calculations:
$F = MS_{SUBJ}/MS_E$
$= 140.5/36.3$
$= 3.87$
P-value $= .000$ [from Minitab]

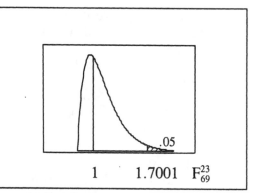

conclusion:
Reject H_0; there is sufficient evidence to
conclude that the choice of subject has an effect on the hearing test score.
This makes practical sense, because all people do not have the same level of hearing.

13. While it is assumed the ANOVA table will be obtained using a statistical software
package, the actual calculations necessary to construct the table by hand are given below.
There are n=4 observations in each of the indicated age-gender cells.

```
                              AGE
               under 20      20-40        over 40
               Σx  = 288     Σx  = 288    Σx  = 288
G              Σx² = 21536   Σx² = 21120  Σx² = 21152
E    male      x̄ = 72        x̄ = 72       x̄ = 72        x̄ = 72
N              s² = 266.667  s² = 128     s² = 138.667
D
E              Σx  = 284     Σx  = 300    Σx  = 264
R              Σx² = 20272   Σx² = 22736  Σx² = 17504
     female    x̄ = 71        x̄ = 75       x̄ = 66        x̄ = 70.667
               s² = 36       s² = 78.667  s² = 26.667

               x̄ = 71.5      x̄ = 73.5     x̄ = 69      Σx  = 1712    x̿ = 71.333
                                                       Σx² = 124320  s² = 95.536
```

$SS_{Gender} = \sum n_i(\bar{x}_i - \bar{\bar{x}})^2 = 12(72-71.333)^2 + 12(70.667-71.333)^2 = 10.667$
$SS_{Age} = \sum n_j(\bar{x}_j - \bar{\bar{x}})^2 = 8(71.5-71.333)^2 + 8(73.5-71.333)^2 + 8(69-71.333)^2 = 81.333$
$SS_{Tot} = \sum(x-\bar{\bar{x}})^2 = df \cdot s^2 = 23(95.536) = 2197.333$
$SS_{Err} = \sum(x-\bar{x}_{ij})^2 = \sum df_{ij}s_{ij}^2 = 3(266.667) + 3(128) + 3(138.667) +$
$\qquad\qquad\qquad\qquad\qquad\qquad 3(36) + 3(78.667) + 3(26.667) = 2024$
$SS_{Int} = SS_{Tot} - SS_{Gender} - SS_{Age} - SS_{Err}$
$\qquad = 2197.333 - 10.667 - 81.333 - 2024 = 81.333$

The resulting ANOVA table is used for the test of hypothesis in this exercise.

Source	df	SS	MS	F
Gender	1	10.667	10.667	.0949
Age	2	81.333	40.667	.3617
Interaction	2	81.333	40.667	.3617
Error	18	2024.000	112.444	
Total	23	2197.333		

H_0: there is no gender-age interaction effect
H_1: there is a gender-age interaction effect
$\alpha = .05$
C.R. $F > F^2_{18,.05} = 3.5546$
calculations:
$F = MS_{INT}/MS_E$
$= 40.667/112.444$
$= .3617$

conclusion:
Do not reject H_0; there is not sufficient
evidence to conclude that there is any
interaction between a person's gender and age (in determining pulse rate).

H_o: $\mu_M = \mu_F$ [there is no gender effect]
H_1: at least one μ_i is different
$\alpha = .05$
C.R. $F > F_{18,.05}^1 = 4.4139$
calculations:
$F = MS_{GENDER}/MS_E$
$\quad = 10.667/112.444$
$\quad = .0949$

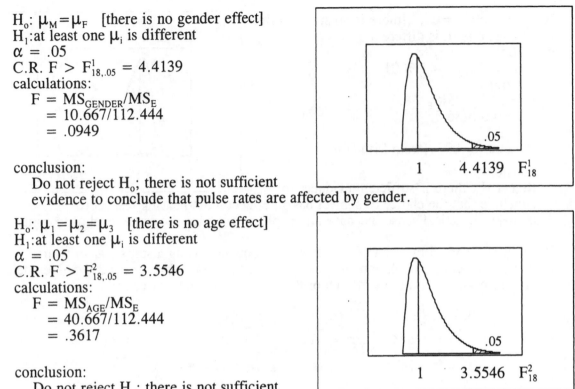

conclusion:
Do not reject H_o; there is not sufficient
evidence to conclude that pulse rates are affected by gender.

H_o: $\mu_1 = \mu_2 = \mu_3$ [there is no age effect]
H_1: at least one μ_i is different
$\alpha = .05$
C.R. $F > F_{18,.05}^2 = 3.5546$
calculations:
$F = MS_{AGE}/MS_E$
$\quad = 40.667/112.444$
$\quad = .3617$

conclusion:
Do not reject H_o; there is not sufficient
evidence to conclude that pulse rates are affected by age.

15. a. No change. The ANOVA calculated statistics are ratios of variances. Since adding the same value to each score does not affect the variances, the ANOVA table will not change.
   b. No change. The ANOVA calculated statistics are ratios of variances. Since multiplying each score by the same nonzero constant will multiply the all the variances by the square of that constant, the numerators and denominators of each ANOVA ratio will be multiplied by the same constant and the ratios will not change.
   c. No change. The same values will appear in different positions, but referring to the same factors as before.
   d. Depends. The change will affect the mean for row 1, the mean for column 1, and the variability within the first cell. In general, all calculated ANOVA statistics will change.

Review Exercises

1. H_o: $\mu_A = \mu_B = \mu_C$
    H_1: at least one $= \mu_i$ is different
    $\alpha = .05$
    C.R. $F > F_{14,.05}^2 = 3.7389$
    calculations:
    $F = s_B^2/s_p^2$
    $\quad = .0038286/.0000816$
    $\quad = 46.90$
    P-value $= .000$ [from Minitab]

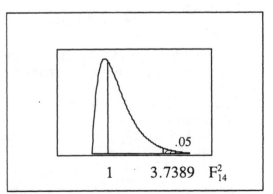

conclusion:
Reject H_o; there is not sufficient evidence to reject the claim that the three groups have the same mean and to conclude that at least one mean is different.

2. Since each group has equal size n, the simplified forms can be used.
    The following preliminary values are identified and/or calculated.

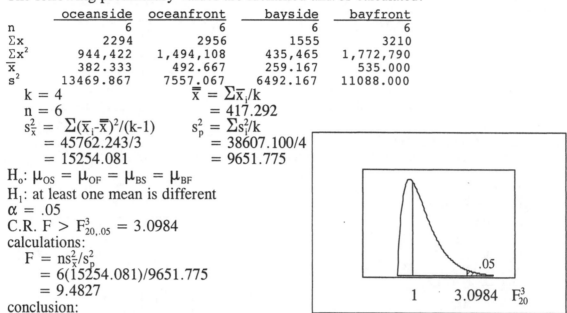

	oceanside	oceanfront	bayside	bayfront
n	6	6	6	6
Σx	2294	2956	1555	3210
Σx^2	944,422	1,494,108	435,465	1,772,790
$\bar{x}$	382.333	492.667	259.167	535.000
s^2	13469.867	7557.067	6492.167	11088.000

$k = 4$ $\bar{\bar{x}} = \Sigma \bar{x}_i/k$
$n = 6$ $= 417.292$
$s_{\bar{x}}^2 = \Sigma(\bar{x}_i-\bar{\bar{x}})^2/(k-1)$ $s_p^2 = \Sigma s_i^2/k$
     $= 45762.243/3$ $= 38607.100/4$
     $= 15254.081$ $= 9651.775$

$H_0: \mu_{OS} = \mu_{OF} = \mu_{BS} = \mu_{BF}$
$H_1:$ at least one mean is different
$\alpha = .05$
C.R. $F > F_{20,.05}^3 = 3.0984$
calculations:
    $F = ns_{\bar{x}}^2/s_p^2$
      $= 6(15254.081)/9651.775$
      $= 9.4827$
conclusion:
    Reject H_0; there is sufficient evidence to conclude that at least one mean is different.

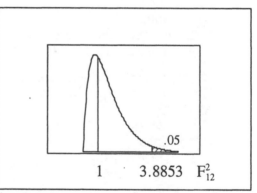

3. $H_0:$ there is no gender-major interaction effect
    $H_1:$ there is a gender-major interaction effect
    $\alpha = .05$
    C.R. $F > F_{12,.05}^2 = 3.8853$
    calculations:
        $F = MS_{INT}/MS_E$
          $= 7.1/37.8$
          $= .19$
        P-value $= .832$ [from Minitab]
    conclusion:
        Do not reject H_0; there is not sufficient
    evidence to conclude that there is any interaction between a person's gender and major
    (in estimating length).

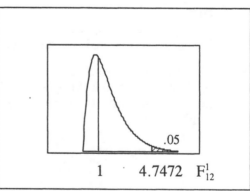

4. $H_0: \mu_M = \mu_F$ [there is no gender effect]
    $H_1:$ the gender means are different
    $\alpha = .05$
    C.R. $F > F_{12,.05}^1 = 4.7472$
    calculations:
        $F = MS_{GENDER}/MS_E$
          $= 29.4/37.8$
          $= .78$
        P-value $= .395$ [from Minitab]
    conclusion:
        Do not Reject H_0; there is not sufficient
    evidence to conclude that estimated length is affected by gender.

5. H_o: $\mu_M = \mu_B = \mu_L$ [there is no major effect]
     H_1: the major means are different
     $\alpha = .05$
     C.R. $F > F^2_{12,.05} = 3.8853$
     calculations:
        $F = MS_{MAJOR}/MS_E$
          $= 5.1/37.8$
          $= .13$
        P-value $= .876$ [from Minitab]

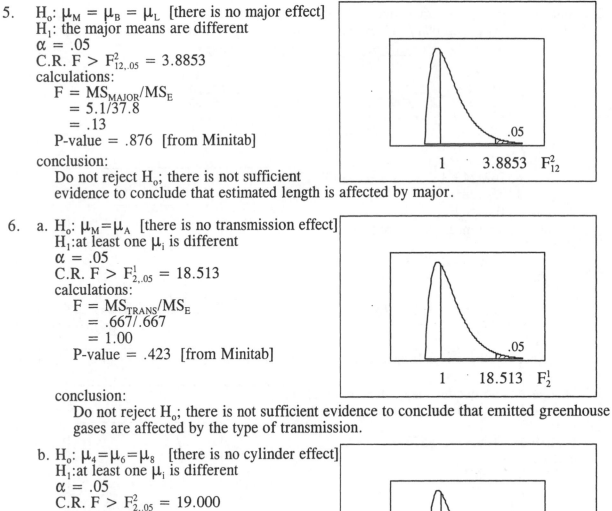

     conclusion:
        Do not reject H_o; there is not sufficient
     evidence to conclude that estimated length is affected by major.

6.   a. H_o: $\mu_M = \mu_A$ [there is no transmission effect]
        H_1: at least one μ_i is different
        $\alpha = .05$
        C.R. $F > F^1_{2,.05} = 18.513$
        calculations:
           $F = MS_{TRANS}/MS_E$
             $= .667/.667$
             $= 1.00$
           P-value $= .423$ [from Minitab]

     conclusion:
        Do not reject H_o; there is not sufficient evidence to conclude that emitted greenhouse
        gases are affected by the type of transmission.

     b. H_o: $\mu_4 = \mu_6 = \mu_8$ [there is no cylinder effect]
        H_1: at least one μ_i is different
        $\alpha = .05$
        C.R. $F > F^2_{2,.05} = 19.000$
        calculations:
           $F = MS_{CYL}/MS_E$
             $= 4.667/.667$
             $= 7.000$
           P-value $= .125$ [from Minitab]

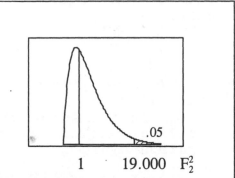

     conclusion:
        Do not reject H_o; there is not sufficient evidence to conclude that emitted greenhouse
        gases are affected by the number of cylinders.

     c. No; we cannot conclude there in no effect, but only that there is not enough evidence to
        be 95% certain that there is an effect. The very small sample sizes make it unlikely that
        the data would provide enough evidence to support any claim, no matter how valid it
        might be.

Cumulative Review Exercises

1. summary statistics: $n = 52$ $\Sigma x = 5.22$ $\Sigma x^2 = 4.0416$

a. $\bar{x} = (\Sigma x)/n$
$= 5.22/52 = .100$

b. $s^2 = [n(\Sigma x^2) - (\Sigma x)^2]/[n(n-1)]$
$= [52(4.0416) - (5.22)^2]/[52(51)] = .06897$
$s = .263$

c. from the ordered list:
$x_1 = 0$
$Q_1 = P_{25} = (x_{13}+x_{14})/2 = (0+0)/2 = 0$
$\tilde{x} = P_{50} = (x_{26}+x_{27})/2 = (0+0)/2 = 0$
$Q_3 = P_{75} = (x_{39}+x_{40})/2 = (.01+.01)/2 = .01$
$x_{52} = 1.41$

d. $\bar{x} + 2s = .100 + 2(.263) = .626$
The outliers are the two points above .626: .92 and 1.41

e. One possible histogram is given below, with a class width of .25 inches.

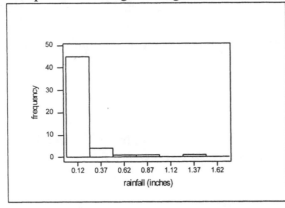

f. No; the distributions are not approximately normal.

g. let R = rain falls on a selected Monday in Boston
Since there was rain on 19 of the 52 observed Mondays, estimate $P(R)=19/52=.365$.

2. For each color, the scores are listed in order at the right.
The following preliminary values are determined.

	Red	Green	Blue
n	20	20	20
Σx	19210	19600	20920
Σx^2	19030374	20298724	22859414
$\bar{x}$	960.5	980.0	1046.0
s^2	30482.579	57406.526	51426.000

a. $\bar{x}_R = 960.5$, $\bar{x}_G = 980.5$, $\bar{x}_B = 1046.0$
The means are fairly close, but the one for the blue
M&M's may be significantly higher – a statistical
test is needed to decide for sure.

b. $\tilde{x}_R = (908+921)/2 = 914.5$
$\tilde{x}_G = (996+1025)/2 = 1010.5$
$\tilde{x}_B = (1004+1013)/2 = 1008.5$
The medians are fairly close, but the one for the red
M&M's may be significantly lower – a statistical
test is needed to decide for sure.

#	Red	Green	Blue
01	621	499	706
02	699	583	793
03	743	630	821
04	813	780	848
05	855	793	866
06	858	828	892
07	896	907	915
08	896	916	939
09	898	993	996
10	908	996	1004
11	921	1025	1013
12	996	1071	1039
13	1030	1111	1068
14	1092	1121	1097
15	1095	1147	1131
16	1130	1153	1159
17	1133	1180	1244
18	1179	1188	1370
19	1190	1229	1408
20	1257	1450	1611

c. $s_R = 174.6$
   $s_G = 239.6$
   $s_B = 226.8$
   The standard deviations are fairly close, but the one for the red M&M's may be significantly lower – a statistical test is needed to decide for sure.

d. Let the scores from the red M&M's be group 1.
   original claim: $\mu_1-\mu_2 = 0$
   $\bar{x}_1-\bar{x}_2 = 960.5 - 980.0 = -19.5$
   $H_o: \mu_1-\mu_2 = 0$
   $H_1: \mu_1-\mu_2 \neq 0$
   $\alpha = .05$ [assumed]
   C.R. $t < -t_{19,.025} = -2.093$
   $t > t_{19,.025} = 2.093$
   calculations:

   $t_{\bar{x}_1-\bar{x}_2} = (\bar{x}_1-\bar{x}_2 - \mu_{\bar{x}_1-\bar{x}_2})/s_{\bar{x}_1-\bar{x}_2}$
   $= (-19.5 - 0)/\sqrt{(30483)/20 + (57407)/20}$
   $= -19.6/66.291$
   $= -.294$
   P-value $= 2 \cdot P(t_{19} < -.29) > .20$

   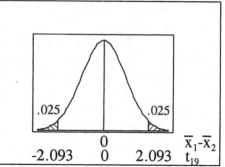

   conclusion:
   Do not reject H_o; there is not sufficient evidence to reject the claim that $\mu_1-\mu_2 = 0$.

e. $\bar{x} \pm t_{19,.025} \cdot s/\sqrt{n}$
   $960.5 \pm 2.093 \cdot (174.593)/\sqrt{20}$
   960.5 ± 81.7
   $878.8 < \mu < 1042.2$

f. Since each group has equal size n, the simplified forms can be used. In addition to the summary statistics above, the following preliminary values are noted.
   $k = 3$ $\bar{\bar{x}} = \Sigma\bar{x}_i/k$
   $n = 20$ $= 995.5$
   $s_{\bar{x}}^2 = \Sigma(\bar{x}_i-\bar{\bar{x}})^2/(k-1)$ $s_p^2 = \Sigma s_i^2/k$
   $= 4015.5/2$ $= 139315.105/3$
   $= 2007.75$ $= 46438.368$

   $H_o: \mu_R = \mu_G = \mu_B$
   $H_1:$ at least one mean is different
   $\alpha = .05$
   C.R. $F > F_{57,.05}^2 = 3.1504$
   calculations:
   $F = ns_{\bar{x}}^2/s_p^2$
   $= 20(2007.75)/46438.368$
   $= .8647$

   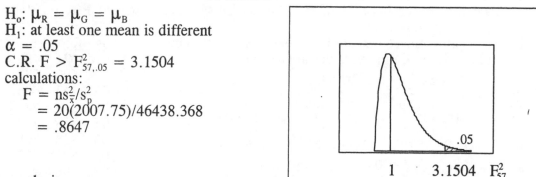

   conclusion:
   Do not reject H_o; there is not sufficient evidence to conclude that at least one mean is different.

3. a. Let x = the weight of a newborn baby.
       normal distribution
       $\mu = 7.54$
       $\sigma = 1.09$
       $P(x > 8.00)$
           $= P(z > .42)$
           $= 1 - P(z < .42)$
           $= 1 - .6628$
           $= .3372$

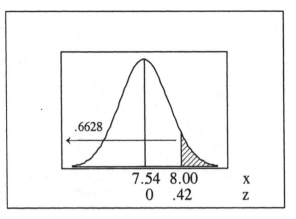

   b. normal distribution,
         since the original distribution is so
       $\mu_{\bar{x}} = \mu = 7.54$
       $\sigma_{\bar{x}} = \sigma/\sqrt{n} = 1.09/\sqrt{16} = .2725$
       $P(\bar{x} > 8.00)$
           $= P(z > 1.69)$
           $= 1 - P(z < 1.69)$
           $= 1 - .9545$
           $= .0455$

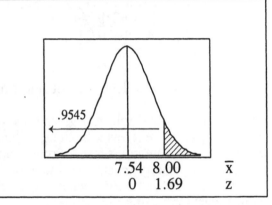

   c. Let B = a baby weighs more than 7.54 lbs.
       $P(B) = .5000$, for each birth
       $P(B_1 \text{ and } B_2 \text{ and } B_3) = P(B_1) \cdot P(B_2) \cdot P(B_3)$
       $\qquad\qquad\qquad\qquad = (.5000) \cdot (.5000) \cdot (.5000)$
       $\qquad\qquad\qquad\qquad = .125$

Chapter 12

Nonparametric Statistics

12-2 Sign Test

1. 10 +'s and 5 −'s
   $n = 15 \leq 25$; use C.V. = 3, from Table A-7
   Since $x = 5 > 3$, do not reject the null hypothesis of no difference.

3. 50 +'s and 40 −'s
   $n = 90 > 25$; use C.V. = −1.96, from the z table
   $z = [(x+.5) - (n/2)]/[\sqrt{n}/2]$
     $= [40.5 - 45]/[\sqrt{90}/2]$
     $= -4.5/4.743$
     $= -.95$
   Since $z = -.95 > 1.96$, do not reject the null hypothesis of no difference.

NOTE for $n \leq 25$: Table A-7 gives only x_L, the <u>lower</u> critical value for the sign test. Accordingly, the text lets x be the <u>smaller</u> of the number of +'s or the number of −'s and warns the user to use common sense to avoid concluding the reverse of what the data indicates. But the problem's symmetry means that the upper critical value is $x_U = n - x_L$ and that $\mu_x = n/2$, the natural expected value for x when H_o is true. For completeness, this manual indicates those values whenever using the sign test.

    Letting x <u>always</u> be the number of +'s is an alternative approach that maintains the natural agreement between the alternative hypothesis and the critical region and is consistent with the logic and notation of parametric tests.

NOTE for $n > 25$: The correction for continuity is a conservative adjustment intending to make less likely a false rejection of H_o by shifting the x value .5 units toward the middle. When x is the smaller of the number of +'s or the number of −'s, this always involves replacing x with $x+.5$. In the alternative approach suggested above, x is replaced with either $x+.5$ or $x-.5$ according to which one shifts the value toward the middle
      − i.e., with $x+.5$ when $x < \mu_x = (n/2)$,
        and with $x-.5$ when $x > \mu_x = (n/2)$.
    The formula given is the usual one for converting a score into its standard score, using the correction for continuity, where the mean and standard deviation of the x's are (n/2) and $\sqrt{n}/2$ respectively.

$$z = \frac{x - \mu_x}{\sigma_x} = \frac{(x \pm .5) - (n/2)}{\sqrt{n}/2}$$

NOTE: The manual follows the lower tail method of the text, with the addition of upper tail references as indicated above. Exercises that are naturally upper-tailed (see #9, #10, #11 in this section) are worked both ways.

5. Let the reported heights be group 1.
    claim: median difference ≠ 0

male	1	2	3	4	5	6	7	8	9	10	11	12
R-M	+	+	-	+	+	-	+	-	-	-	+	+

    n = 12: 7+'s and 5-'s

    H_o: median difference = 0
    H_1: median difference ≠ 0
    α = .05
    C.R. x ≤ $x_{L,12,.025}$ = 2
         x ≥ $x_{U,12,.025}$ = 12-2 = 10
    calculations:
        x = 5 (using less frequent count)
        x = 7 (using + count)

    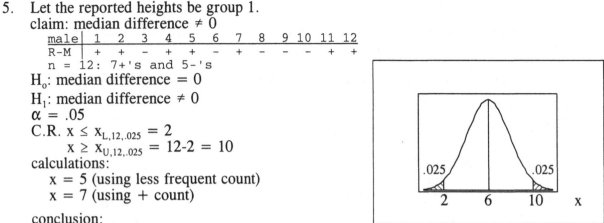

    conclusion:
        Do not reject H_o; there is not sufficient evidence to conclude that the median difference
        is different from 0.

NOTE: Compare the above results to those obtained in exercise #6 of section 8-4, where a
parametric test was used on the same data. Although the text does not so indicate, several
exercises in this chapter are re-tests of data that was previously analyzed parametrically;
comparisons of the two results are usually informative and provide statistical insights.

7. claim: median < 98.6

temp	1	2	3	4	5	6	7	8	9	10	11	12
t-98.6	-	-	0	-	-	+	-	-	-	-	-	-

    n = 11: 1+'s and 10-'s

    H_o: median = 98.6
    H_1: median < 98.6
    α = .05
    C.R. x ≤ $x_{L,11,.05}$ = 2
    calculations:
        x = 1

    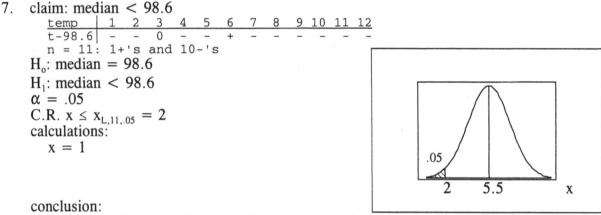

    conclusion:
        Reject H_o; there is sufficient evidence to conclude that the median is less than 98.6.

9. Let x = the number who did not vote.
    claim: p < .5
        701 +'s (voted)   301 -'s (did not vote)
        n = 1002 +'s or -'s
    Since n > 25, use z with
        μ_x = n/2 = 1002/2 = 501
        σ_x = $\sqrt{n}$ /2 = $\sqrt{1002}$ /2 = 15.827
    H_o: p = .5
    H_1: p < .5
    α = .05 [assumed]
    C.R. z < $-z_{.05}$ = -1.645
    calculations:
        x = 301
        z_x = [(x+.5)-μ_x]/σ_x
            = [301.5 - 501]/15.827
            = -199.5/15.827
            = -12.605

    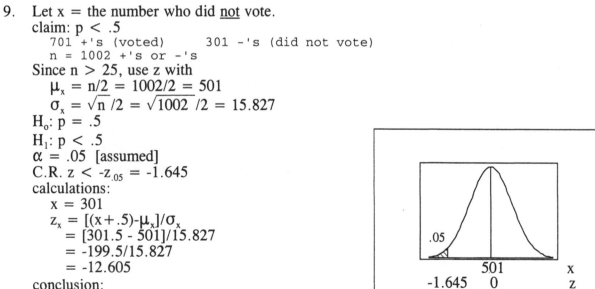

    conclusion:
        Reject H_o; there is sufficient evidence to conclude that p < .5 (i.e., that the majority of

the people say they <u>did</u> vote in the election).

NOTE: Exercises #9 is a one-tailed test naturally worded toward the upper tail. For such an exercise, working with only the lower critical region necessitates notation contrary to the its natural wording. The exercise may be worked more naturally in the manner in which it is stated as follows. See the NOTES preceding exercise #5 of this section.

Let x = the number who voted.
claim: $p > .5$

```
  701 +'s (voted)      301 -'s (did not vote)
  n = 1002 +'s or -'s
```

Since $n > 25$, use z with
$\quad \mu_x = n/2 = 1002/2 = 501$
$\quad \sigma_x = \sqrt{n}/2 = \sqrt{1002}/2 = 15.827$
H_o: $p = .5$
H_1: $p > .5$
$\alpha = .05$ [assumed]
C.R. $z > z_{.05} = 1.645$
calculations:
$\quad x = 701$
$\quad z_x = [(x-.5)-\mu_x]/\sigma_x$
$\quad\quad = [700.5 - 501]/15.827$
$\quad\quad = 199.5/15.827$
$\quad\quad = 12.605$
conclusion:

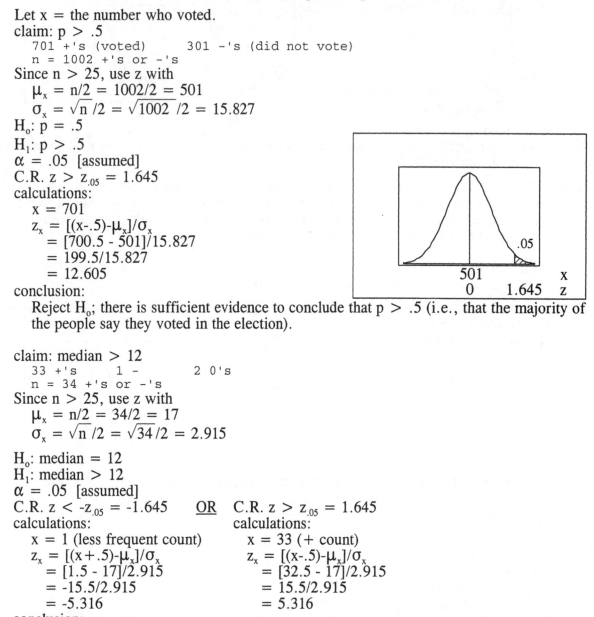

Reject H_o; there is sufficient evidence to conclude that $p > .5$ (i.e., that the majority of the people say they voted in the election).

11.   claim: median > 12

```
      33 +'s        1 -        2 0's
      n = 34 +'s or -'s
```

Since $n > 25$, use z with
$\quad \mu_x = n/2 = 34/2 = 17$
$\quad \sigma_x = \sqrt{n}/2 = \sqrt{34}/2 = 2.915$

H_o: median $= 12$
H_1: median > 12
$\alpha = .05$ [assumed]
C.R. $z < -z_{.05} = -1.645$ <u>OR</u>   C.R. $z > z_{.05} = 1.645$
calculations: calculations:
$\quad x = 1$ (less frequent count) $x = 33$ (+ count)
$\quad z_x = [(x+.5)-\mu_x]/\sigma_x$ $z_x = [(x-.5)-\mu_x]/\sigma_x$
$\quad\quad = [1.5 - 17]/2.915$ $= [32.5 - 17]/2.915$
$\quad\quad = -15.5/2.915$ $= 15.5/2.915$
$\quad\quad = -5.316$ $= 5.316$
conclusion:

Reject H_o; there is sufficient evidence to conclude that median > 12.
Yes; assuming that the intent of the company is to produce a product with a median weight slightly larger than the stated weight of 12 ounces, it appears that the cans are being filled correctly.

13. claim: median > 77

    37 +'s   13 -'s   no 0's
    n = 50 +'s or -'s

Since n > 25, use z with

$\mu_x = n/2 = 50/2 = 25$

$\sigma_x = \sqrt{n}/2 = \sqrt{25}/2 = 3.536$

H_o: median = 77

H_1: median > 77

$\alpha = .05$ [assumed]

C.R. $z < -z_{.05} = -1.645$ <u>OR</u>   C.R. $z > z_{.05} = 1.645$

calculations: calculations:

  x = 13 (less frequent count)   x = 37 (+ count)

  $z_x = [(x+.5)-\mu_x]/\sigma_x$ $z_x = [(x-.5)-\mu_x]/\sigma_x$

    $= [13.5 - 25]/3.536$ $= [36.5 - 25]/3.536$

    $= -11.5/3.536$ $= 11.5/3.536$

    $= -3.253$ $= 3.253$

conclusion:

  Reject H_o; there is sufficient evidence to conclude that median > 77.

15. claim: median < 100

    40 +'s   60 -'s   21 0's

a. Using the usual method: discard the 0's.

    n = 100 +'s or -'s

Since n > 25, use z with

  $\mu_x = n/2 = 100/2 = 50$

  $\sigma_x = \sqrt{n}/2 = \sqrt{100}/2 = 5.000$

H_o: median = 100

H_1: median < 100

$\alpha = .05$

C.R. $z < -z_{.05} = -1.645$

calculations:

  x = 40

  $z_x = [(x+.5)-\mu_x]/\sigma_x$

    $= [40.5 - 50]/5.000$

    $= -9.5/5.000$

    $= -1.900$

conclusion:

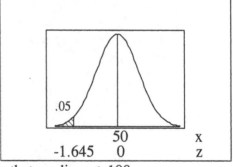

  Reject H_o; there is sufficient evidence to conclude that median < 100.

b. Using the second method: count half the zeros each way (and discarding the odd zero).

    50 +'s   70 -'s
    n = 120 +'s or -'s

Since n > 25, use z with

  $\mu_x = n/2 = 120/2 = 60$

  $\sigma_x = \sqrt{n}/2 = \sqrt{120}/2 = 5.477$

H_o: median = 0

H_1: median < 0

$\alpha = .05$

C.R. $z < -z_{.05} = -1.645$

calculations:

  x = 50

  $z_x = [(x+.5)-\mu_x]/\sigma_x$

    $= [50.5 - 60]/5.477$

    $= -9.5/5.477$

    $= -1.734$

conclusion:

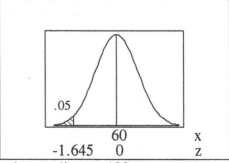

  Reject H_o; there is sufficient evidence to conclude that median < 100.

c. Using the third method: count the zeros in a one-tailed test to favor H_o.

```
   61 +'s        60 -'s
     n = 161 +'s or -'s
```
Since n > 25, use z with
$\mu_x = n/2 = 121/2 = 60.5$
$\sigma_x = \sqrt{n}/2 = \sqrt{121}/2 = 5.500$
H_o: median ≥ 0
H_1: median < 0
$\alpha = .05$
C.R. $z < -z_{.05} = -1.645$
calculations:
  x = 60
$z_x = [(x+.5)-\mu_x]/\sigma_x$
   $= [60.5 - 60.5]/5.500$
   $= 0/5.500 = 0$
conclusion:

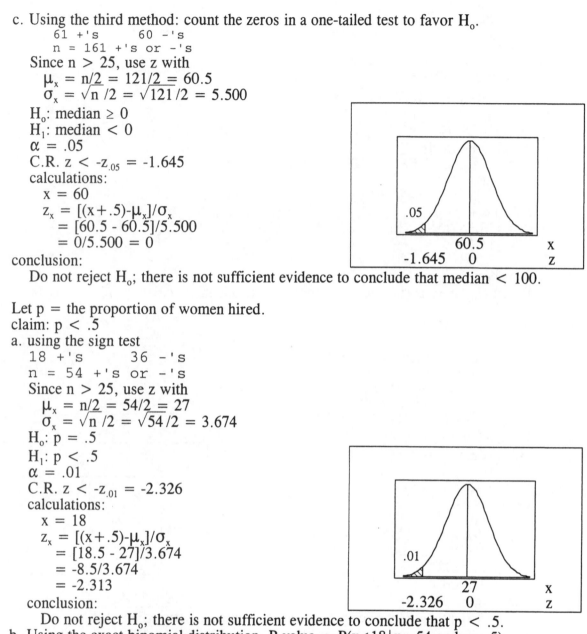

Do not reject H_o; there is not sufficient evidence to conclude that median < 100.

17.   Let p = the proportion of women hired.
claim: p < .5
a. using the sign test
```
     18 +'s         36 -'s
     n = 54 +'s or -'s
```
Since n > 25, use z with
$\mu_x = n/2 = 54/2 = 27$
$\sigma_x = \sqrt{n}/2 = \sqrt{54}/2 = 3.674$
H_o: p = .5
H_1: p < .5
$\alpha = .01$
C.R. $z < -z_{.01} = -2.326$
calculations:
  x = 18
$z_x = [(x+.5)-\mu_x]/\sigma_x$
   $= [18.5 - 27]/3.674$
   $= -8.5/3.674$
   $= -2.313$
conclusion:
Do not reject H_o; there is not sufficient evidence to conclude that p < .5.
b. Using the exact binomial distribution, P-value = $P(x \leq 18 \mid n=54$ and $p=.5)$.

The test in part (a) uses the normal approximation
$P(x \leq 18) = P_c(x < 18.5) = P(z < -2.313) \approx .0104 > .01$ [do not reject H_o].
Using the exact binomial requires using the binomial formula 19 times.
$P(x) = [n!/x!(n-x)!] \cdot p^x \cdot (1-p)^{n-x}$
$P(x) = [54!/x!54!] \cdot (.5)^x \cdot (.5)^{54-x}$ for $x = 0,1,2,...,18$
The statDisk program indicates
$P(x \leq 18) = P(x=0) + P(x=1) + ... + P(x=18) = .00992 < .01$ [reject H_o].

12-3 Wilcoxon Signed-Ranks Test for Matched Pairs

NOTE: Table A-8 gives only T_L, the <u>lower</u> critical value for the signed-ranks test. Accordingly, the text lets T be the <u>smaller</u> of the sum of positive ranks or the sum of the negative ranks and warns the user to use common sense to avoid concluding the reverse of what the data indicates. But the problem's symmetry means that the upper critical value is $T_U = \Sigma R \text{-} T_L$ and that $\mu_T = \Sigma R/2$, the natural expected value for T when H_0 is true.

Letting T <u>always</u> be the sum of the positive ranks is an alternative approach that maintains the natural agreement between the alternative hypothesis and the critical region and is consistent with the logic and notation of parametric tests. The manual follows the lower tail method of the text, with the addition of upper tail references as indicated above.

NOTE 2: This manual follows the text and the directions to the exercises of this section by using "the populations have the same distribution" as the null hypothesis. To be more precise, the signed-rank test doesn't test "distributions" but tests the "location" (i.e., central tendency – as opposed to variation) of distributions. The test discerns whether one group taken as a whole tends to have higher or lower scores than another group taken as a whole. The test does not discern whether one group is more variable than another. This distinction is reflected in the wording of the conclusion when rejecting H_0. Notice also that the signed-rank test measures overall differences between the groups and <u>not</u> whether the two groups give the same results for individuals subjects. If the group 1 scores, for example, were higher for half the subjects and lower for the other half of the subjects by the same amounts, then $\Sigma R\text{-}$ would equal $\Sigma R+$ (so we could not reject H_0), but the distributions would be very different.

NOTE 3: This manual uses a minus sign preceding ranks associated with negative differences. The ranks themselves are not negative, but the use of the minus sign helps to organize the information.

1. claim: the populations have the same distribution

```
x-y |  0   -1   2    5    8    9   10   10
R   |  -   -1   2    3    4    5  6.5  6.5
ΣR- =  1              n = 7 non-zero ranks
ΣR+ = 27
ΣR  = 28
```
check: $\Sigma R = n(n+1)/2 = 7(8)/2 = 28$

H_0: the populations have the same distribution
H_1: the populations have different distributions
$\alpha = .05$
C.R. $T \le T_{L,7,.025} = 2$
　　　$T \ge T_{U,7,.025} = 28\text{-}2 = 26$
calculations
　　$T = 1$ (using the smaller ranks)
　　$T = 27$ (using the positive ranks)

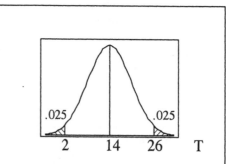

conclusion:
　　Reject H_0; there is sufficient evidence to reject the claim that the populations have the same distribution and to conclude that they have different distributions (in fact, that the x scores are greater).

3. Let the reported heights be group 1.
    claim: the populations have the same distribution

```
R-M | .1   1.1  -1.9  1.7   .7  -.6   .5  -3.0  -1.6  -11.2  1.0  1.2
R   |  1    6   -10    9    4   -3    2   -11   -8    -12    5    7
ΣR- = 44                  n = 12 non-zero ranks
ΣR+ = 34
ΣR  = 78
```
check: $\Sigma R = n(n+1)/2 = 12(13)/6 = 78$

H_o: the populations have the same distribution
H_1: the populations have different distributions
$\alpha = .05$
C.R. $T \le T_{L,12,.025} = 14$
   $T \ge T_{U,12,.025} = 78\text{-}14 = 64$
calculations
   $T = 34$

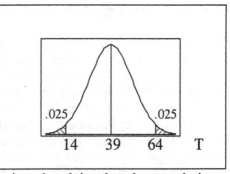

conclusion:
   Do not reject H_o; there is not sufficient evidence to reject the claim that the populations have the same distribution.

5.   Let Sitting be group 1.
   claim: the populations have the same distribution

pair	1	2	3	4	5	6	7	8	9	10
Si-Su	.99	1.60	.98	.82	1.01	1.54	.21	.70	1.67	1.32
R	5	9	4	3	6	8	1	2	10	7

   $\Sigma R- = 0$ n = 10 non-zero ranks
   $\Sigma R+ = 55$
   $\Sigma R = 55$
check: $\Sigma R = n(n+1)/2 = 10(11)/2 = 55$

H_o: the populations have the same distribution
H_1: the populations have different distributions
$\alpha = .05$
C.R. $T \le T_{L,10,.025} = 8$
   $T \ge T_{U,10,.025} = 55\text{-}8 = 47$
calculations:
   $T = 0$ (using the smaller ranks)
   $T = 55$ (using the positive ranks)

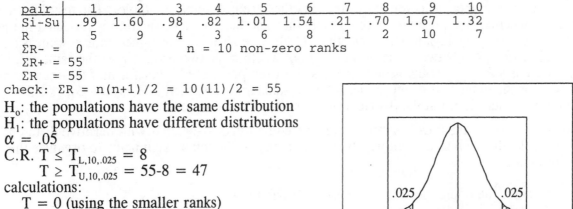

conclusion:
   Reject H_o; there is sufficient evidence to reject the claim that the populations have the same distribution and to conclude that they have different distributions (in fact, the sitting positions have larger capacity scores).

7.   The 31" actual minus predicted" differences and their signed ranks are as follows.

```
31 (actual-predicted) values
    0    -1     1     1    -6     0    -2     2    -1    -6     6
    0     5    -7    -4     2     3    -4    -4     5     1     2
   -5    -2     6    -1     2    -1    -6    -3     4
------------------------------------------------------------------
31 Ra-p values
    -    -4     4     4   -25     -  -10.5  10.5    -4   -25    25
    -    21   -28 -17.5  10.5  14.5 -17.5 -17.5    21     4  10.5
  -21 -10.5    25    -4  10.5    -4   -25 -14.5  17.5
```

   $\Sigma R- = 228$ n = 28 non-zero ranks
   $\Sigma R+ = 178$
   $\Sigma R = 406$
check: $\Sigma R = n(n+1)/2 = 28(29)/2 = 406$

H_o: the populations have the same distribution
H_1: the populations have different distributions
$\alpha = .05$ [assumed]
C.R. $T \le T_{L,28,.025} = 117$
   $T \ge T_{U,28,.025} = 406\text{-}117 = 289$
calculations:
   $T = 178$

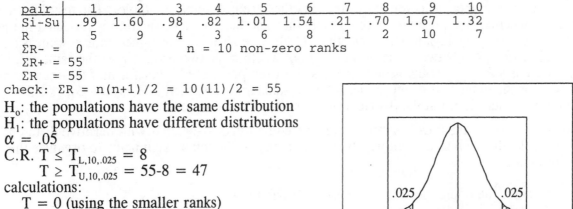

conclusion:
    Do not reject H_0; there is not sufficient evidence to reject the claim that the populations have the same distribution.

9. The 106 differences obtained by subtracting 98.6 from each temperature, and the signed ranks of the differences are as follows.

```
    106 (x-98.6) values
    0       0     -.6    -.6     .4    -.2    -.2    -.2    -.2      0      0
    .2      0    -1.6   -1.6    0.2   -1.0    -.9     .2    -.6    -.6    -.3
   -.1    -1.3     .1   -1.2     .3      0     .9   -1.1   -1.3   -1.0    -.4
   1.0     .1     .8    -.4    -.6      0      0   -1.4    -.2      0    -.4
   -.6    -.8    -.6    -.2      0      0    -.8     .4   -2.1   -1.0    -.6
  -1.7   -1.0   -1.5    -.7    -.2   -1.3    -.6   -1.1   -1.0    -.4    -.1
    .2     .1    -.8    -.6   -1.5   -1.2     .8    -.2      0    -.2    -.1
    0    -.3     .1     .2     .5      0    -.7     .2    -.6     .1    -.1
    .3    -.2      0   -1.5    -.7     .2     .1   -1.0    -.4     .6    -.8
   -.6    -.2    -.8    -.2   -1.2    -.6   -1.6
```

```
    106 R_x-98.6 values
     -      -   -48.5  -48.5   37.0  -16.5  -16.5  -16.5  -16.5     -      -
   26.0     -   -88.0  -88.0   26.0  -71.0  -66.5   26.0  -48.5  -48.5  -30.5
   -5.5  -81.0    5.5  -78.0   32.5     -    66.5  -75.5  -81.0  -71.0  -37.0
   71.0    5.5   64.5  -37.0  -48.5     -      -   -83.0  -16.5     -   -37.0
  -48.5  -61.0  -48.5  -16.5     -      -   -61.0   37.0  -91.0  -71.0  -48.5
  -90.0  -71.0  -85.0  -57.0  -16.5  -81.0  -48.5  -75.5  -71.0  -37.0   -5.5
   26.0    5.5  -61.0  -48.5  -85.0  -78.0   64.5  -16.5     -    -16.5   -5.5
     -   -30.5    5.5   26.0   41.0     -   -57.0   26.0  -48.5    5.5   -5.5
   32.5  -16.5     -   -85.0  -57.0   26.0    5.5  -71.0  -37.0   48.5  -61.0
  -48.5  -16.5  -61.0  -16.5  -78.0  -48.5  -88.0
```

    $\Sigma R- = 3476$ n = 91 non-zero's to be ranked
    $\Sigma R+ = 710$
    $\Sigma R = 4186$
    check: $\Sigma R = n(n+1)/2 = (91)(92)/2 = 4186$

For n = 91 ranks, use the z approximation with
$$\mu_T = n(n+1)/4$$
$$= 91(92)/4 = 2093$$
$$\sigma_T = \sqrt{[n(n+1)(2n+1)/24]}$$
$$= \sqrt{[91(92)(183)/24]} = 252.66$$

H_0: median $= 98.6$
H_1: median $\neq 98.6$
$\alpha = .05$
C.R. $z < -z_{.025} = -1.960$
    $z > z_{.025} = 1.960$
calculations:
    T = 710
    $z = (T - \mu_T)/\sigma_T$
      $= (710 - 2093)/252.66$
      $= -1383/252.66$
      $= -5.473$
conclusion:
    Reject H_0, there is sufficient evidence reject the claim that median $= 98.6$ and to conclude that median $\neq 98.6$ (in fact, median < 98.6).

12-4 Wilcoxon Rank-Sum Test for Two Independent Samples

NOTE: As in the previous section, the manual follows the wording in the text and tests the hypothesis that "the populations have the same distribution" with the understanding that the test detects only differences in location and not differences in variability. In addition, always letting $R = \Sigma R_1$ guarantees agreement and consistency between the directions of H_1 and the C.R. as in the previous chapters.

1. Below are the ordered scores for each group.
    claim: the populations have the same distribution

grp 1	R	grp 2	R
1	1	2	2
3	3	5	5
4	4	7	7
6	6	9	9
8	8	11	10
12	11	13	12
15	14	14	13
16	15	18	17
17	16	19	18
22	20	20	19
26	22.5	25	21
	120.5	26	22.5
			155.5

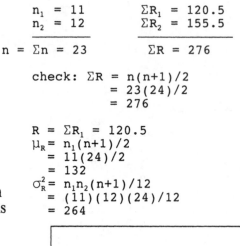

$n_1 = 11$ $\Sigma R_1 = 120.5$
$n_2 = 12$ $\Sigma R_2 = 155.5$
$n = \Sigma n = 23$ $\Sigma R = 276$

check: $\Sigma R = n(n+1)/2$
$= 23(24)/2$
$= 276$

$R = \Sigma R_1 = 120.5$
$\mu_R = n_1(n+1)/2$
$= 11(24)/2$
$= 132$
$\sigma_R^2 = n_1 n_2 (n+1)/12$
$= (11)(12)(24)/12$
$= 264$

H_0: the populations have the same distribution
H_1: the populations have different distributions
$\alpha = .05$
C.R. $z < -z_{.025} = -1.96$
     $z > z_{.025} = 1.96$
calculations:
   $z_R = (R - \mu_R)/\sigma_R$
     $= (120.5 - 132)/\sqrt{264}$
     $= -11.5/16.248 = -.708$
conclusion:
   Do not reject H_0; there is not sufficient evidence to reject the claim that the populations have the same distribution.

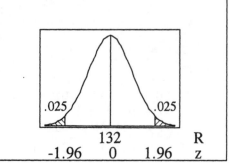

3. Below are the ordered scores for each group. The group listed first is considered group 1.
    claim: the populations have the same distribution

O-C	R	Con	R
.210	1	.334	7.5
.287	2	.349	11
.288	3	.402	12
.304	4	.413	14
.305	5	.429	15
.308	6	.445	16
.334	7.5	.460	18.5
.340	9	.476	20.5
.344	10	.483	21
.407	13	.501	22
.455	17	.519	23
.463	19	.594	24
	96.5		203.5

$n_1 = 12$ $\Sigma R_1 = 96.5$
$n_2 = 12$ $\Sigma R_2 = 203.5$
$n = \Sigma n = 24$ $\Sigma R = 300.0$

check: $\Sigma R = n(n+1)/2$
$= 24(25)/2$
$= 300$

$R = \Sigma R_1 = 96.5$
$\mu_R = n_1(n+1)/2$
$= 12(25)/2$
$= 150$
$\sigma_R^2 = n_1 n_2 (n+1)/12$
$= (12)(12)(25)/12$
$= 300$

H_o: the populations have the same distribution
H_1: the populations have different distributions
$\alpha = .01$
C.R. $z < -z_{.005} = -2.575$
$\quad\quad z > z_{.005} = 2.575$
calculations:
$$z_R = (R - \mu_R)/\sigma_R$$
$$\quad = (96.5 - 150)/\sqrt{300}$$
$$\quad = -53.5/17.321$$
$$\quad = -3.089$$

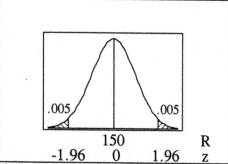

conclusion:
Reject H_o; there is sufficient evidence to reject the claim that the populations have the same distribution and to conclude that they have different distributions (in fact, that population 1 has smaller volumes).

Based on this result, we can be 99% confident that there are biological factors related to obsessive-compulsive disorders.

5. Below are the ordered scores for each group. The group listed first is considered group 1.

E to D	R	D to E	R
7.10	1	26.63	11
16.32	2	26.68	12
20.60	3	27.24	14
21.06	4	27.62	15
21.13	5	29.34	21
21.96	6	29.49	22
24.23	7	30.20	24
24.64	8	30.26	25
25.49	9	32.34	29
26.43	10	32.54	30
26.69	13	33.53	34
27.85	16	33.62	35
28.02	17	34.02	36
28.71	18	35.32	37
28.89	19	35.91	38
28.90	20	42.91	41
30.02	23		424
30.29	26		
30.72	27		
31.73	28		
32.83	31		
32.86	32		
33.31	33		
38.81	39		
39.29	40		
	437		

$n_1 = 25 \quad\quad \Sigma R_1 = 437$
$n_2 = 16 \quad\quad \Sigma R_2 = 424$
_____
$n = \Sigma n = 41 \quad\quad \Sigma R = 861$

check: $\Sigma R = n(n+1)/2$
$$\quad\quad = 41(42)/2$$
$$\quad\quad = 861$$

$R = \Sigma R_1 = 437$

$\mu_R = n_1(n+1)/2$
$$\quad = 25(42)/2$$
$$\quad = 525$$

$\sigma_R^2 = n_1 n_2(n+1)/12$
$$\quad = (25)(16)(42)/12$$
$$\quad = 1400$$

H_o: the populations have the same distribution
H_1: the populations have different distributions
$\alpha = .05$
C.R. $z < -z_{.025} = -1.96$
$\quad\quad z > z_{.025} = 1.96$
calculations:
$$z_R = (R - \mu_R)/\sigma_R$$
$$\quad = (437 - 525)/\sqrt{1400}$$
$$\quad = -88/37.417$$
$$\quad = -2.352$$

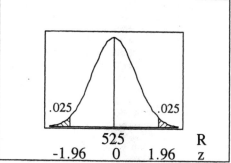

conclusion:
Reject H_o; there is sufficient evidence to reject the claim that the populations have the same distribution and to conclude that they have different distributions (in fact, that population 1 has lower scores).

Below are the ordered scores for exercises #7, #8, #9 and #10. The group listed first is considered group 1.

#	Rowl	R	Tols	R	McGw	R	Bond	R	w	R	e	R	male	R	fema	R
01	3.2	1.0	5.9	10.5	340	3.0	320	1.5	15	4.5	12	1.0	19.6	6.0	17.7	1.0
02	3.7	2.0	5.9	10.5	341	4.0	320	1.5	16	9.5	15	4.5	19.9	9.0	18.3	2.0
03	4.1	3.0	6.1	12.5	350	7.0	347	5.0	16	9.5	15	4.5	20.7	12.5	19.2	3.0
04	4.4	4.5	7.2	16.0	350	7.0	350	7.0	16	9.5	15	4.5	21.5	16.0	19.3	4.0
95	4.4	4.5	7.7	17.0	360	11.5	360	11.5	16	9.5	15	4.5	21.6	17.0	19.6	6.0
06	4.9	6.0	8.2	18.0	360	11.5	360	11.5	17	13.5	15	4.5	22.7	21.0	19.6	6.0
07	5.2	7.0	8.4	19.0	360	11.5	360	11.5	17	13.5	17	13.5	23.2	24.0	19.8	8.0
08	5.6	8.0	8.6	20.0	369	17.0	361	15.0	18	21.0	17	13.5	23.3	25.0	20.5	10.0
09	5.7	9.0	9.8	21.0	370	20.5	365	16.0	18	21.0	18	21.0	23.4	26.0	20.6	11.0
10	6.1	12.5	10.9	22.0	370	20.5	370	20.5	18	21.0	18	21.0	23.5	27.5	20.7	12.5
11	6.7	14.0	11.0	23.0	370	20.5	370	20.5	18	21.0	18	21.0	23.8	30.5	21.2	14.0
12	6.9	15.0	11.5	24.0	370	20.5	375	25.5	18	21.0	18	21.0	23.8	30.5	21.4	15.0
13		86.5		213.5	377	28.0	375	25.5	18	21.0	18	21.0	24.2	34.0	21.9	18.5
14					380	33.5	375	25.5	19	33.0	19	33.0	24.5	35.0	21.9	18.5
15					380	33.5	375	25.5	19	33.0	19	33.0	24.6	36.5	22.0	20.0
16					380	33.5	380	33.5	19	33.0	19	33.0	24.6	36.5	22.8	22.5
17					380	33.5	380	33.5	19	33.0	19	33.0	25.2	39.5	22.8	22.5
18					380	33.5	380	33.5	20	45.0	19	33.0	25.5	41.0	23.5	27.5
19					385	40.0	380	33.5	21	53.0	19	33.0	25.6	42.0	23.8	30.5
20					385	40.0	380	33.5	21	53.0	19	33.0	26.2	44.5	23.8	30.5
21					388	42.0	385	40.0	21	53.0	19	33.0	26.2	44.5	24.0	33.0
22					390	45.5	390	45.5	22	59.5	19	33.0	26.3	46.0	25.1	38.0
23					390	45.5	390	45.5	22	59.5	20	45.0	26.4	47.5	25.2	39.5
24					390	45.5	391	49.0	22	59.5	20	45.0	26.4	47.5	26.0	43.0
25					390	45.5	394	50.0	23	66.5	20	45.0	26.6	50.0	26.5	49.0
26					398	52.0	396	51.0	23	66.5	20	45.0	26.7	51.0	27.5	56.0
27					400	55.5	400	55.5	24	73.5	20	45.0	26.9	52.0	28.5	60.0
28					400	55.5	400	55.5	24	73.5	20	45.0	27.0	53.0	28.7	61.5
29					409	61.0	400	55.5	24	73.5	20	45.0	27.1	54.0	28.9	63.0
30					410	69.0	400	55.5	24	73.5	20	45.0	27.4	55.0	29.1	64.0
31					410	69.0	404	59.0	24	73.5	20	45.0	27.8	57.0	29.7	65.0
32					410	69.0	405	60.0	25	80.0	20	45.0	28.1	58.0	29.8	66.0
33					410	69.0	410	69.0	25	80.0	21	53.0	28.3	59.0	29.9	67.0
34					410	69.0	410	69.0	26	85.0	21	53.0	28.7	61.5	30.9	68.5
35					420	89.0	410	69.0	26	85.0	22	59.5	30.9	68.5	31.0	70.0
36					420	89.0	410	69.0	27	88.5	22	59.5	31.4	71.0	31.7	72.0
37					420	89.0	410	69.0	29	92.5	22	59.5	31.9	73.0	33.5	77.0
38					420	89.0	410	69.0	29	92.5	22	59.5	32.1	74.0	37.7	78.0
39					420	89.0	410	69.0	30	95.5	22	59.5	33.1	75.0	40.6	79.0
40					423	96.0	410	69.0	30	95.5	23	66.5	33.2	76.0	44.9	80.0
41					425	97.0	410	69.0	31	100.0	23	66.5		1727.5		1512.5
42					430	104.5	410	69.0	31	100.0	23	66.5				
43					430	104.5	411	77.0	32	104.0	23	66.5				
44					430	104.5	415	78.5	33	106.5	24	73.5				
45					430	104.5	415	78.5	33	106.5	24	73.5				
46					430	104.5	416	80.0	34	109.0	24	73.5				
47					430	104.5	417	81.5	38	116.5	25	80.0				
48					430	104.5	417	81.5	39	118.0	25	80.0				
49					440	117.0	420	89.0	40	119.0	25	80.0				
50	(McGw/Bond continued)				440	117.0	420	89.0	41	121.0	26	85.0	(east continued)			
51	470	134.0	430	104.5	440	117.0	420	89.0	41	121.0	26	85.0	32	104.0		
52	470	134.0	430	104.5	450	124.0	420	89.0	42	123.5	26	85.0	32	104.0		
53	470	134.0	435	111.5	450	124.0	420	89.0	43	125.0	27	88.5	34	109.0		
54	478	136.0	435	111.5	450	124.0	420	89.0	45	126.0	28	90.5	34	109.0		
55	480	137.0	436	113.0	450	124.0	420	89.0	48	129.0	28	90.5	35	111.5		
56	500	139.0	440	117.0	452	127.0	420	89.0	66	131.0	30	95.5	35	111.5		
57	510	140.5	440	117.0	458	129.0	429	98.0		3861.0	30	95.5	36	113.0		
58	510	140.5	440	117.0	460	130.5	430	104.5			31	100.0	37	114.5		
59	527	142.0	440	117.0	460	130.5	430	104.5			31	100.0	37	114.5		
60	550	143.0	442	121.0	461	132.0	430	104.5			31	100.0	38	116.5		
		5468.5			450	124.0	←			→			41	121.0		
					454	128.0	←						42	123.5		
					488	138.0							47	127.0		
						4827.5							48	129.0		
													48	129.0		
														4785.0		

7. See the data summary on the data page.
    claim: the populations have different distributions
    preliminary calculations:

$n_1 = 12$		$\Sigma R_1 = 86.5$
$n_2 = 12$		$\Sigma R_2 = 213.5$
$n = \Sigma n = 24$		$\Sigma R = 300.0$

$R = \Sigma R_1 = 86.5$
$\mu_R = n_1(n+1)/2 = 12(25)/2 = 150$
$\sigma_R^2 = n_1 n_2(n+1)/12 = (12)(12)(25)/12 = 300$

check: $\Sigma R = n(n+1)/2$
            $= 24(25)/2$
            $= 300$

H_0: the populations have the same distribution
H_1: the populations have different distributions
$\alpha = .05$
C.R. $z < -z_{.025} = -1.96$
       $z > z_{.025} = 1.96$
calculations:

$$z_R = (R - \mu_R)/\sigma_R$$
$$= (86.5 - 150)/\sqrt{300}$$
$$= -63.5/17.321 = -3.666$$

conclusion:

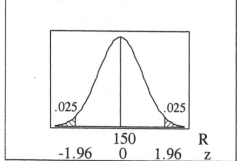

Reject H_0; there is sufficient evidence to conclude the populations have different distributions (in fact, that the Tolstoy scores are higher).

9. See the data summary on the data page.
    claim: the populations have different distributions
    preliminary calculations:

$n_1 = 56$		$\Sigma R_1 = 3861$
$n_2 = 75$		$\Sigma R_2 = 4785$
$n = \Sigma n = 131$		$\Sigma R = 8646$

$R = \Sigma R_1 = 3861$
$\mu_R = n_1(n+1)/2 = 56(132)/2 = 3696$
$\sigma_R^2 = n_1 n_2(n+1)/12 = (56)(75)(132)/12 = 46200$

check: $\Sigma R = n(n+1)/2$
            $= 131(132)/2$
            $= 8646$

H_0: the populations have the same distribution
H_1: the populations have different distributions
$\alpha = .05$
C.R. $z < -z_{.025} = -1.96$
       $z > z_{.025} = 1.96$
calculations:

$$z_R = (R - \mu_R)/\sigma_R$$
$$= (3861 - 3696)/\sqrt{46200}$$
$$= 165/214.942 = .768$$

conclusion:

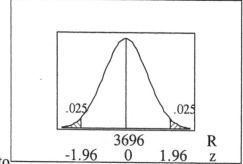

Do not reject H_0; there is not sufficient evidence to reject the claim that the ages come from populations with the same distribution.

11. The denominators of the Mann-Whitney and Wilcoxon statistics are the same.
    The numerator of the Mann-Whitney statistic, assuming $R = \Sigma R_1$, is

$$U - n_1 n_2/2 = [n_1 n_2 + n_1(n_1+1)/2 - R] - n_1 n_2/2$$
$$= n_1 n_2/2 + n_1(n_1+1)/2 - R$$
$$= (n_1/2)(n_2+n_1+1) - R$$
$$= -[R - n_1(n_1+n_2+1)/2] = -[\text{numerator of Wilcoxon statistic}]$$

For the readability example in this section,

$$U = n_1 n_2 + n_1(n_1+1)/2 - R$$
$$= (13)(12) + (13)(14)/2 - 236.5$$
$$= 10.5$$
$$\mu_U = n_1 n_2/2 = (13)(12)/2 = 78$$
$$\sigma_U^2 = \sigma_R^2 = n_1 n_2(n+1)/12 = (13)(12)(26)/12 = 338$$
$$z_U = (U - \mu_U)/\sigma_U$$
$$= (10.5 - 78)/\sqrt{338}$$
$$= -67.5/18.385$$
$$= -3.67$$

As predicted, z_U has the same absolute value but the opposite sign of z_R.

12-5 Kruskal-Wallis Test

NOTE: As in the previous sections, the manual follows the wording in the text and tests the hypothesis that "the populations have the same distribution" with the understanding that the test detects only differences in location and not differences in variability.

1. H_o: the populations have the same distribution
    H_1: the populations have different distributions
    $\alpha = .05$
    C.R. $H > \chi^2_{2,.05} = 5.991$
    calculations:
    $$H = [12/n(n+1)] \cdot [\Sigma(R_i^2/n_i)] - 3(n+1)$$
    $$= .58 \text{ [from Minitab]}$$
    P-value $= P(\chi^2_2 > .58)$
    $\qquad = .747$ [from Minitab]

    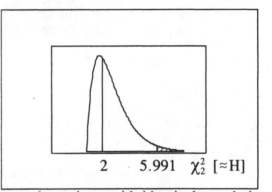

    conclusion:
    Do not reject H_o; there is not sufficient evidence to reject the claim that the different ages have times with identical population distributions.

    NOTE: Compare these results to those of the parametric test in section 11-2, exercise #3.

3. Below are the ordered scores for each group. The group listed first is group 1, etc.

    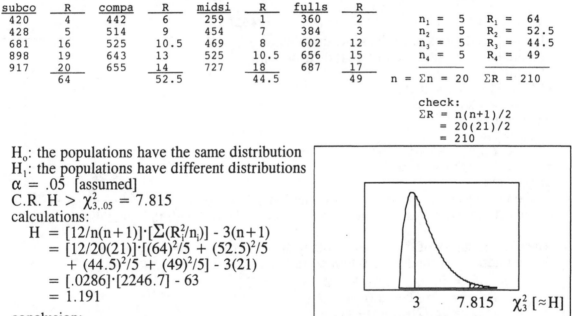

subco	R	compa	R	midsi	R	fulls	R						
420	4	442	6	259	1	360	2	$n_1 =$	5	$R_1 =$	64
428	5	514	9	454	7	384	3	$n_2 =$	5	$R_2 =$	52.5
681	16	525	10.5	469	8	602	12	$n_3 =$	5	$R_3 =$	44.5
898	19	643	13	525	10.5	656	15	$n_4 =$	5	$R_4 =$	49
917	20	655	14	727	18	687	17				
	64		52.5		44.5		49	$n = \Sigma n = 20$		$\Sigma R = 210$	

    check:
    $\Sigma R = n(n+1)/2$
    $\qquad = 20(21)/2$
    $\qquad = 210$

    H_o: the populations have the same distribution
    H_1: the populations have different distributions
    $\alpha = .05$ [assumed]
    C.R. $H > \chi^2_{3,.05} = 7.815$
    calculations:
    $$H = [12/n(n+1)] \cdot [\Sigma(R_i^2/n_i)] - 3(n+1)$$
    $$= [12/20(21)] \cdot [(64)^2/5 + (52.5)^2/5 + (44.5)^2/5 + (49)^2/5] - 3(21)$$
    $$= [.0286] \cdot [2246.7] - 63$$
    $$= 1.191$$

    conclusion:
    Do not reject H_o; there is not sufficient evidence to support the claim that the head injury measurements have different population distributions.

    No; for this particular measurement, the data do not show that the heavier cars are safer in a crash. While the subcompact head injury values might seem higher, they are not statistically significant and could have occurred by chance when sampling from identical populations.

5.   Below are the ordered scores for each group. The group listed first is group 1, etc.

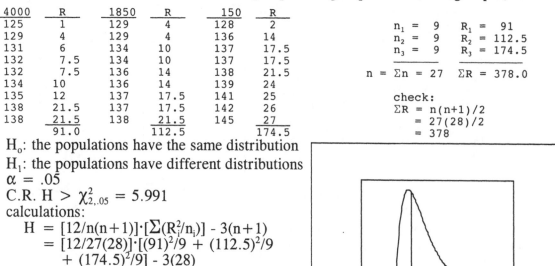

4000	R	1850	R	150	R
125	1	129	4	128	2
129	4	129	4	136	14
131	6	134	10	137	17.5
132	7.5	134	10	137	17.5
132	7.5	136	14	138	21.5
134	10	136	14	139	24
135	12	137	17.5	141	25
138	21.5	137	17.5	142	26
138	21.5	138	21.5	145	27
	91.0		112.5		174.5

$n_1 = 9$ $R_1 = 91$
$n_2 = 9$ $R_2 = 112.5$
$n_3 = 9$ $R_3 = 174.5$

$n = \Sigma n = 27$ $\Sigma R = 378.0$

check:
$\Sigma R = n(n+1)/2$
$= 27(28)/2$
$= 378$

H_0: the populations have the same distribution
H_1: the populations have different distributions
$\alpha = .05$
C.R. $H > \chi^2_{2,.05} = 5.991$
calculations:
$$H = [12/n(n+1)] \cdot [\Sigma(R_i^2/n_i)] - 3(n+1)$$
$$= [12/27(28)] \cdot [(91)^2/9 + (112.5)^2/9 + (174.5)^2/9] - 3(28)$$
$$= [.0159] \cdot [5709.7] - 84$$
$$= 6.631$$

2   5.991 $\chi^2_2 \, [\approx H]$

conclusion:
   Reject H_0; there is sufficient evidence to reject the claim that the populations have the same distribution and to conclude that they populations have different distributions.
NOTE: Compare these results to those of the parametric test in section 11-2, exercise #7.

7.   Below are the ordered weights (in thousandths of a gram) for each group. The group listed first is group 1, etc.

red	R	ora	R	yel	R	bro	R	blu	R	gre	R
870	9.5	861	5	868	8	856	2	838	1	890	26
872	12	897	31	876	16.5	858	3	870	9.5	902	39.5
874	13	898	34	877	18	860	4	875	14.5	902	39.5
882	21	903	42	879	19.5	866	6	956	87	911	52
888	24	920	61	879	19.5	867	7	968	90.5	930	72
891	27	942	82	886	22.5	871	11		202.5	949	83.5
897	31	971	92	886	22.5	875	14.5			1002	98
898	34	1009	99	892	28	876	16.5				410.5
908	46		446	893	29	889	25				
908	46			900	36.5	897	31				
908	46			906	44	898	34				
911	52			910	50	900	36.5				
912	54			911	52	902	39.5				
913	55			917	58	902	39.5				
920	61			921	63.5	904	43				
924	67			924	67	909	48.5				
924	67			926	69	909	48.5				
933	75			934	76	914	56.5				
936	77.5			939	79	914	56.5				
952	85			940	80	919	59				
983	95			941	81	920	61				
	998.0			949	83.5	921	63.5				
				960	88	923	65				
				968	90.5	928	70				
				978	94	930	72				
				989	97	930	72				
					1392.5	932	74				
						936	77.5				
						955	86				
						965	89				
						976	93				
						988	96				
						1033	100				
							1600.5				

$n_1 = 21$ $R_1 = 998.0$
$n_2 = 8$ $R_2 = 446.0$
$n_3 = 26$ $R_3 = 1392.5$
$n_4 = 33$ $R_4 = 1600.5$
$n_5 = 5$ $R_5 = 202.5$
$n_6 = 7$ $R_6 = 410.5$

$n = \Sigma n = 100$ $\Sigma R = 5050.0$

check:
$\Sigma R = n(n+1)/2$
$= 100(101)/2$
$= 5050$

H_o: the populations have the same distribution
H_1: the populations have different distributions
$\alpha = .05$ [assumed]
C.R. $H > \chi^2_{5,.05} = 11.071$
calculations:

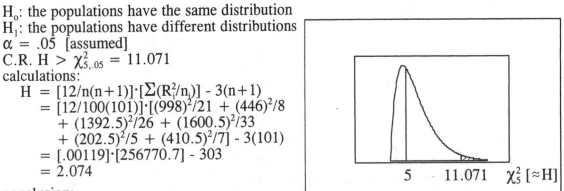

$$H = [12/n(n+1)]\cdot[\Sigma(R_i^2/n_j)] - 3(n+1)$$
$$= [12/100(101)]\cdot[(998)^2/21 + (446)^2/8$$
$$+ (1392.5)^2/26 + (1600.5)^2/33$$
$$+ (202.5)^2/5 + (410.5)^2/7] - 3(101)$$
$$= [.00119]\cdot[256770.7] - 303$$
$$= 2.074$$

conclusion:

Do not reject H_o; there is not sufficient evidence to reject the claim that the populations have the same distribution.

No; the results do not indicate a problem that requires corrective action.

NOTE: Compare these results to those of the parametric test in section 11-2, exercise #9.

9. $H = [12/n(n+1)]\cdot[\Sigma(R_i^2/n_j)] - 3(n+1)$, which depends only on the rank of each score and the number of scores within each group.

a. Adding or subtracting a constant to each score does not change the number of scores within each group. Since adding or subtracting a constant to each score does not affect the order of the scores, their ranks and the calculated H statistic are not affected.

b. Multiplying or dividing each score by a positive constant does not change the number of scores within each group. Since multiplying or dividing each score by a positive constant does not affect the order of the scores, their ranks and the calculated H statistic are not affected.

c. Changing a single sample value so that it becomes an outlier will not change the overall values of the ranks (i.e., they will still vary from 1 to n), but it may change individual ranks by ± 1 and affect how the ranks are distributed among the groups. In general, the effect on the value of the test statistic H will be minimal unless the change places the largest (or smallest) rank in the group which previously had the smallest (or largest) ranks.

11.

rank	t	T = $t^3 - t$
4.5	2	6
9.5	2	6
12	3	24
		36

correction factor:
$$1 - \Sigma T/(n^3-n) = 1 - 36/(18^3-18)$$
$$= 1 - 36/5814$$
$$= .99380805$$

The original calculated test statistic is $H = 14.74853801$
The corrected calculated test statistic is $H = 14.74853801/.99380805 = 14.8404$
No; the corrected value of H does not differ substantially from the original one.

NOTE: Be careful when counting the number of tied ranks; in addition to the easily recognized ".5's" that result from an even number of ties, there may be multiple whole number ranks resulting from an odd number of ties.

12-6 Rank Correlation

NOTE: This manual calculates $d = R_x - R_y$, thus preserving the sign of d. This convention means Σd must equal 0 and provides a check for the assigning and differencing of the ranks. In addition, it must always be true that $\Sigma R_x = \Sigma R_y = n(n+1)/2$.

1. In each case the n=5 pairs are pairs of ranks, called R_x and R_y below to stress that fact.

a.

R_x	R_y	d	d^2
1	1	0	0
3	3	0	0
5	5	0	0
4	4	0	0
2	2	0	0
15	15	0	0

$r_s = 1 - [6(\Sigma d^2)]/[n(n^2-1)]$
$= 1 - [6(0)]/[5(24)]$
$= 1 - 0$
$= 1$

Yes; there appears to be a perfect positive correlation between R_x and R_y.

b.

R_x	R_y	d	d^2
1	5	-4	16
2	4	-2	4
3	3	0	0
4	2	2	4
5	1	4	16
15	15	0	40

$r_s = 1 - [6(\Sigma d^2)]/[n(n^2-1)]$
$= 1 - [6(40)]/[5(24)]$
$= 1 - 2$
$= -1$

Yes; there appears to be a perfect negative correlation between R_x and R_y.

c.

R_x	R_y	d	d^2
1	2	-1	1
2	5	-3	9
3	3	0	0
4	1	3	9
5	4	1	1
15	15	0	20

$r_s = 1 - [6(\Sigma d^2)]/[n(n^2-1)]$
$= 1 - [6(20)]/[5(24)]$
$= 1 - 1$
$= 0$

No; there does not appear to be any correlation between R_x and R_y.

3. The following table summarizes the calculations.

R_x	R_y	d	d^2
2	2	0	0
6	7	-1	1
3	7	-3	9
5	4	1	1
7	5	2	4
10	8	2	4
9	9	0	0
8	10	-2	4
4	3	1	1
1	1	0	0
55	55	0	24

$r_s = 1 - [6(\Sigma d^2)]/[n(n^2-1)]$
$= 1 - [6(24)]/[10(99)]$
$= 1 - .145$
$= .855$

$H_0: \rho_s = 0$
$H_1: \rho_s \neq 0$
$\alpha = .05$
C.R. $r_s < -.648$
     $r_s > .648$
calculations:
     $r_s = .855$
conclusion:

Reject H_0; there is sufficient evidence to reject the claim that $\rho_s = 0$ and to conclude that $\rho_s \neq 0$ (in fact, that $\rho > 0$).

Yes; it does appear that salary increases as stress increases.

IMPORTANT NOTE: The rank correlation is correctly calculated using the ranks in the Pearson product moment correlation formula 9-1 of chapter 9 to produce

$$r_s = [\Sigma R_x R_y - (\Sigma R_x)(\Sigma R_y)]/[\sqrt{\Sigma R_x^2 - (\Sigma R_x)^2} \cdot \sqrt{\Sigma R_y^2 - (\Sigma R_y)^2}]$$

Since $\Sigma R_x = \Sigma R_y = 1+2+\ldots+n = n(n+1)/2$ [always]

and $\Sigma R_x^2 = \Sigma R_y^2 = 1^2+2^2+\ldots+n^2 = n(n+1)(2n+1)/6$ [when there are ties in the ranks],

it can be shown by algebra that the above formula can be shortened to

$$r_s = 1 - [6(\Sigma d^2)]/[n(n^2-1)]$$ when there are no ties in the ranks.

As ties typically make no difference in the first 3 decimals of r_s, this manual uses the shortened formula exclusively and notes when use of the longer formula gives a slightly different result.

5.   The following table summarizes the calculations.

R_x	R_y	d	d^2
1	3	-2	4
2	5	-3	9
4	4	0	0
5	1	4	16
3	10	-7	49
6	7	-1	1
8	6	2	4
7	8	-1	1
10	2	8	64
9	9	0	0
55	55	0	148

$$\begin{aligned} r_s &= 1 - [6(\Sigma d^2)]/[n(n^2-1)] \\ &= 1 - [6(148)]/[10(99)] \\ &= 1 - .897 \\ &= .103 \end{aligned}$$

H_o: $\rho_s = 0$
H_1: $\rho_s \neq 0$
$\alpha = .05$
C.R. $r_s < -.648$
        $r_s > .648$
calculations:
    $r_s = .261$
conclusion:

    Do not reject H_o; there is not sufficient evidence to reject the claim that $\rho_s = 0$.
No; there does not appear to be a relationship between the two rankings.

7.   The following table summarizes the calculations.

x	R_x	y	R_y	d	d^2
71	7.0	125	6	1	1
70.5	5	119	3.5	1.5	2.25
71	7.0	128	8.5	-1.5	2.25
72	9	128	8.5	.5	.25
70	3.0	119	3.5	-.5	.25
70	3.0	127	7	-4	16
66.5	1	105	1	0	0
70	3.0	123	5	-2	4
71.0	7	115	2	5	25
	45.0		45.0	0.0	51.00

$$\begin{aligned} r_s &= 1 - [6(\Sigma d^2)]/[n(n^2-1)] \\ &= 1 - [6(51.00)]/[9(80)] \\ &= 1 - .425 \\ &= .575 \end{aligned}$$

H_o: $\rho_s = 0$

H_1: $\rho_s \neq 0$
$\alpha = .05$
C.R. $r_s < -.683$
        $r_s > .683$
calculations:
    $r_s = .575$
conclusion:

    Do not reject H_o; there is not sufficient evidence to conclude that there is a correlation between heights and weights of supermodels.

NOTE: Using formula 9-1 (since there are ties) yields $r_s = .557$. Compare this to the parametric hypothesis test using Pearson's correlation in section 9-2 exercise #10.

9. The following table summarizes the calculations, where grams of fat is denoted x.

x	R_x	y	R_y
.07	16	3.7	7.5
.02	8.0	3.6	4.5
.01	6	3.6	4.5
.03	12.0	4.0	14.0
.03	12.0	4.0	14.0
.00	3.0	3.6	4.5
.03	12.0	3.8	9
.03	12.0	3.7	7.5
.06	15	4.1	16
.00	3.0	3.9	11.0
.02	8.0	3.9	11.0
.02	8.0	3.3	1
.00	3.0	3.5	2
.00	3.0	3.6	4.5
.00	3.0	3.9	11.0
.03	12.0	4.0	14.0
	136.0		136.0

Since there are so many ties, use formula 9-1 on the ranks.

$n = 16$
$\Sigma R_x = 136$
$\Sigma R_x^2 = 1474$
$\Sigma R_y = 136$
$\Sigma R_y^2 = 1486.5$
$\Sigma R_x R_y = 1320$

$n(\Sigma R_x^2) - (\Sigma R_x)^2 = 16(1474) - (136)^2$
$\qquad = 5088$
$n(\Sigma R_y^2) - (\Sigma R_y)^2 = 16(1486.5) - (136)^2$
$\qquad = 5288$
$n(\Sigma R_x R_y) - (\Sigma R_x)(\Sigma R_y) = 16(1320) - (136)(136)$
$\qquad = 2624$
$r_s = 2624/[\sqrt{5088}\sqrt{5288}]$
$\qquad = .5058$

$H_0: \rho_s = 0$
$H_1: \rho_s \neq 0$
$\alpha = .05$
C.R. $r_s < -.507$
$\qquad r_s > .507$
calculations:
$\qquad r_s = .506$
conclusion:

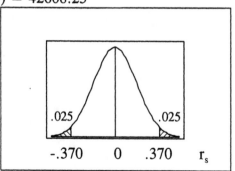

Do not reject H_0; there is not sufficient evidence to conclude that there is a correlation between grams of fat and calorie count.

NOTE: Compare this to the parametric hypothesis test using Pearson's correlation in section 9-2 exercise #9.

11. Refer to the data page for exercises #10, #11 and #12.
Since there are many ties, use formula 9-1 applied to the ranks.
The critical values are $\pm z/\sqrt{n-1} = \pm 1.96/\sqrt{28} \ \pm.370$.

$n(\Sigma R_n^2) - (\Sigma R_n)^2 = 29(8509) - (435)^2 = 57536$
$n(\Sigma R_t^2) - (\Sigma R_t)^2 = 29(8530) - (435)^2 = 58145$
$n(\Sigma R_c^2) - (\Sigma R_c)^2 = 29(8519) - (435)^2 = 57826$
$n(\Sigma R_n R_t) - (\Sigma R_n)(\Sigma R_t) = 29(8355.50) - (435)(435) = 53084.50$
$n(\Sigma R_n R_c) - (\Sigma R_n)(\Sigma R_c) = 29(7994.25) - (435)(435) = 42608.25$

a. Use the nicotine (n) & tar (t) values.
$H_0: \rho_s = 0$
$H_1: \rho_s \neq 0$
$\alpha = .05$
C.R. $r_s < -.370$
$\qquad r_s > .370$
calculations:
$\qquad r_s = 53084.50/[\sqrt{57536}\sqrt{58145}]$
$\qquad = .9177$
conclusion:

Reject H_0; there is sufficient evidence to conclude that the nicotine content is (positively) correlated with the amount of tar.

Below are the relevant values for exercises #10, #11 and #12.

	exercise #10				exercise #11					exercise #12				
#	cho	R_i	BMI	R_d	nic	R_n	tar	R_t	car	R_c	act	R_a	1dp	R_1	5dp	R_5
01	264	25	19.6	5.5	1.2	24.0	16	23.5	15	19.0	30	4.0	28	5.0	28	3.5
12	181	20	23.8	19.5	1.2	24.0	16	23.5	15	19.0	25	1	29	7.0	27	2
03	267	26	19.6	5.5	1.0	15.0	16	23.5	17	26.0	31	6.0	32	13.5	28	3.5
04	384	34	29.1	30	0.8	8.5	9	8.0	6	3	33	11.0	29	7.0	30	5.5
05	98	8.5	35.2	23	0.1	1	1	1	1	1	29	3	30	10.5	26	1
06	62	4.5	21.4	12	0.8	8.5	8	5.5	8	6	36	19.0	35	20	35	16.0
07	126	13	22.0	15	0.8	8.5	10	10	10	8.0	39	19.0	35	20.0	34	13.0
08	89	6	27.5	26	1.0	15.0	16	23.5	17	26.0	37	22.5	32	13.5	34	13.0
09	531	37	33.5	37	1.0	15.0	14	16.5	13	14.0	32	8	27	3.0	33	9.5
10	130	14	20.6	9	1.0	15.0	13	14.0	13	14.0	28	2	25	1	35	16.0
11	175	19	29.9	33	1.1	20.0	13	14.0	13	14.0	43	30	41	29.5	38	26.5
12	44	3	17.7	1	1.2	24.0	15	19.0	15	19.0	37	22.5	30	10.5	37	24.5
13	8	2	24.0	21	1.2	24.0	16	23.5	15	19.0	36	19.0	33	15.5	36	20.5
14	112	10	28.9	29	0.7	5.5	9	8.0	11	10.5	37	22.5	40	27.5	36	20.5
15	462	36	37.7	38	0.9	11	11	11	15	19.0	34	14	34	17.5	45	31
16	62	4.5	18.3	2	0.2	2	2	2	3	2	41	29	38	25.5	36	20.5
17	98	8.5	19.8	7	1.4	28.5	18	28.5	18	28.5	40	28	33	15.5	33	9.5
18	447	35	29.8	32	1.2	24.0	15	19.0	15	19.0	33	11.0	35	20.0	34	13.0
19	125	12	29.7	31	1.1	20.0	13	14.0	12	12	35	16.0	40	27.5	36	20.5
20	318	32	31.7	36	1.0	15.0	15	19.0	16	23.5	33	11.0	27	3.0	33	9.5
21	325	33	23.8	19.5	1.3	27	17	27	16	23.5	31	6.0	27	3.0	31	7
22	600	39	44.9	40	0.8	8.5	9	8.0	10	8.0	33	11.0	30	10.5	30	5.5
23	237	23	19.2	3	1.0	15.0	12	12	10	8.0	35	16.0	37	24	38	26.5
24	173	18	28.7	28	1.0	15.0	14	16.5	17	26.0	38	25.5	38	25.5	39	28
25	309	31	28.5	27	0.5	3	5	3	7	4.5	37	22.5	29	7.0	33	9.5
26	94	7	19.3	4	0.6	4	6	4	7	4.5	31	6.0	36	22.5	36	20.5
27	280	28	31.0	35	0.7	5.5	8	5.5	11	10.5	38	25.5	34	17.5	37	24.5
28	254	24	25.1	22	1.4	28.5	18	28.5	15	19.0	35	16.0	30	10.5	35	16.0
29	123	11	22.8	16.5	1.1	20.0	16	23.5	18	28.5	33	11.0	36	22.5	40	29.5
30	596	38	30.9	34		435.0		435.0		435.0	39	27	41	29.5	36	20.5
31	301	30	26.5	25							46	31	42	31	40	29.5
32	223	22	21.2	11								496.0		496.0		496.0
33	293	29	40.6	39												
34	146	15	21.9	13.5												
35	149	16.5	26.0	24												
36	149	16.5	23.5	18												
37	920	40	22.8	16.5												
38	271	27	20.7	10												
39	207	21	20.5	8												
40	2	1	21.9	13.5												
		820.0		820.0												

exercise #10

$\Sigma R_c^2 = 22138.50$

$\Sigma R_B^2 = 22138.00$

$\Sigma R_c R_B = 19581.50$

exercise #11

$\Sigma R_n^2 = 8509.00$

$\Sigma R_t^2 = 8530.00$

$\Sigma R_c^2 = 8519.00$

$\Sigma R_n R_t = 8355.50$

$\Sigma R_n R_c = 7994.25$

exercise #12

$\Sigma R_a^2 = 10394.50$

$\Sigma R_1^2 = 10401.50$

$\Sigma R_5^2 = 10887.00$

$\Sigma R_a R_1 = 9531.75$

$\Sigma R_a R_5 = 9369.50$

b. Use the nicotine (n) & carbon monoxide (c) values.

H_0: $\rho_s = 0$
H_1: $\rho_s \neq 0$
$\alpha = .05$
C.R. $r_s < -.370$
$\qquad r_s > .370$
calculations:
$\qquad r_s = 42608.25/[\sqrt{57536}\sqrt{57826}]$
$\qquad = .7387$
conclusion:

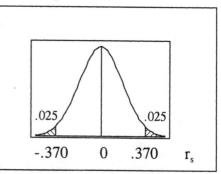

Reject H_0; there is sufficient evidence to conclude that the nicotine content is (positively) correlated with the amount of carbon monoxide.

c. Tar is a better predictor than carbon monoxide, because it has a higher correlation with nicotine.

NOTE: Compare this to the parametric hypothesis test using Pearson's correlation in section 9-2 exercise #20.

13. a. $t_{6,.025} = 2.365$; $r_s^2 = (2.447)^2/[(2.447)^2 + 6] = .499$, $r_s = \pm.707$
     b. $t_{13,.025} = 2.160$; $r_s^2 = (2.160)^2/[(2.160)^2 + 13] = .264$, $r_s = \pm.514$
     c. $t_{28,.025} = 2.048$; $r_s^2 = (2.048)^2/[(2.048)^2 + 28] = .130$, $r_s = \pm.361$
     d. $t_{28,.005} = 2.763$; $r_s^2 = (2.763)^2/[(2.763)^2 + 28] = .214$, $r_s = \pm.463$
     e. $t_{6,.005} = 3.707$; $r_s^2 = (3.707)^2/[(3.707)^2 + 6] = .696$, $r_s = \pm.834$

12-7 Runs Test for Randomness

NOTE: In each exercise, the item that appears first in the sequence is considered to be of the first type and its count is designated by n_1.

1. $n_1 = 11$ (# of H's)   G = 3 (# of runs)
    $n_2 = 7$ (# of M's)   CV: 5,14 (from Table A-10)
    No; the sequence does not appear to be random – fewer runs than expected by chance.

3. $n_1 = 10$ (# of A's)   G = 10 (# of runs)
    $n_2 = 10$ (# of B's)   CV: 6,16 (from Table A-10)
    Yes; the sequence appears to be random.
    NOTE: This example illustrates an obvious shortcoming of this test – it considers only the number of runs, and not the pattern within the runs. Here there seems to be an obvious pattern of 2 A's followed by 2 B's – but the <u>number</u> of runs is not considered unusual.

5. Since $n_1 = 12$ and $n_2 = 8$, use Table A-10.
    H_0: the sequence is random
    H_1: the sequence is not random
    $\alpha = .05$
    C.R. $G \leq 6$
    $\qquad G \geq 16$
    calculations:
    $\qquad G = 10$
    conclusion:

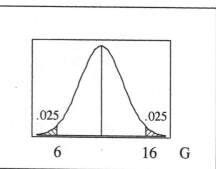

Do not reject H_0; there is not sufficient evidence to reject the claim that the values occur in a random sequence.
A lack of randomness would mean there was a pattern. Recognition of that pattern would give the bettor an advantage and put the casino at a disadvantage.

7. Since $n_1 = 19$ and $n_2 = 13$, use Table A-10.
    H_o: the sequence is random
    H_1: the sequence is not random
    $\alpha = .05$
    C.R. $G \le 10$
          $G \ge 23$
    calculations:
        $G = 6$
    conclusion:

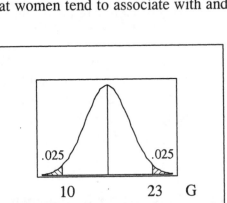

    Reject H_o; there is sufficient evidence to reject the
    claim that the values occur in a random sequence.
    Fred should conclude that there is not randomness in the marital status of the women he
    selects. There are fewer runs than expected by chance alone – i.e., the married and single
    women tend to come in bunches.   Perhaps his "random" selection process includes input
    from the present date in selecting the next date – and that women tend to associate with and
    recommend friends of their same marital status.

9. Since $n_1 = 18$ and $n_2 = 14$, use Table A-10.
    H_o: the sequence is random
    H_1: the sequence is not random
    $\alpha = .05$
    C.R. $G \le 10$
          $G \ge 23$
    calculations
        $G = 15$
    conclusion:

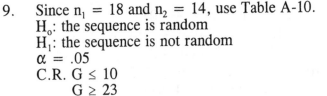

    Do not reject H_o; there is not sufficient evidence to
    reject the claim that the values occur in a random sequence.
    Yes; it appears that we elect Democrat and Republican candidates in a sequence that is
    random.

11. The median of the 21 values (arranged in numerical order from $x_1 = 1000$ to $x_{21} = 11568$)
     is $x_{11} = 3000$. Passing through the values chronologically and assigning A's and B's as
     directed (and ignoring values equal to the median), yields the sequence
     B B B B B B B B B B B A A A A A A A A A A
     Since $n_1 = 10$ and $n_2 = 10$, use Table A-10.
     H_o: B&A values occur in a random sequence
     H_1: B&A values do not occur in a random sequence
     $\alpha = .05$
     C.R. $G \le 6$
           $G \ge 16$
     calculations:
         $G = 2$
     conclusion:

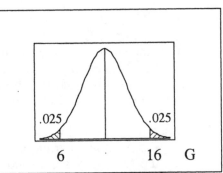

     Reject H_o; there is sufficient evidence to reject the
     claim that values below and above the median occur in a random sequence and to
     conclude that the sequence is not random (in fact, that values on the same side of the
     median tend to occur in groups).
     The results suggest that the values in the stock market are steadily increasing and not
     randomly fluctuating around a long-term typical value.

13. The sequence in O's and E's is as follows.

```
O E O O O E E O O O E O O O O E O E E E
E E E O O E O E O O O E E E E E O O O O
E O O O O O O O O E O E E E O O E O E E
O O E O E O E O E E E E E E E E E E O O
E E E E E O E E E O O E E O O O E E O O
```

Since $n_1 = 49$ and $n_2 = 51$, use the normal approximation.

$\mu_G = 2n_1n_2/(n_1+n_2) + 1$
 $= 2(49)(51)/100 + 1 = 50.98$
$\sigma_G^2 = [2n_1n_2(2n_1n_2-n_1-n_2)]/[(n_1+n_2)^2(n_1+n_2-1)]$
 $= [2(49)(51)(4898)]/[(100)^2(99)] = 24.727$

H_o: the sequence is random
H_1: the sequence is not random
$\alpha = .05$ [assumed]
C.R. $z < -z_{.025} = -1.96$
  $z > z_{.025} = 1.96$
calculations:
  $G = 43$
  $z_G = (G - \mu_G)/\sigma_G$
   $= (43 - 50.98)/\sqrt{24.73}$
   $= -7.98/4.973 = -1.605$
conclusion:

  Do not reject H_o; there is not sufficient evidence to conclude that the sequence is not random.

15. The 150 runners finished in the following order by gender.

```
16M  F 2M  F 10M F  M  F 4M F 7M 2F 11M F 9M 2F 5M 3F
 4M  F 3M 2F  M F 4M 3F  M F 2M  F   M F 4M 2F 4M  F
 M 2F 4M 2F   M F 4M  F  M F 2M  F  3M F 2M 2F 4M 2F
```

There are 111M's, 39F's and 54 runs.

Since $n_1 = 111$ and $n_2 = 39$, use the normal approximation.

$\mu_G = 2n_1n_2/(n_1+n_2) + 1$
 $= 2(111)(39)/150 + 1 = 58.72$
$\sigma_G^2 = [2n_1n_2(2n_1n_2-n_1-n_2)]/[(n_1+n_2)^2(n_1+n_2-1)]$
 $= [2(111)(39)(8508)]/[(150)^2(149)] = 21.965$

H_o: the M&F sequence is random
H_1: the M&F sequence is not random
$\alpha = .05$ [assumed]
C.R. $z < -z_{.025} = -1.96$
  $z > z_{.025} = 1.96$
calculations:
  $G = 54$
  $z_G = (G - \mu_G)/\sigma_G$
   $= (54 - 58.72)/\sqrt{21.96}$
   $= -4.72/4.687 = -1.007$
conclusion:

  Do not reject H_o; there is not sufficient evidence to conclude that the sequence is not random.

No, there is not sufficient evidence to support the claim that male runners tend to finish before female runners – the hypothesis that finishing order was gender-random cannot be rejected.

17. The minimum possible number of runs is $G = 2$ and occurs when all the A's are together and all the B's are together (e.g., A A B B).

The maximum possible number of runs is $G = 4$ and occurs when the A's and B's alternate (e.g., A B A B).

Because the critical region for $n_1 = n_2 = 2$ is

C.R. $G \leq 1$

$G \geq 6$

the null hypothesis of the sequence being random can never be rejected at the .05 level. Very simply, this means that it is not possible for such a small sample to provide 95% certainty that a non-random phenomenon is occurring.

Review Exercises

1. Let the after-course scores be group 1.

claim: median difference $\neq 0$

subj	A	B	C	D	E	F	G	H	I	J
R-M	+	0	-	+	+	+	-	+	+	+

$n = 9$: 7+'s and 2-'s

H_0: median difference $= 0$

H_1: median difference $\neq 0$

$\alpha = .05$

C.R. $x \leq x_{L,9,.025} = 1$

$x \geq x_{U,9,.025} = 9-1 = 8$

calculations:

$x = 2$ (using less frequent count)

$x = 7$ (using + count)

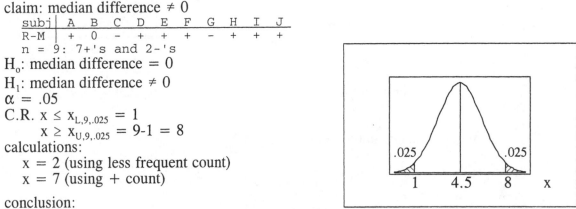

conclusion:

Do not reject H_0; there is not sufficient evidence to conclude that the median difference is different from 0 – i.e., there is not sufficient evidence to reject the claim that the course has no effect.

2. Let the after-course scores be group 1.

claim: the populations have the same distribution

A-B	20	0	-10	40	30	10	-30	20	20	10
R	5	–	-2	9	7.5	2	-7.5	5	5	2

$\Sigma R- = 9.5$

$\Sigma R+ = 35.5$ $n = 9$ non-zero ranks

$\Sigma R = 45.0$

check: $\Sigma R = n(n+1)/2 = 9(10)/2 = 45$

H_0: the populations have the same distribution

H_1: the populations have different distributions

$\alpha = .05$

C.R. $T \leq T_{L,9,.025} = 6$

$T \geq T_{U,9,.025} = 45-6 = 39$

calculations

$T = 9.5$ (using the smaller ranks)

$T = 35.5$ (using the positive ranks)

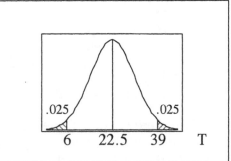

conclusion:

Do not reject H_0; there is not sufficient evidence to conclude that the populations have different distributions – i.e., there is not sufficient evidence to reject the claim that the course has no effect.

3. Let x = the number of women hired.
    claim: p < .5
    ```
    22 +'s (women)      44 -'s (men)
    n = 66 +'s or -'s
    ```
    Since n > 25, use z with
    $\mu_x = n/2 = 66/2 = 33$
    $\sigma_x = \sqrt{n}/2 = \sqrt{66}/2 = 4.062$
    $H_o: p = .5$
    $H_1: p < .5$
    $\alpha = .01$
    C.R. $z < -z_{.01} = -2.326$
    calculations:
    x = 22
    $z_x = [(x+.5)-\mu_x]/\sigma_x$
    $= [22.5 - 33]/4.062$
    $= -10.5/4.062$
    $= -2.585$
    conclusion:

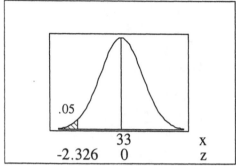

    Reject H_o; there is sufficient evidence to conclude that p < .5 (i.e., that there is bias
    against women in the hiring practices).

4. Below are the ordered scores for each group.
    claim: the populations have the same distribution

beer	R	liquor	R
.129	1	.182	9
.146	2	.185	10
.148	3	.190	12.5
.152	4	.205	15
.154	5	.220	17
.155	6	.224	18
.164	7	.225	19
.165	8	.226	20
.187	11	.227	21
.190	12.5	.234	22
.203	14	.241	23
.212	16	.247	24
	89.5	.253	25
		.257	26
			261.5

    $n_1 = 12$ $\Sigma R_1 = 89.5$
    $n_2 = 14$ $\Sigma R_2 = 261.5$

    $n = \Sigma n = 26$ $\Sigma R = 351.0$

    check: $\Sigma R = n(n+1)/2$
    $= 26(27)/2$
    $= 351$
    $R = \Sigma R_1 = 89.5$

    $\mu_R = n_1(n+1)/2$
    $= 12(27)/2$
    $= 162$
    $\sigma_R^2 = n_1 n_2 (n+1)/12$
    $= (12)(14)(27)/12 = 378$

    H_o: the populations have the same distribution
    H_1: the populations have different distributions
    $\alpha = .05$
    C.R. $z < -z_{.025} = -1.96$
    $\quad\ z > z_{.025} = 1.96$
    calculations:
    $z_R = (R - \mu_R)/\sigma_R$
    $= (89.5 - 162)/\sqrt{378}$
    $= -72.5/19.442$
    $= -3.729$
    conclusion:

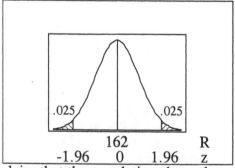

    Reject H_o; there is sufficient evidence to reject the claim that the populations have the
    same distribution and to conclude that the distributions are different (in fact, that the
    beer scores are lower).
    It appears that the liquor drinkers are more dangerous.

5. The following table summarizes the calculations.

x	R_x	y	R_y	d	d^2
29	5	31	8.5	-3.5	12.25
35	9	27	2	7	49
28	3.5	29	6,5	-3	9
44	10	25	1	9	81
25	2	31	8.5	-6.5	42.25
34	8	29	6.5	1.5	2.25
30	6	28	4.0	2	4
33	7	28	4.0	3	9
28	3.5	28	4.0	-0.5	0.25
24	1	33	10	-9	81
	55.0		55.0	0.0	290.00

$r_s = 1 - [6(\Sigma d^2)]/[n(n^2-1)]$
$= 1 - [6(290)]/[10(99)]$
$= 1 - 1.758$
$= -.758$

$H_o: \rho_s = 0$
$H_1: \rho_s \neq 0$
$\alpha = .05$
C.R. $r_s < -.648$
    $r_s > .648$
calculations:
    $r_s = -.758$

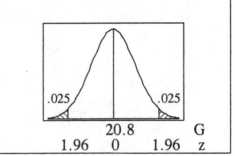

conclusion:
    Reject H_o; there is sufficient evidence to reject the claim that $\rho_s = 0$ conclude that $\rho_s \neq 0$ (in fact, that $\rho_s < 0$).
Yes; based on these results you can expect to pay more for gas if you buy a heavier car. Since the calculations are based entirely on ranks, and the ranks would not change whether the weights are given in pounds or 100-pounds, such a modification would not change the results at all.
NOTE: Using formula 9-1 (since there are ties) yields $r_s = -.796$.

6. The sequence in O's and E's is as follows.
   O O E O O O O O E E E O E E O E O E E O
   O O O E E E E O O E O O O O O E O E E E
Since $n_1 = 22$ and $n_2 = 18$, use the normal approximation.
   $\mu_G = 2n_1n_2/(n_1+n_2) + 1$
   $= 2(22)(18)/40 + 1 = 20.8$
   $\sigma_G^2 = [2n_1n_2(2n_1n_2-n_1-n_2)]/[(n_1+n_2)^2(n_1+n_2-1)]$
   $= [2(22)(18)(752)]/[(40)^2(39)] = 9.545$

H_o: the O&E sequence is random
H_1: the O&E sequence is not random
$\alpha = .05$
C.R. $z < -z_{.025} = -1.96$
    $z > z_{.025} = 1.96$
calculations:
   $G = 18$
   $z_G = (G - \mu_G)/\sigma_G$
   $= (18 - 20.8)/\sqrt{9.545}$
   $= -2.8/3.089$
   $= -.906$

conclusion:
   Do not reject H_o; there is not sufficient evidence to conclude that the sequence is not random.

7. Below are the scores for each group. The group listed first is group 1, etc.

subco	R	compa	R	midsi	R	fulls	R
595	5	1051	15	629	6	1985	17
1063	16	1193	18	1686	20	971	12
885	10	946	11	880	9	996	14
519	3	984	13	181	1	804	8
422	2	584	4	645	7	1376	19
	36		61		43		70

$n_1 = 5 \quad R_1 = 36$
$n_2 = 5 \quad R_2 = 61$
$n_3 = 5 \quad R_3 = 43$
$n_4 = 5 \quad R_4 = 70$

$n = \Sigma n = 20 \quad \Sigma R = 210$

H_o: the populations have the same distribution
H_1: the populations have different distributions
$\alpha = .05$
C.R. $H > \chi^2_{3,.05} = 7.815$
calculations:

check:
$\Sigma R = n(n+1)/2$
$= 20(21)/2$
$= 210$

$$H = [12/n(n+1)]\cdot[\Sigma(R_i^2/n_i)] - 3(n+1)$$
$$= [12/20(21)]\cdot[(36)^2/5 + (61)^2/5$$
$$+ (43)^2/5 + (70)^2/5] - 3(21)$$
$$= [.0286]\cdot[2353.2] - 63 = 4.234$$

conclusion:

Do not reject H_o; there is not sufficient evidence to conclude that leg injury measurements for the four weight categories are not all the same.

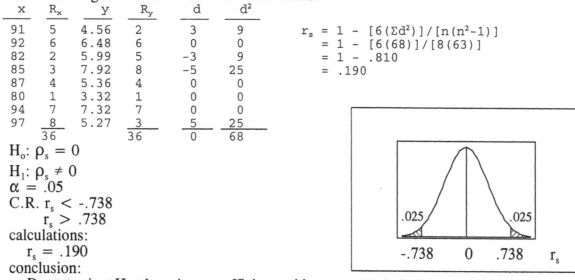

No; for this particular measurement, the data do not show that the heavier cars are safer in a crash.

8. The following table summarizes the calculations.

x	R_x	y	R_y	d	d^2
91	5	4.56	2	3	9
92	6	6.48	6	0	0
82	2	5.99	5	-3	9
85	3	7.92	8	-5	25
87	4	5.36	4	0	0
80	1	3.32	1	0	0
94	7	7.32	7	0	0
97	8	5.27	3	5	25
	36		36	0	68

$r_s = 1 - [6(\Sigma d^2)]/[n(n^2-1)]$
$= 1 - [6(68)]/[8(63)]$
$= 1 - .810$
$= .190$

H_o: $\rho_s = 0$

H_1: $\rho_s \neq 0$

$\alpha = .05$

C.R. $r_s < -.738$
$\quad\quad r_s > .738$

calculations:
$r_s = .190$

conclusion:

Do not reject H_o; there is not sufficient evidence to conclude that $\rho_s \neq 0$.
The results suggest that the higher priced tapes do not necessarily have higher performance ratings. When buying tapes, choose less expensive ones.

Cumulative Review Exercises

1. a. Since $n_1 = 20$ and $n_2 = 7$, use Table A-10.
      H_o: the M&F sequence is random
      H_1: the M&F sequence is not random
      $\alpha = .05$
      C.R. $G \leq 6$
          $G \geq 16$
      calculations:
          $G = 15$

      conclusion:
          Do not reject H_o; there is not sufficient
      evidence to reject the claim that the sequence is random with respect to gender

    b. Let p be the proportion of women: $\hat{p} = x/n = 7/27 = .259$
       claim: $p \neq .50$
       H_o: $p = .50$
       H_1: $p \neq .50$
       $\alpha = .05$
       C.R. $z < -z_{.025} = -1.96$
           $z > z_{.025} = 1.96$
       calculations:
           $z_{\hat{p}} = (\hat{p} - \mu_{\hat{p}})/\sigma_{\hat{p}}$
           $= (.259 - .50)/\sqrt{(.50)(.50)/27}$
           $= -.241/.0962$
           $= -2.502$
           P-value $= 2 \cdot P(z < -2.50) = 2 \cdot (.0062) = .0124$
       conclusion:
           Reject H_o; there is sufficient evidence to conclude that $p \neq .50$ (in fact, that $p < .50$).

    c. Let x = the number of women.
       claim: $p \neq .50$
           7 +'s (women)   20 -'s (men)
           n = 27 +'s or -'s
       Since n > 25, use z with
           $\mu_x = n/2 = 27/2 = 13.5$
           $\sigma_x = \sqrt{n}/2 = \sqrt{27}/2 = 2.598$
       H_o: $p = .50$
       H_1: $p \neq .50$
       $\alpha = .05$
       C.R. $z < -z_{.025} = -1.96$
           $z > z_{.025} = 1.96$
       calculations:
           $x = 7$
           $z_x = [(x + .5) - \mu_x]/\sigma_x$
           $= [7.5 - 13.5]/2.598$
           $= -6/2.598$
           $= -2.309$
           P-value $= 2 \cdot P(z < -2.31) = 2 \cdot (.0104) = .0208$
       conclusion:
           Reject H_o; there is sufficient evidence to conclude that $p \neq .50$ (in fact, that $p < .50$).
       NOTE: The smaller P-value in part (b) [$.0124 < .0208$] indicates that the parametric tests are generally better than their non-parametric counterparts.

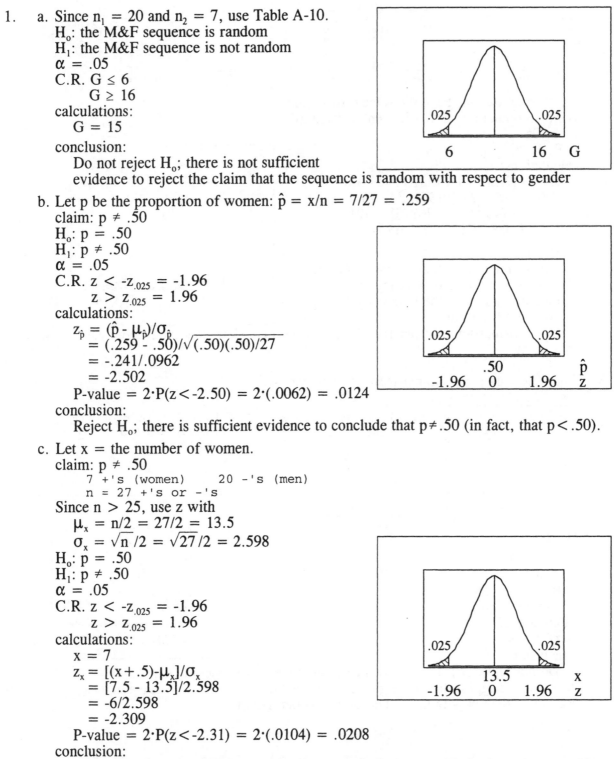

d. Let p be the proportion of women: $\hat{p} = x/n = 7/27 = .259$

$\hat{p} \pm z_{.025}\cdot\sqrt{\hat{p}\hat{q}/n}$

$.259 \pm 1.96\cdot\sqrt{(.259)(.741)/27}$

$.259 \pm .165$

$.094 < p < .425$

e. There is no evidence that the passengers were not sampled in random sequence according to gender. But the above results cannot address whether or not the sample is biased against either gender without knowing the proportion of each gender in the population of passengers. If the sample contains significantly more of one gender than another, that might reflect the gender distribution of the passengers and not gender bias by the pollster. Otherwise, there are no problems with the survey in these areas.

2. The following table summarizes the preliminary calculations necessary for all parts of this exercise. Since there are ties, the shortcut formula for r_s using $d = R_x - R_y$ will not be used.

Let x = height of the winner.

Let y = height of the runner-up.

raw scores			sign	signed	ranks	
x	y	d=x-y	d	rank d	R_x	R_y
76	64	12	+	7	8	1
66	71	-5	-	-5	1	4.5
70	72	-2	-	-2.5	2.5	6.5
70	72	-2	-	-2.5	2.5	6.5
74	68	6	+	6	6.5	2
71.5	71	0.5	+	1	4	4.5
73	69.5	3.5	+	4	5	3
74	74	0	0	-	6.5	8
Σv 574.5	561.5	13.0			36	36
Σv^2 41325.25	39476.25	225.50			203.00	203.00
Σuv 40288.00					143.00	

a. $n(\Sigma xy) - (\Sigma x)(\Sigma y) = 8(40288.00) - (574.1)(561.5) = -277.75$

$n(\Sigma x^2) - (\Sigma x)^2 = 8(41325.25) - (574.5)^2 = 551.75$

$n(\Sigma y^2) - (\Sigma y)^2 = 8(39476.25) - (561.5)^2 = 527.75$

$r = [n(\Sigma xy) - (\Sigma x)(\Sigma y)]/[\sqrt{n(\Sigma x^2) - (\Sigma x)^2}\cdot\sqrt{n(\Sigma y^2) - (\Sigma y)^2}]$

$= -277.75/[\sqrt{551.75}\cdot\sqrt{527.75}]$

$= -.5147$

$H_o: \rho = 0$

$H_1: \rho \neq 0$

$\alpha = .05$

C.R. $r < -.707$ OR   C.R. $t < -t_{6,.025} = -2.447$

   $r > .707$ $t > t_{6,.025} = 2.447$

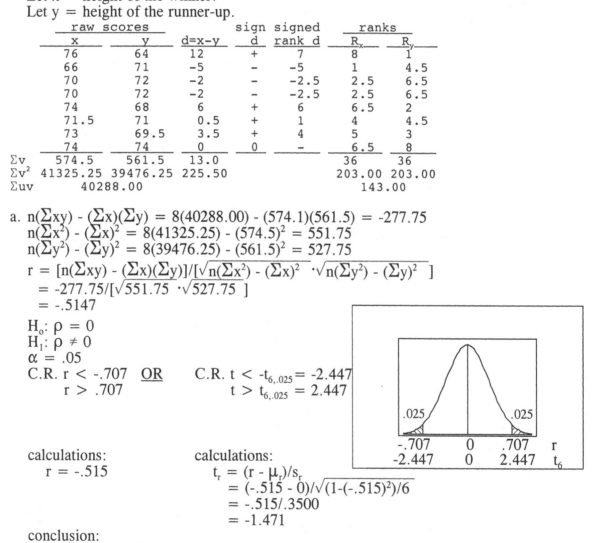

calculations: calculations:

   $r = -.515$ $t_r = (r - \mu_r)/s_r$

                         $= (-.515 - 0)/\sqrt{(1-(-.515)^2)/6}$

                         $= -.515/.3500$

                         $= -1.471$

conclusion:

   Do not reject H_o; there is not sufficient evidence to reject the claim that $\rho = 0$.

No; there is no significant correlation between the heights.

b. Refer to the summary calculations at the beginning of the exercise.
   Since there are ties, use R_x for x and R_y for y in formula 9-1.
   $n(\Sigma xy) - (\Sigma x)(\Sigma y) = 8(143.00) - (36)(36) = -152.00$
   $n(\Sigma x^2) - (\Sigma x)^2 = 8(203.00) - (36)^2 = 328.00$
   $n(\Sigma y^2) - (\Sigma y)^2 = 8(203.00) - (36)^2 = 328.00$
   $r = -152.00/[\sqrt{328.00} \cdot \sqrt{328.00}\ ]$
   $= -.4634$
   $H_o: \rho_s = 0$
   $H_1: \rho_s \neq 0$
   $\alpha = .05$
   C.R. $r_s < -.738$
   $\quad r_s > .738$
   calculations:
   $\quad r_s = -.463$

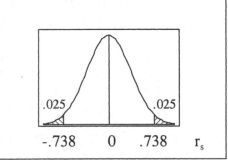

conclusion:
   Do not reject H_o; there is not sufficient evidence to conclude that there is any
   correlation between the heights.
   NOTE: The smaller (in absolute value) r_s in part (b) [$|-.463| < |-.515|$] indicates that
   non-parametric tests are generally weaker than their parametric counterparts.

c. Refer to the summary calculations at the beginning of the exercise.
   claim: median difference $\neq 0$
   $\quad\quad$ n = 7: 4+'s and 3-'s
   H_o: median difference $= 0$
   H_1: median difference $\neq 0$
   $\alpha = .05+$
   C.R. $x \leq x_{L,7,.025} = 0$
   $\quad\quad x \geq x_{U,7,.025} = 7-0 = 7$
   calculations:
   $\quad x = 3$ (using less frequent count)
   $\quad x = 4$ (using + count)
   conclusion:

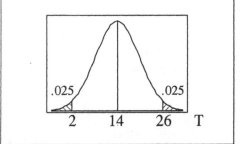

   Do not reject H_o; there is not sufficient evidence to conclude that the median
   difference is different from 0.

d. Refer to the summary calculations at the beginning of the exercise.
   claim: the populations have different distributions
   $\quad\quad \Sigma R- = 10 \quad\quad\quad$ n = 7 non-zero ranks
   $\quad\quad \Sigma R+ = 18$
   $\quad\quad \Sigma R\ \ = 28$
   check: $\Sigma R = n(n+1)/2 = 7(8)/2 = 28$
   H_o: the populations have the same distribution
   H_1: the populations have different distributions
   $\alpha = .05$
   C.R. $T \leq T_{L,7,.025} = 2$
   $\quad\quad T \geq T_{U,7,.025} = 28-2 = 26$
   calculations:
   $\quad T = 10$ (using the smaller ranks)
   $\quad T = 18$ (using the positive ranks)
   conclusion:
   Do not reject H_o; there is not sufficient evidence to conclude there is a difference
   between the heights of the winning and losing candidates.

e. Refer to the summary calculations at the beginning of the exercise.

$n = 8$ $\Sigma d = 13.0$ $\bar{d} = 1.625$
                 $\Sigma d^2 = 225.50$ $s_d = 5.403$

original claim: $\mu_d \neq 0$
$H_o: \mu_d = 0$
$H_1: \mu_d \neq 0$
$\alpha = .05$
C.R. $t < -t_{7,.025} = -2.365$
      $t > t_{7,.025} = 2.365$
calculations:

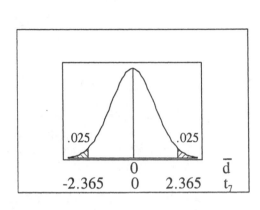

$t_{\bar{d}} = (\bar{d} - \mu_{\bar{d}})/s_{\bar{d}}$
      $= (1.625 - 0)/(5.403/\sqrt{8})$
      $= 1.625/1.910$
      $= .851$
P-value $= 2 \cdot P(t_7 > .851) > .20$
conclusion:
    Do not reject H_o; there is not sufficient evidence to conclude that $\mu_d \neq 0$.

f. The is not sufficient evidence to conclude there is any correlation between the heights of the winning and losing candidates, nor that there is any difference between the heights.

Chapter 13

Statistical Process Control

13-2 Control Charts for Variation and Mean

NOTE: In this section, k = number of sample subgroups
n = number of observations per sample subgroup
Although the $\bar{\bar{x}}$ line is not specified in the text as a necessary part of the run chart, such a line is included in the Minitab examples and is included in this manual.

1. a. Process data are data arranged in sequence according to time.
   b. Process data are out of statistical control when they exhibit more than natural variation – i.e., when the data include patterns, cycles, or unusual points.
   c. A process is out of statistical control whenever any of the following 3 criteria are met:
      (1) The data exhibit a non-random trend, pattern, or cycle.
      (2) The is a point beyond the upper or lower control limit.
      (3) The are 8 or more consecutive points on the same side of the center line.
   d. Random variation is due to chance; assignable variation is the result of identifiable causes.
   e. An R chart monitors the ranges of consecutive samples on size n and is used to determine whether the process variation is within statistical control. An $\bar{x}$ chart monitors the means of consecutive sample of size n and is used to determine whether the process average is within statistical control.

The following chart summarizes the information necessary for exercises #2, #3 and #4.

sample	consumption			$\bar{x}$	R
1a	4762	3875	2657	3764.7	2105
1b	4358	2201	3187	3248.7	2157
2a	4504	3237	2198	3313.0	2306
2b	2511	3020	2857	2796.0	509
3a	3952	2785	2118	2951.7	1834
3b	2658	2139	3071	2622.7	932
4a	3863	3013	2023	2966.3	1840
4b	2953	3456	2647	3018.7	809

3. $\bar{R} = \Sigma R/k = 12492/8 = 1561.5$ $LCL = D_3\bar{R} = 0.000(1561.5) = 0.0$
   $UCL = D_4\bar{R} = 2.574(1561.5) = 4019.3$

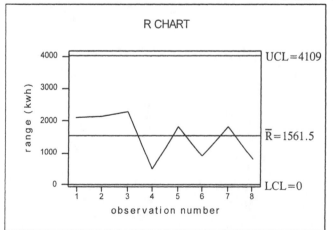

The process variation is within statistical control.

5. $\bar{R} = \Sigma R/k = 1374/25 = 54.96$ $LCL = D_3\bar{R} = 0.076(54.96) = 4.18$
$UCL = D_4\bar{R} = 1.924(54.96) = 105.74$

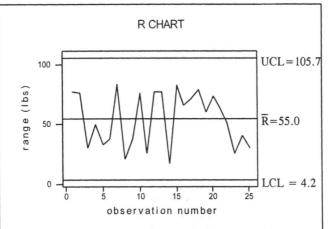

The process variation appears to be within statistical control.

7.   There are $k \cdot n$ = (20 samples)·(5 obs/sample) = 100 observations on the run chart.
For convenience, the problem is worked in milligrams (i.e., the original weights x 1000).
$\bar{\bar{x}} = \Sigma\bar{x}/k = \Sigma x/(k \cdot n) = 570980/100 = 5709.8$

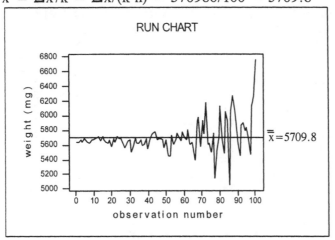

Yes, there appears to be a pattern that suggests the process is not within statistical control
– viz., the variability is increasing. In practical terms, this suggests that the equipment is
not holding its tolerance and that the process needs to be shut down for an adjustment.

9.   For convenience, the problem is worked in milligrams (i.e., the original weights x 1000).
$\bar{\bar{x}} = \Sigma\bar{x}/k = 114196.0/20 = 5709.8$
$\bar{R} = \Sigma R/k = 7292/20 = 364.6$
$LCL = \bar{\bar{x}} - A_2\bar{R} = 5709.8 - (.577)(364.6) = 5709.8 - 210.4 = 5499.4$
$UCL = \bar{\bar{x}} + A_2\bar{R} = 5709.8 + (.577)(364.6) = 5709.8 + 210.4 = 5920.2$

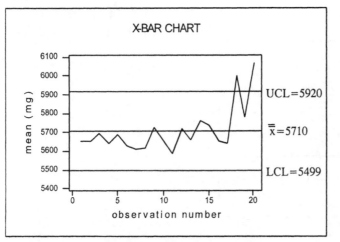

The process mean is not within statistical control. To be out of control, the process must meet at least one of the three out-of-control criteria. This process actually meets all three. (1) There is a pattern to the $\bar{x}$ values – in this case, an upward drift. (2) There is a point beyond the upper or lower control limit – in this case, there are 2 points above the UCL. (3) There are at least eight consecutive points on the same side of the center line – in this case, the first 8 points are below the center line. Yes, corrective action is needed.

The following chart summarizes the information necessary for exercises #11 and #12.

week	$\bar{x}$	R	week	$\bar{x}$	R	week	$\bar{x}$	R	week	$\bar{x}$	R
1	.019	.05	14	.000	.00	27	.036	.14	40	.067	.24
2	.027	.10	15	.097	.40	28	.027	.11	41	.003	.02
3	.101	.71	16	.237	.87	29	.010	.05	42	.003	.02
4	.186	.64	17	.067	.47	30	.017	.12	43	.181	.68
5	.016	.05	18	.054	.24	31	.082	.44	44	.243	1.48
6	.091	.64	19	.036	.14	32	.044	.18	45	.290	1.28
7	.051	.30	20	.237	.92	33	.164	.64	46	.139	.96
8	.001	.01	21	.049	.27	34	.134	.85	47	.116	.79
9	.039	.16	22	.003	.01	35	.009	.03	48	.109	.41
10	.121	.39	23	.000	.00	36	.017	.12	49	.017	.08
11	.181	.78	24	.101	.71	37	.050	.26	50	.000	.00
12	.029	.17	25	.097	.33	38	.057	.40	51	.106	.74
13	.416	1.41	26	.000	.00	39	.017	.12	52	.146	.43

11. $\bar{R} = \Sigma R/k = 20.36/52 = .3915$ $LCL = D_3\bar{R} = 0.076(.3915) = .030$
$UCL = D_4\bar{R} = 1.924(.3915) = .753$

R CHART

range (in)

1.5

1.0

0.5   UCL=.753

                                                 $\bar{R}$=.392

0.0   LCL=.030

   0   10   20   30   40   50

week number

The process variation is not within statistical control. Since there are points above the upper control limit and points below the lower control limit, reject the claim that rainfall amounts exhibit stable variation. While only one point above or below the appropriate control limits is sufficient evidence of unstable variation, these data have multiple points beyond each limit.

13. $\overline{s} = \Sigma s/k = 193.85/20 = 9.6925$

$$LCL = B_3\overline{s} = 0(9.6925)$$
$$= 0$$
$$UCL = B_4\overline{s} = 2.266(9.6925)$$
$$= 21.963$$

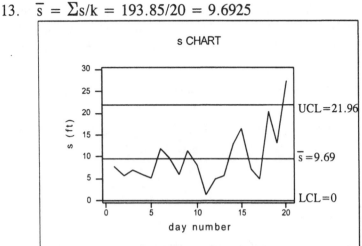

s CHART

This is very similar to the R chart given in this section, and both charts indicate the process variation is not within statistical control. For small n's in the sub-samples (n=4 in this example), there will be a very high correlation between the values of R and s – and so the two charts will be almost identical, but with different labels on the vertical axis.

13-3 Control Charts for Attributes

1.   This process is within statistical control. Since the first third of the sample means are generally less than the overall mean, the middle third are generally more than the overall mean, and the final third are generally less than the overall mean, however, one may wish to check future analyses to see whether such a pattern tends to repeat itself.

3.   This process is out of statistical control. There is an upward trend, and there is a point above the upper control limit.

5. $\overline{p} = (\Sigma x)/(\Sigma n) = (30+29+...+23)/(13)(100,000) = 332/1,300,000 = .000255$
    $\sqrt{\overline{p}\cdot\overline{q}/n} = \sqrt{(.000255)(.999745)/100000} = .0000505$
    $LCL = \overline{p} - 3\sqrt{\overline{p}\cdot\overline{q}/n} = .000255 - 3(.0000505) = .000255 - .000152 = .000104$
    $UCL = \overline{p} + 3\sqrt{\overline{p}\cdot\overline{q}/n} = .000255 + 3(.0000505) = .000255 + .000152 = .000407$

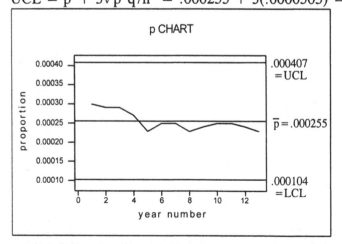

p CHART

NOTE: The 13 sample proportions are .00030, .00029, .00029,..., .00023.
There are three out-of-control criteria, and meeting any one of them means that the process is not within statistical control. This process meets two of those criteria. (1) There is a

pattern – in this case, a downward trend. (3) The run of 8 rule applies – in this case, the last 9 points are all below the center line. By definition, therefore, process is statistically unstable. In this instance, however, that is good and means that the age 0-4 death rate for infectious diseases is decreasing.

7. $\bar{p} = (\Sigma x)/(\Sigma n) = (3+3+1...+5)/(52)(7) = 127/364 = .3489$
$\sqrt{\bar{p}\cdot\bar{q}/n} = \sqrt{(.3489)(.6511)/7} = .1801$
LCL $= \bar{p} - 3\sqrt{\bar{p}\cdot\bar{q}/n} = .3489 - 3(.1801) = .3489 - .5404 = -.192$ [truncated at 0.000]
UCL $= \bar{p} + 3\sqrt{\bar{p}\cdot\bar{q}/n} = .3489 + 3(.1801) = .3489 + .5404 = .889$

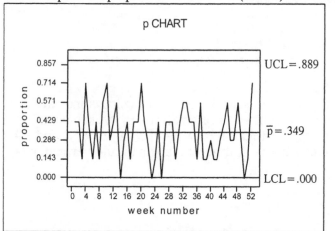

NOTE: The 52 sample proportions are: 3/7, 3/7, 1/7,...
The vertical labels correspond to 0/7=.000, 1/7=.143, 2/7=.286, 3/7=.429,...
The process is within statistical control. Viewing the proportion of days each week with measurable rainfall as a statistical process, it does not need correction. Whether the proportions are too high or too low for individual preferences or other needs is another issue.

9. $\bar{p} = (\Sigma x)/(\Sigma n) = (25+24+...+31)/(13)(100,000) = 375/1,300,000 = .000288$
$n\cdot\bar{p} = (100,000)(.000288) = 28.846$ [$= \bar{x}$]
$\sqrt{n\cdot\bar{p}\cdot\bar{q}} = \sqrt{(100000)(.000288)(.999712)} = 5.370$
LCL $= n\cdot\bar{p} - 3\sqrt{n\cdot\bar{p}\cdot\bar{q}} = 28.846 - 3(5.370) = 28.846 - 16.110 = 12.736$
UCL $= n\cdot\bar{p} + 3\sqrt{n\cdot\bar{p}\cdot\bar{q}} = 28.846 + 3(5.370) = 28.846 + 16.110 = 44.956$

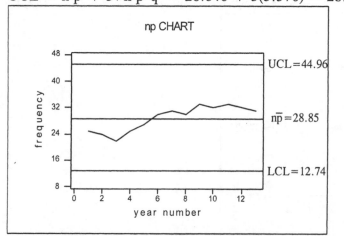

The charts are identical except for the labels. Since the labels on the np chart are n=100,000 times the labels on the chart for p, the center line is 100,000(.000288) = 28.486 and the upper control limit is 100,000(.000450) = 44.956 and the lower control limit is 100,000(.000127) = 12.736. The placement of the points does not change.

Review Exercises

The chart at the right summarizes the information necessary for exercises #1, #2 and #3.

$$\overline{\overline{x}} = \Sigma\overline{x}/k$$
$$= 126.922/11$$
$$= 11.538$$

$$\overline{R} = \Sigma R/k$$
$$= 55.86/11$$
$$= 5.078$$

year	$\overline{x}$	R
1980	12.480	9.33
1981	12.106	2.99
1982	10.707	6.26
1983	9.430	5.11
1984	13.372	7.09
1985	10.468	4.32
1986	15.084	6.34
1987	11.684	4.50
1988	10.240	3.99
1989	10.734	4.50
1990	10.706	1.34

1.

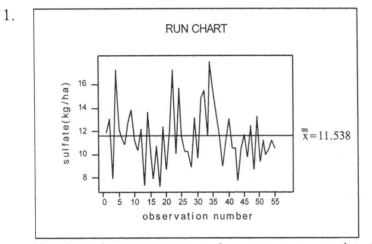

No; there does not appear to be a pattern suggesting that the process is not within statistical control.

2. $\overline{R} = \Sigma R/k = 55.86/11 = 5.078$ $LCL = D_3\overline{R} = 0(5.078) = 0$
                                                              $UCL = D_4\overline{R} = 2.114(5.078) = 10.73$

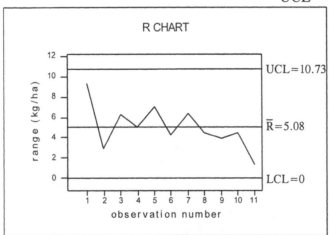

The process variation is within statistical control.

3. $\bar{\bar{x}} = \Sigma\bar{x}/k = 126.922/11 = 11.538$
   $\bar{R} = \Sigma R/k = 55.86/11 = 5.078$
   $LCL = \bar{\bar{x}} - A_2\bar{R} = 11.538 - (.577)(5.078) = 11.538 - 2.930 = 8.608$
   $UCL = \bar{\bar{x}} + A_2\bar{R} = 11.538 + (.577)(5.078) = 11.538 + 2.930 = 14.468$

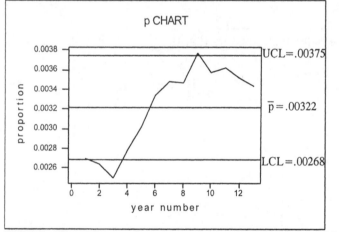

The process mean is not within statistical control because there is a point above the UCL.

4. $\bar{p} = (\Sigma x)/(\Sigma n) = (270+264+...+343)/(13)(100,000) = 4183/1,300,000 = .00322$
   $\sqrt{\bar{p}\cdot\bar{q}/n} = \sqrt{(.00322)(.99678)/100000} = .000179$
   $LCL = \bar{p} - 3\sqrt{\bar{p}\cdot\bar{q}/n} = .00322 - 3(.000179) = .00322 - .00053 = .00268$
   $UCL = \bar{p} + 3\sqrt{\bar{p}\cdot\bar{q}/n} = .00322 + 3(.000179) = .00322 + .00053 = .00375$

NOTE: The 13 sample proportions are: .00270 .00264 .00250 .00278 ... .00343
There are three out-of-control criteria, and meeting any one of them means that the process is not within statistical control. This process meets all three criteria. (1) There is a pattern – in this case, an upward trend. (2) There is a point outside the lower or upper control limits – in this case, there are two points below the LCL and one point above the UCL. (3) The run of 8 rule applies – in this case, the last 8 points are all above the center line. By definition, therefore, process is statistically unstable. In this instance, that means that the age 65+ death rate for infectious diseases is increasing.

5. $\bar{p} = (\Sigma x)/(\Sigma n) = (608+466+...+491)/(15)(1,000) = 7031/15000 = .4687$
    $\sqrt{\bar{p}\cdot\bar{q}/n} = \sqrt{(.4687)(.5313)/1000} = .01578$
    LCL $= \bar{p} - 3\sqrt{\bar{p}\cdot\bar{q}/n} = .4687 - 3(.01578) = .4687 - .0473 = .4214$
    UCL $= \bar{p} + 3\sqrt{\bar{p}\cdot\bar{q}/n} = .4687 + 3(.01578) = .4687 + .0473 = .5161$

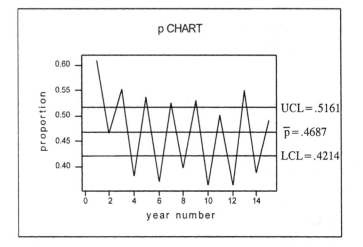

NOTE: The 15 sample proportions are: .608 .466 .552 .382 ... .491
There are three out-of-control criteria, and meeting any one of them means that the process
is not within statistical control. This process meets the first two criteria. (1) There is a
pattern – in this case, an alternating trend. (2) There is a point outside the lower or upper
control limits – in this case, most of the points are beyond the control limits. By
definition, therefore, process is statistically unstable. The pattern is caused by the fact that
the national elections occur every two years, every other national election involving a
presidential election and bring out the voters in larger numbers.

Cumulative Review Exercises

1. a. $\bar{p} = (\Sigma x)/(\Sigma n) = (10+8+...+11)/(20)(400) = 150/8000 = .01875$
    $\sqrt{\bar{p}\cdot\bar{q}/n} = \sqrt{(.01875)(.98125)/400} = .006782$
    $LCL = \bar{p} - 3\sqrt{\bar{p}\cdot\bar{q}/n} = .01875 - 3(.006782) = .01875 - .02035 = 0$ [truncate at 0]
    $UCL = \bar{p} + 3\sqrt{\bar{p}\cdot\bar{q}/n} = .01875 + 3(.006782) = .01875 + .02035 = .0391$
    NOTE: The LCL is truncated at 0 because of the story problem – since $\bar{p}$ is the proportion of defectives in the sample, which cannot be negative.

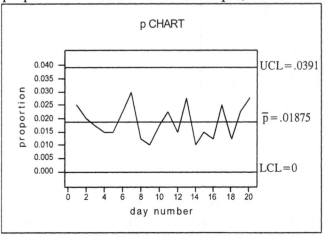

    NOTE: The 20 sample proportions are: $10/400 = .025$, $8/400 = .020, ..., 11/400 = .0275$
    The process is within statistical control, and so the data can be treated as coming from a binomial population with a fixed mean and variance.

    b. Using all the data combined, $\hat{p} = x/n = 150/8000$
    $\hat{p} \pm z_{.025}\cdot\sqrt{\hat{p}\hat{q}/n}$
    $.01875 \pm 1.960\cdot\sqrt{(.01875)(.98125)/8000}$
    $.01875 \pm .00297$
    $.0158 < p < .0217$

    c. original claim $p > .01$
    $H_o: p = .01$
    $H_1: p > .01$
    $\alpha = .05$
    C.R. $z > 1.645$
    calculations:
    $z_{\hat{p}} = (\hat{p} - \mu_{\hat{p}})/\sigma_{\hat{p}}$
    $= (.01875 - .01)/\sqrt{(.01)(.99)/8000}$
    $= .00875/.001112$
    $= 7.866$
    P-value $= P(z > 7.87) = 1 - P(z < 7.87) = 1 - .9999 = .0001$
    conclusion:
    Reject H_o; there is sufficient evidence to conclude that $p > .01$.

2. a. let A_i = the ith point is above the center line
    $P(A_i) = .5$ for each value of i
    $P(A) = P(\text{all 8 above}) = P(A_1)\cdot P(A_2)\cdot...\cdot P(A_8) = (.5)^8 = .00391$
    b. let B_i = the ith point is below the center line
    $P(B_i) = .5$ for each value of i
    $P(B) = P(\text{all 8 below}) = P(B_1)\cdot P(B_2)\cdot...\cdot P(B_8) = (.5)^8 = .00391$
    c. Notice that the events in parts (a) and (b) are mutually exclusive.
    $P(A \text{ or } B) = P(A) + P(B) - P(A \text{ and } B)$
    $= .00391 + .00391 - 0 = .00781$

3. The chart at the right summarizes
    the information in the same manner used for
    the kwh consumption data in section 13-2
    exercises #2, #3 and #4. The analysis also
    follows the format of those exercises.

sample	temperature			$\overline{x}$	R
1a	32	35	59	42.0	27
1b	76	66	42	61.3	34
2a	22	33	56	37.0	34
2b	70	63	42	58.3	28
3a	30	38	55	41.0	25
3b	71	61	38	56.7	33
4a	32	40	57	43.0	25
4b	72	65	45	60.7	27

a. Run Chart: There are $k \cdot n$ = (8 samples)·(3 observations/sample) = 24 observations.
   $\overline{\overline{x}} = \Sigma\overline{x}/k = 400/8 = 50.00$

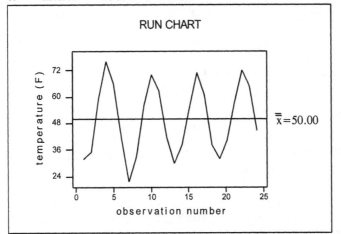

The process is not within statistical control because there is a pattern of cycles
indicating something other than random variation is at work.

b. R Chart: $\overline{R} = \Sigma R/k = 233/8 = 29.125$ $LCL = D_3\overline{R} = 0(29.125) = 0$
                                                              $UCL = D_4\overline{R} = 2.544(29.125) = 74.97$

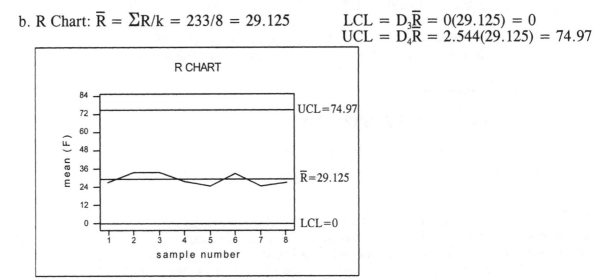

The process variation is within statistical control.

c. $\bar{x}$ Chart: $\bar{\bar{x}} = \Sigma\bar{x}/k = 400/8 = 50.00$
   $\bar{R} = \Sigma R/k = 233/8 = 29.125$
   $LCL = \bar{\bar{x}} - A_2\bar{R} = 50.00 - (1.023)(29.125) = 50.00 - 29.79 = 20.21$
   $UCL = \bar{\bar{x}} + A_2\bar{R} = 50.00 + (1.023)(29.125) = 50.00 + 29.79 = 79.79$

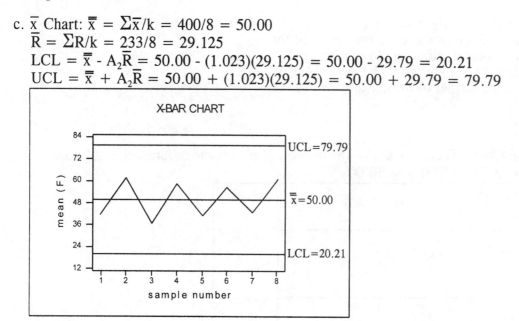

The process mean is not within statistical control because there is an alternating pattern of points above and below the overall mean.

4. Letting x represent temperature (°F) and y represent consumption (kwh), the following numerical summary contains all the necessary information for the n=24 paired values.
   $\Sigma x = 1200$ $\Sigma y = 74045$
   $\Sigma x^2 = 65850$ $\Sigma y^2 = 242624810$ $\Sigma xy = 3562714$
   $n(\Sigma x^2)-(\Sigma x)^2 = 140400$ $n(\Sigma y^2)-(\Sigma y)^2 = 340333415$ $n(\Sigma xy)-(\Sigma x)(\Sigma y) = -3348864$
   a. $r = -3348864/[\sqrt{140400} \cdot \sqrt{340333415}\ ]$
      $= -.484$
      The critical values being $r = \pm.396$ (for n=25 at the $\alpha=.05$ level), we conclude there is a significant negative linear correlation between outside temperature and energy consumption.
   b. $b_1 = -3348864/140400$
      $= -23.852$
      $b_0 = (74045/24) - (-23.852)(1200/24)$
      $= 4277.824$
      The regression line is $\hat{y} = 4277.824 - 23.852x$
   c. $\hat{y}_{60} = 4277.824 - 23.852(60) = 2847$ kwh

FINAL NOTE: Congratulations! You have completed statistics – the course that everybody likes to hate. I hope that this manual has helped to make the course a little more understandable for you – and that you leave the course with an appreciation of broad principles, and not memories of merely manipulating formulas. I wish you well in your continued studies, and that you achieve your full potential wherever your journey of life may lead.